STOCHASTIC ANALYSIS

ACADEMIC PRESS RAPID MANUSCRIPT REPRODUCTION

Proceedings of the International Conference
on Stochastic Analysis, April 10–14, 1978,
Northwestern University, Evanston, Illinois

STOCHASTIC ANALYSIS

edited by

AVNER FRIEDMAN
MARK PINSKY

Department of Mathematics
Northwestern University
Evanston, Illinois

ACADEMIC PRESS New York San Francisco London 1978
A Subsidiary of Harcourt Brace Jovanovich, Publishers

ACADEMIC PRESS, INC.
111 Fifth Avenue, New York, New York 10003

United Kingdom Edition published by
ACADEMIC PRESS, INC. (LONDON) LTD.
24/28 Oval Road, London NW1 7DX

Library of Congress Cataloging in Publication Data

International Conference on Stochastic Analysis,
Northwestern University, 1978.
Stochastic analysis.

1. Stochastic analysis—Congresses.
I. Friedman, Avner. II. Pinsky, Mark A.,
DATE III. Title.
QA274.2.I57 1978 519.2 78-14932
ISBN 0-12-268380-3

PRINTED IN THE UNITED STATES OF AMERICA

Contents

List of Contributors

Numbers in parentheses indicate the pages on which authors' contributions begin.

Robert F. Anderson (1), University of Pittsburgh, Pittsburgh, Pennsylvania
A. Bensoussan (25), IRIA-Laboria, University of Paris IX
M. D. Donsker (57), Courant Institute, New York University, New York
E. B. Dynkin (63), Department of Mathematics, Cornell University, Ithaca, New York
David Elworthy (79), University of Warwick, Coventry, England
W. H. Fleming (97), Division of Applied Mathematics, Brown University, Providence, Rhode Island
Avner Friedman (109), Department of Mathematics, Northwestern University, Evanston, Illinois
Bert Fristedt (127), Department of Mathematics, University of Minnesota, Minneapolis, Minnesota
Charles J. Holland (139), Courant Institute of Mathematical Sciences, New York University, New York
R. Holley (153), Department of Mathematics, University of Colorado, Boulder, Colorado
Nobuyuki Ikeda (175), Department of Mathematics, Osaka University, Toyonaka, Japan
Kiyosi Itô (187), Research Institute for Mathematical Sciences, Kyoto University, Kyoto, Japan
Paul Malliavin (199, 327), Institut Henri Poincaré, Paris, France
Shojiro Manabe (175), Department of Mathematics, Osaka University, Toyonaka, Japan
Steven Orey (127), Department of Mathematics, University of Minnesota, Minneapolis, Minnesota

George C. Papanicolaou (215), Courant Institute, New York University, New York

Etienne Pardoux (239), IRIA(CNRS)-Laboria, Le Chesnay, France

Mark Pinsky (271), Department of Mathematics, Northwestern University, Evanston, Illinois

Maurice Robin (285), IRIA-Laboria, Rocquencourt, France

D. W. Stroock (153), Department of Mathematics, University of Colorado, Boulder, Colorado

Hiroshi Tanaka (301), Department of Mathematics, Hiroshima University, Hiroshima, Japan

S. R. S. Varadhan (57), Courant Institute, New York University, New York

Michel Viot (97), IRIA-Laboria, Rocquencourt, France

Shinzo Watanabe (315), Department of Mathematics Kyoto University, Kyoto, Japan

Preface

The International Conference on Stochastic Analysis was held April 10–14, 1978 in the Louis Room of Norris Center at Northwestern University. The conference was sponsored by a grant from the National Science Foundation and supplemented by Northwestern University.

The purpose of this conference was to bring together leading workers in stochastic differential equations and related areas. The specific areas include foundations, stochastic optimization, stochastic differential geometry, limit theorems, and martingale methods.

The opening ceremonies included welcoming remarks by Robert Strotz, President of Northwestern University. They were followed by a brief commentary from Professor Joseph Doob of the University of Illinois. Doob's illuminating remarks on the early history of probability theory in the United States served as an appropriate prelude to the specialized talks that followed.

In addition to the invited speakers there were twelve contributed papers. A complete list of the participants appears on the following pages.

Participants

Mohamed Abdel-Hameed, University of North Carolina at Charlotte
A. N. Al-Hussaini, University of Alberta
[1] *Robert Anderson,* University of Pittsburgh
Richard Arratia, University of Wisconsin
Alexandra Bellow, Northwestern University
[1] *Alain Bensoussan,* IRIA
[2] *Marc Berger,* University of Wisconsin
Mircea Bradescu, Northwestern University
[2] *Maury Bramson,* Courant–NYU
Neil Bromberg, Rochester Institute of Technology
Donald Burkholder, University of Illinois–Urbana
Russel Caflisch, Courant–NYU
Joshua Chover, University of Wisconsin
Paul Chow, Wayne State University
[2] *Erhan Cinlar,* Northwestern University
Michael Cranston, University of Minnesota
[2] *Donald Dawson,* Carleton University
Joseph Doob, University of Illinois–Urbana
James Droser, Northwestern University
Tyrone Duncan, University of Kansas
[1] *Eugene Dynkin,* Cornell University
[1] *David Elworthy,* University of Warwick
Stewart Ethier, Michigan State University–East Lansing
[1] *Wendell Fleming,* Brown University
[1] *Avner Friedman,* Northwestern University
Ronald Getoor, University of California–San Diego
[2] *Tepper Gill,* Howard University
Steven Goldstein, University of Wisconsin
Victor Goodman, Indiana University
[2] *Luis Gorostiza,* Centro de Investigacion Del IPN

Priscilla Greenwood, University of British Columbia–Vancouver
[2] *Richard Griego,* University of New Mexico, Albuquerque
David Griffeath, University of Wisconsin
Alberto Grunbaum, University of California–Berkeley
Theodore Harris, University of Southern California
Kevin Hastings, Northwestern University
H. Hering, University of Wisconsin
[2] *Kenneth Hochberg,* Carleton University
[1] *Charles Holland,* Courant–NYU
[1] *Richard Holley,* University of Colorado
Steel Huang, University of Cincinnati
Jurg Husler, University of Pittsburgh
[1] *Nobuyuki Ikeda,* Osaka University
[1] *Kiyosi Ito,* Kyoto University
Jeffrey Joel, University of Michigan–Ann Arbor
A. Joffe, University of Wisconsin
Manuel Keepler, South Carolina State College–Orangeburg
[2] *Jerome Keisler,* University of Wisconsin
Ronald Kirk, Southern Illinois University
Frank Knight, University of Illinois–Urbana
Arthur Krener, University of California–Davis
William Kuchenbecker, Northwestern University
Alan Lambert, University of North Carolina at Charlotte
Arthur Lubin, Illinois Institute of Technology
[2] *Michael Magill,* Northwestern University
[1] *Paul Malliavin,* Institut Henri Poincaré, Paris, France
Robert Marcus, City University of New York
Benjamin Melamed, Northwestern University
Joanna Mitro, University of Illinois–Urbana
Victor Mizel, Carnegie–Mellon University
Itrel Monroe, University of Arkansas
[2] *A. G. Mucci,* University of Maryland
Naresh Jain, University of Minnesota
Peter Ney, University of Wisconsin
[1] *Steven Orey,* University of Minnesota
Suleyman Ozekici, Northwestern University
Thomas Paine, Southern Illinois University
[1] *George Papanicolaou,* Courant–NYU
[1] *Etienne Pardoux,* IRIA
[1] *Mark Pinsky,* Northwestern University
Stan Pliska, Northwestern University
[2] *Paul Polansky,* Northwestern University
Zoran Pop-Stojanovic, University of Florida
[2] *Philip Protter,* The Institute for Advanced Study

Joseph Quinn, University of North Carolina at Charlotte
[1] *Maurice Robin,* IRIA
Millu Rosenblatt-Roth, State University of New York at Buffalo
Steven Rosencrans, Tulane University
Uwe Rosler, University of Minnesota
Michael Rubinovitch, Northwestern University
Thomas Savits, University of Pittsburgh
Marty Silverstein, Washington University
[1] *Daniel Stroock,* University of Colorado
Robert Strotz, Northwestern University
Keith Stroyan, University of Iowa, Iowa City
David Sze, University of Chicago
[1] *Hiroshi Tanaka,* Hiroshima University
David Tanny, University of Rochester
Murad Taqqu, Cornell University
Charles Tier, University of Illinois–Chicago Circle
George Tsien, University of New Mexico
[1] *Srinivasa Varadhan,* Courant–NYU
Thomas Van Haagen, Southern Illinois University
Albert Wang, University of Tennessee, Knoxville
[1] *Shinzo Watanabe,* Kyoto University
Robert Wolpert, Duke University
Wobjun Woyczynski, Cleveland State University

[1] *Invited Paper*
[2] *Contributed Paper*

OPTIMAL STOPPING IN A RELIABILITY PROBLEM

Robert F. Anderson

University of Pittsburgh
Pittsburgh, PA

I. INTRODUCTION

This is a report on joint work of the author, part with J. K. Ghosh [1] and part with W. E. Winkler [2].

Let $\Omega = D([0,\infty): R^1_+ x\{0,1\}^n)$ and define λ: $[0,\infty)x\Omega \to R^1_+$ by $\lambda(t)(\omega) = \omega_0(t)$ and X: $[0,\infty)x\Omega \to \{0,1\}^n$ by

$$X(t)(\omega) = (\omega_1(t),\omega_2(t),\ldots, \omega_n(t))$$

where for $\omega \in \Omega$ we have use the notation $\omega = (\omega_0,\omega_1,\ldots, \omega_n)$.

Let

$$M_t = \sigma(\lambda(s),X(s): 0\leq s\leq t) \qquad \mathcal{F}_t = \sigma(X(s): 0\leq s\leq t)$$

$$M = \sigma(M_t: t\geq 0) \qquad \text{and} \qquad \mathcal{F} = \sigma(\mathcal{F}_t: t\geq 0).$$

For each $\lambda \in R^1_+$ and $X \in \{0,1\}^n$, define on (Ω,M), $P^{\lambda,X}$ by its finite dimensional distributions as follows: Let $A_1,A_2,\ldots, A_j \in \mathcal{B}(R^1_+)$, $B_1,B_2,\ldots, B_j \in \mathcal{B}(\{0,1\}^n)$ and $0 \leq t_1 \leq t_2 \leq\ldots\leq t_j$,

$$P^{\lambda,X}(\lambda(t_1)\varepsilon A_1,X(t_1)\varepsilon B_1,\lambda(t_2)\varepsilon A_2,X(t_2)\varepsilon B_2,\ldots, \lambda(t_j)\varepsilon A_j,X(t_j)\varepsilon B_j) \tag{1.1}$$

$$= I_{\bigcap\limits_{1\leq i\leq j} A_i}(\lambda)\int_{B_1} p^\lambda(t_1,X,dX_1)\int_{B_2} p^\lambda(t_2-t_1,X_1,dX_2)\ldots\int_{B_j} p^\lambda(t_j-t_{j-1},X_{j-1},dX_j)$$

where for $s \geq 0$ and $X,Y \in \{0,1\}^n$

$$p^\lambda(s,X,Y) = \begin{cases} 0 & X > Y \\ (e^{-\lambda s})^{n-k(Y)}(1-e^{-\lambda s})^{k(Y)-k(X)} & X \leq Y, \end{cases}$$

ISBN 0-12-268380-3

$k(X) = \#$ of entries of X which are equal to 1.

Note that $P^{\lambda,X}(\lambda(t)=\lambda:\ t\geq 0) = 1$ and $\{P^{\lambda,X}\}_{\lambda\varepsilon R^1_+, X\varepsilon\{0,1\}^n}$ is just a representation of n independent random variables each taking values 0 or 1, which if start at 0 jump to 1 according to an exponential distribution with parameter λ and if at 1 remain there. That is, we have n independent Poisson processes with parameter λ each of which has 1 as an absorbing state.

Define for $\alpha, \beta > 0$ the gamma density

$$f(\lambda,\alpha,\beta) = \frac{\beta^\alpha}{\Gamma(\alpha)} e^{-\beta\lambda} \lambda^{\alpha-1} \qquad \lambda > 0$$

and let

$$P^{X,\alpha,\beta} = \int_0^\infty d\lambda\ f(\lambda,\alpha,\beta)P^{\lambda,X}$$

We use the notation $\tau \varepsilon (\mathfrak{F}_t)_{t\geq 0}$ to mean τ is an $(\mathfrak{F}_t)_{t\geq 0}$ stopping time. Let c_1 and $c(\lambda,k)$ be positive and $c(\lambda,k)$ and $g(\lambda)$ be Borel measurable in λ. For $\tau \varepsilon (\mathfrak{F}_t)_{t\geq 0}$, define

$$J(\tau)=E^{0,\alpha,\beta}[c_1 k(X(\tau))+(g(\lambda(\tau))-E^{0,\alpha,\beta}[g(\lambda(\tau))|\mathfrak{F}_\tau])^2 + \int_0^\tau c(\lambda(s),k(X(s))ds] \tag{1.2}$$

Here $0 = (0,0,\ldots, 0)$. We wish to find $\bar{\tau} \varepsilon (\mathfrak{F}_t)_{t\geq 0}$ so that

$$J(\bar{\tau}) = \inf_{\tau\varepsilon(\mathfrak{F}_t)_{t\geq 0}} J(\tau).$$

The situation we are modeling is that of estimating the reliability $g(\lambda)$ of a machine by taking a sample of n and observing them at a cost c_1 per machine failure and a cost $c(\lambda,k(X))$ per unit of time.

The problem at hand is a partial observation optimal stopping problem. $X(t)$ by itself is not Markov, but $\lambda(t),X(t)$ is Markov. We are allowed only to observe $\mathfrak{F}_t = \sigma(X(s):\ 0\leq s\leq t)$.

In Section 2, we embed our problem into an optimal stopping problem for a Markov process by use of a sufficient statistic for $\lambda(t)$ which is $\mathfrak{F}_t$ measurable. We translate the new problem into an analytic problem of solving a variational inequality.

In Section 3, we solve the variational inequality explicitly for the two cases $g(\lambda) = \lambda$ and $g(\lambda) = 1/\lambda$. We solve the variational inequality in general in Section 4.

II. PRELIMINARIES

We first show the posterior distribution of $\lambda(t)$ given $\mathfrak{F}_t$ is also gamma with parameter $\alpha + k(t)$ and $\beta + w(t)$ where

$$k(t) = k(X(t))$$

and

$$w(t) = \int_0^t \sum_1^n I_{\{0\}}(x_i(s))ds, \quad x_i(s) = \omega_i(s) \qquad i = 1,2,\ldots, n.$$

More precisely, for $B \in \mathcal{B}(R_+^1)$ and $k = k(X)$

$$P^{X,\alpha+k,\beta}(\lambda(t)\in B|\mathfrak{F}_t) = \int_B d\lambda\ f(\lambda,\alpha+k(t),\beta+w(t)) \tag{2.1}$$

A simple exercise will show for $\lambda,\mu > 0$,

$$\frac{dP^{\mu,X}}{dP^{\lambda,X}}\Big|\mathfrak{F}_t = \prod_1^n \frac{\mu^{x_i(t)-x_i} e^{-\mu t_i(t)}}{\lambda^{x_i(t)-x_i} e^{-\lambda t_i(t)}}, \quad X = (x_1,x_2,\ldots, x_n),$$

where $t_i(t) = \inf\{s\colon x_i(s)=1\}\wedge t$. Note that

$$\frac{dP^{\mu,X}}{dP^{\lambda,X}}\Big|\mathfrak{F}_t = \frac{\mu^{k(t)-k(X)} e^{-\mu w(t)}}{\lambda^{k(t)-k(X)} e^{-\lambda w(t)}}.$$

Also

$$\int_0^\infty d\mu\ f(\mu,\alpha+k(t),\beta+w(t)) = 1.$$

Thus for $A \in \mathfrak{F}_t$, $B \in \mathcal{B}(R_+^1)$ and $k(X) = k$,

$$P^{X,\alpha+k,\beta}(\lambda(t)\in B,A) = \int_B d\lambda\ f(\lambda,\alpha+k,\beta)P^{\lambda,X}(A)$$

$$= \int_B d\lambda\ f(\lambda,\alpha+k,\beta)E^{\lambda,X}[I_A\int_0^\infty d\mu\ f(\mu,\alpha+k(t),\beta+w(t))]$$

$$= \int_B d\mu\ f(\lambda,\alpha+k,\beta)\int_B d\lambda\ E^{\lambda,X}[I_A f(\lambda,\alpha+k(t),\beta+w(t))\Big(\frac{dP^{\mu,X}}{dP^{\lambda,X}}\Big|\mathfrak{F}_t\Big)]$$

$$= \int_B d\mu\ f(\mu,\alpha+k,\beta)E^{\mu,X}[I_A \int_B d\lambda\ f(\lambda,\alpha+k(t),\beta+w(t))]$$

which proves (2.1).

Let $\tau \in (\mathfrak{F}_t)_{t\geq 0}$. A standard approximation argument shows that (2.1) extends to

$$P^{X,\alpha+k,\beta}(\lambda(\tau)\in B|\mathfrak{F}_\tau) = \int_B d\lambda\ f(\lambda,\alpha+k(\tau),\beta+w(\tau)). \tag{2.2}$$

We assume that for $k = 0,1,\ldots, n$

$$\begin{aligned} &c(\lambda,k) \text{ is non-negative, bounded.} \\ &g(\lambda) \text{ is continuous with} \\ &\int_0^\infty d\lambda\ f(\lambda,\alpha,\beta)g^2(\lambda) \leq C(1+\beta)^N,\ N \text{ positive} \end{aligned} \tag{2.3}$$

Define

$$h(k,w) = \int_0^\infty d\lambda\ f(\lambda,\alpha+k,\beta+w)g^2(\lambda) - (\int_0^\infty d\lambda\ f(\lambda,\alpha+k,\beta+w)g(\lambda))^2 \tag{2.4}$$

and

$$b(k,w) = \int_0^\infty d\lambda\ f(\lambda,\alpha+k,\beta+w)c(\lambda,k). \tag{2.5}$$

Clearly $b(k,w)$ and $h(k,w)$ are non-negative.

Theorem 1. For any $\tau \in (\mathfrak{F}_t)_{t\geq 0}$

$$J(\tau) = E^{0,\alpha,\beta}[c_1 k(\tau)+h(k(\tau),w(\tau))+\int_0^\tau b(k(s),w(s))ds],$$

$J(\tau)$ given by (1.2).

Proof. The proof is immediate from (2.2), (2.4) and (2.5).

Define for each $X \in \{0,1\}^n$ and $w \geq 0$ with $k = k(X)$,

$$P^{X,\alpha+k,\beta+w} = \int_0^\infty d\lambda\ f(\lambda,\alpha+k,\beta+w)P^{\lambda,X}$$

Let

$$w_w(t) = w + w(t).$$

We claim for $k = k(X)$,

$$\begin{aligned} &P^{X,\alpha+k,\beta+w}(X(t+s)\in A,k(t+s)\in B,w_w(t+s)\in C|\mathfrak{F}_s) \\ &\quad = P^{X(s),\alpha+k(s),w_w(s)}(X(t)\in A,k(t)\in B,w_{w_w(s)}(t)\in C). \end{aligned}$$

Since

$$P^{X,\alpha+k,\beta+w}(X(t+s)\varepsilon A, k(t+s)\varepsilon B, w_W(t+s)\varepsilon C|M_s)$$

$$= P^{\lambda(s),X(s)}(X(t)\varepsilon A, k(t)\varepsilon B, w_{W_W(s)}(t)\varepsilon C),$$

(2.1) yields

$$E^{X,\alpha+k,\beta+w}[P^{X,\alpha+k,\beta+w}(X(t+s)\varepsilon A, k(t+s)\varepsilon B, w_W(t+s)\varepsilon C|M_s)|\mathfrak{F}_s]$$

$$= \int_0^\infty d\lambda\ f(\lambda,\alpha+k(s),\beta+w_W(s))P^{\lambda,X(s)}(X(t)\varepsilon A, k(t)\varepsilon B, w_{W_W(s)}(t)\varepsilon C)$$

$$= P^{X(s),\alpha+k(s),\beta+w_W(s)}(X(t)\varepsilon A, k(t)\varepsilon B, w_{W_W(s)}(t)\varepsilon C)$$

which establishes the claim and the fact that $\{P^{X,\alpha+k,\beta+w}\}_{\substack{X\varepsilon\{0,1\}^n \\ k=k(X) \\ w\geq 0}}$

is a Markov process. Moreover, it is Feller and strong Markov.

An easy calculation shows

$$AU(X,k,w) = \lim_{s\downarrow 0} \frac{E^{X,\alpha+k,\beta+w}[U(X(s),k(s),w_W(s))]-U(X,k,w)}{s}$$

$$= \sum_1^n \frac{\alpha+k}{\beta+w}(U(X+e_i,k,w)-U(X,k,w)) + (n-k)D_W^+ U(X,k,w)$$

$$+ \frac{(n-k)(\alpha+k)}{\beta+w}(U(X,k+1,w)-U(X,k,w)), \quad k = k(X)$$

where

$$e_i = \begin{cases} 0 & \text{if } x_i = 1 \\ \text{unit vector in ith direction} & \text{if } x_i = 0. \end{cases}$$

Also

$$D_W^+U(X,k,w) = \lim_{h\downarrow 0} \frac{U(X,k,w+h)-U(X,k,w)}{h},$$

the right hand derivative coming about since w(s) increases in s. A is the weak infinitesimal generator for the process.

Define

$$\psi(k,w) = c_1 k + h(k,w)$$

and for $\tau \varepsilon (\mathfrak{F}_t)_{t\geq 0}$ and $k = k(X)$,

$$J(X,k,w;\tau) = E^{X,\alpha+k,\beta+w}[\psi(k(\tau),w_w(\tau))+\int_0^\tau b(k(s),w_w(s))ds].$$

By Theorem 1

$$J(\tau) = J(0,0,0;\tau).$$

We wish to find $\bar{\tau} \varepsilon (\mathfrak{F}_t)_{t\geq 0}$ so that

$$J(X,k,w;\bar{\tau}) = \inf_{\tau\varepsilon(\mathfrak{F}_t)_{t\geq 0}} J(X,k,w;\tau).$$

Theorem 2. Assume that (2.3) holds. Suppose

$$V(n,w) = \psi(n,w) \qquad w \geq 0$$

and for $k = 0,1,\ldots, n-1$ fixed, $V(k,w)$ is continuous in w, $D_w^+ V(k,w)$ exists and is continuous except at possibly for a finite number of jump discontinuities. Suppose also $V(k,w)$ satisfies

$$AV(k,w) + b(k,w) \geq 0 \tag{2.6}$$

$$V(k,w) \leq \psi(k,w) \tag{2.7}$$

$$(AV(k,w)+b(k,w))(\psi(k,w)-V(k,w)) = 0. \tag{2.8}$$

Then for X with $k = k(X)$,

$$V(k,w) = \inf_{\tau\varepsilon(\mathfrak{F}_t)_{t\geq 0}} J(X,k,w;\tau).$$

Moreover, if

$$S = \{(k,w): V(k,w)=\psi(k,w)\}$$

and

$$\bar{\tau}_w = \inf\{t: (k(t),w_w(t))\varepsilon S\}$$

then

$$V(k,w) = J(X,k,w;\bar{\tau}_w), \qquad k(X) = k.$$

We first make a few remarks before proving the result.

Remark 2.1. If $\tau' = \inf\{t: k(t)=n\}$ and τ is any other $(\mathcal{F}_t)_{t\geq 0}$ stopping time, then

$$E^{X,\alpha+k,\beta+w}[\psi(k(\tau\wedge\tau'),w_w(\tau\wedge\tau'))+\int_0^{\tau\wedge\tau'} b(k(s),w_w(s))ds]$$
$$\leq E^{X,\alpha+k,\beta+w}[\psi(k(\tau),w_w(\tau))+\int_0^{\tau} b(k(s),w_w(s))ds], \quad k(X) = k,$$

since for $t \geq \tau'$, $\psi(k(\tau'),w_w(\tau')) = \psi(k(t),w_w(t))$ and $b(k,w) \geq 0$. Thus we need only consider stopping times which are less than or equal to τ'.

2.2. If for each k, $U(k,w)$ is continuous in w and $D_w^+U(k,w)$ is continuous except at a finite number of jump discontinuities, then by Dynkin [3], p. 133,

$$U(k,w) = E^{X,\alpha+k,\beta+w}[U(k(t),w_w(t))-\int_0^t AU(k(s),w_w(s))ds], \quad k = k(X) \tag{2.9}$$

Moreover, if τ is a stopping time such that

$$E^{X,\alpha+k,\beta+w}[\,|U(k(\tau),w_w(\tau))|\,] < \infty$$

and (2.10)

$$E^{X,\alpha+k,\beta+w}[\int_0^{\tau}|AU(k(s),w_w(s))|ds] \leq w$$

Then (2.9) is valid for τ in place of t. (2.9) is Dynkin's formula.

2.3. Conditions (2.3) and (2.4) and (2.5) imply $\psi(k,w)$ is continuous in w, $D_w^+\psi(k,w)$ exists and is continuous. Moreover, if $\tau' = \inf\{t: k(t)=n\}$, $E^{X,\alpha+k,\beta+w}[(\tau')^m] < \infty$, m a positive integer and hence by (2.3) again,

$$E^{X,\alpha+k,\beta+w}[\ \psi(k(\tau'),w_w(\tau'))\ ] < \infty$$

and

$$E^{X,\alpha+k,\beta+w}[\int_0^{\tau'} A\psi(k(s),w_w(s))\ ds] < \infty, \quad k = k(X).$$

Further (2.6)-(2.8) imply

$$|V(k,w)| \leq C(1+w)^N$$

and $|AV(k,w)| \leq C(1+w)^N$, N positive.

Hence if $\tau \leq \tau'$, Dynkin's formula (2.9) is valid for τ.

Proof of Theorem 2: Let $\tau \in (\mathcal{F}_t)_{t\geq 0}$ and $\tau \leq \tau'$,

$\tau' = \inf\{t: k(t)=n\}$. By Remark 2.3, and (2.9)

$$V(k,w) = E^{X,\alpha+k,\beta+w}[V(k(\tau),w_w(\tau))-\int_0^\tau AV(k(s),w_w(s))ds], \quad k = k(X).$$

Thus by (2.6) and (2.7),

$$V(k,w) \leq E^{X,\alpha+k,\beta+w}[\psi(k(\tau),w_w(\tau))+\int_0^\tau b(k(s),w_w(s))ds] \tag{2.11}$$

$$\leq J(x,k,w;\tau) \qquad k = k(X).$$

On the other hand, by (2.8), for $\bar{\tau}_w$ the inequality (2.11) becomes an equality. This completes the proof.

Conditions (2.6)-(2.8) are referred to as a variational inequality.

Remark 2.4. Theorems 1 and 2 are valid for other types of costs. All that is necessary is that V(k,w) satisfy Dynkin's formula, (2.9). In particular, the case of one sided hypothesis testing is covered.

III. TWO EXAMPLES OF ESTIMATION

In this section, we solve explicitly the variational (2.6)-(2.8) for the two examples $g(\lambda) = \lambda$ and $g(\lambda) = 1/\lambda$ with $c(\lambda,k) = c_2 > 0$. We also find the explicit nature of S, $S = \{(k,w): V(k,w)=\psi(k,w)\}$.

A straight forward calculation using (2.2) shows for $\tau \in (\mathcal{F}_t)_{t\geq 0}$

$$E^{0,\alpha,\beta}[(\lambda(\tau)-E^{0,\alpha,\beta}[\lambda(\tau)|\mathcal{F}_\tau])^2|\mathcal{F}_\tau] = \frac{\alpha+k(\tau)}{(\beta+w(\tau))^2}$$

and

$$E^{0,\alpha,\beta}[(\frac{1}{\lambda(\tau)} - E^{0,\alpha,\beta}[\frac{1}{\lambda(\tau)}|\mathcal{F}_\tau])^2|\mathcal{F}_\tau] = \frac{\beta+w(\tau)}{(\alpha+k(\tau)-1)^2(\alpha+k(\tau)-2)}$$

Let

$$\psi_1(k,w) = c_1 k + \frac{\alpha+k}{(\beta+w)^2} \tag{3.1}$$

and

$$\psi_2(k,w) = c_1 k + \frac{(\beta+w)^2}{(\alpha+k-1)^2(\alpha+k-2)}. \tag{3.2}$$

Also

$$b(k,w) = c_2.$$

Note that (2.3) can be verified in either case so Theorem 2 holds if we can solve for $V(k,w)$. Note also for w large,

$$A\psi_1(k,w) + c_2 = \frac{(n-k)(\alpha+k)}{\beta+w}\left(c_1 + \frac{c_2(\beta+w)}{(n-k)(\alpha+k)} - \frac{1}{(\beta+w)^2}\right) > 0 \tag{3.3}$$

and

$$A\psi_2(k,w) + c_2 = \frac{(n-k)(\alpha+k)}{(\beta+w)}\left(c_1 + \frac{c_2(\beta+w)}{(n-k)(\alpha+k)} - \frac{(\beta+w)^2}{(\alpha+k)^2(\alpha+k-1)^2}\right) < 0 \tag{3.4}$$

Theorem 3. Assume $\alpha > 1$ and $\beta > 0$. Set

$$V(n,w) = \psi_1(n,w) \qquad w \geq 0$$

and for $k = 0,1,\ldots, n-1$,

$$V(k,w) = \begin{cases} \psi_1(k,w) & w \geq b_k \\ (\beta+w)^{\alpha+k}\left(\dfrac{\psi_1(k,b_k)}{(\beta+b_k)^{\alpha+k}} + \displaystyle\int_w^{b_k} dz\, \frac{c_2}{(n-k)(\beta+z)^{\alpha+k}} + \int_w^{b_k} dz\, \frac{(\alpha+k)V(k+1,z)}{(\beta+z)^{\alpha+k+1}}\right), & w \leq b_k \end{cases} \tag{3.5}$$

where if

$$R(k,w) = \frac{\beta+w}{(n-k)(\alpha+k)}(A\psi_1(k,w)+c_2) + (V(k+1,w)-\psi_1(k+1,w))$$

then

$$D_w R(k,w) \geq 0 \qquad w \geq 0 \tag{3.6}$$

and

$$b_k \text{ is the unique positive solution of } R(k,b_k) = 0$$
$$\text{or in the case no such root exists, } b_k = 0. \tag{3.7}$$

Then $V(k,w)$ is continuous and bounded, $D_w V(k,w)$ is continuous and bounded, and for $k = 0,1,\ldots, n-1$ satisfies (2.6)-(2.8).

In the case $n \leq \alpha + 1$,

$$b_0 \geq b_1 \geq \ldots \geq b_{n-1} \geq 0 \tag{3.8}$$

and in the case $n > \alpha + 1$, let k_0 be the largest integer such that $n - 2k_0 - \alpha - 1 \geq 0$, then

$$0 \leq b_0 \leq b_1 \leq \ldots \leq b_{k_0+1}$$
$$b_{k_0+1} \geq b_{k_0+2} \geq \ldots \geq b_{n-1} \geq 0. \tag{3.9}$$

Proof: We first show that if b_k is defined by (3.7) and $V(k,w)$ by (3.5) then the variational inequality (2.6)-(2.8) is satisfied. If (3.6) and (3.7) hold, then it follows from (3.3) and (3.5), for $0 \leq w \leq b_k$

$$V(k,w) - \psi_1(k,w) = (\beta+w)^{\alpha+k} \int_w^{b_k} dz \, \frac{(\alpha+k)R(k,z)}{(\beta+z)^{\alpha+k+1}} < 0 \tag{3.10}$$

so

$$V(k,w) \leq \psi_1(k,w) \qquad w \geq 0$$

For $0 \leq w \leq b_k$, by (3.5)

$$AV(k,w) + c_2 = 0$$

and for $w \geq b_k$, by (3.3), (3.6), and (3.7)

$$AV(k,w) + c_2 = \frac{(n-k)(\alpha+k)}{\beta+w} R(k,w) \geq 0.$$

From (3.7) and (3.10) it follows that if $b_k > 0$, $D_w V(k,b_k) = D_w \psi_1(k,b_k)$ and hence $D_w V(k,w)$ is continuous and bounded.

What we must do is to establish (3.6). We do this by backwards induction on k. Let

$$H(k,w) = \frac{\beta+w}{(n-k)(\alpha+k)} (A\psi_1(k,w)+c_2) \tag{3.11}$$

By (3.3)

$$D_w H(k,w) > 0 \qquad w \geq 0 \tag{3.12}$$

Hence for n - 1,

$$D_w R(n-1,w) = D_w H(n-1,w) > 0 \qquad w \geq 0$$

We assume (3.6) for $k + 1$. Now

$$D_w R(k,w) = D_w H(k,w) + D_w (V(k+1,w)-\psi_1(k+1,w)).$$

If $b_{k+1} = 0$, then $D_w R(k,w) > 0$, $w \geq 0$ by (3.12). If $b_{k+1} > 0$, then again for $w > b_{k+1}$, $D_w R(k,w) > 0$ by (3.12). For $0 \leq w \leq b_{k+1}$ from (3.5) and (3.10) with $k+1$ in place of k, it follows after differentiation and an integration by parts that

$$D_w(V(k+1,w)-\psi_1(k+1,w)) = (\alpha+k+1)(\beta+w)^{\alpha+k} \int_w^{b_{k+1}} dz \frac{D_z R(k+1,z)}{(\beta+z)^{\alpha+k+1}} - \frac{R(k+1,b_{k+1})}{(\beta+b_{k+1})^{\alpha+k+1}} \geq 0$$

by (3.7) and the inductive hypothesis. This establishes (3.6).

Finally, we prove (3.8) and (3.9). Let $n \leq \alpha + 1$. In this case $H(k,w)$ is increasing in both w and k. Since

$$R(n-1,w) = H(n-1,w)$$

by (3.7), b_{n-1} either satisfies $H(n-1,b_{n-1}) = 0$ or $H(n-1,w) > 0$ for $w \geq 0$, i.e. $b_{n-1} = 0$. Now

$$R(n-2,w) \leq H(n-2,w) < H(n-1,w)$$

so either $b_{n-2} > b_{n-1}$ or $b_{n-2} = 0$. The argument for (3.8) is completed by repeated applications of this argument.

Let $n > \alpha + 1$ and k_0 the largest integer such that $n- 2k_0 - \alpha - 1 \geq 0$. Now (3.3) and (3.11) yield

$$H(k,w) \leq H(k+1,w) \qquad k \geq k_0 + 1$$

$$H(k,w) \geq H(k+1\ w) \qquad 0 \leq k \leq k_0$$

For $k \geq k_0 + 1$, the previous argument shows $b_k \geq b_{k+1}$. For $k \leq k_0$,

$$R(k,b_{k+1}) = H(k,b_{k+1}) \geq R(k+1,b_{k+1}) - (V(k+2,b_{k+1}) - \psi_1(k+2,b_{k+1})) \geq 0$$

which shows $b_k \leq b_{k+1}$ proving (3.9) and the theorem.

Theorem 4. Assume $\alpha > 2$ and $\beta > 0$. Define $U(n,w) = \psi_2(n,w)$, $w \geq 0$. For X with $k(X) = k$ define inductively for $k = 0,1,\ldots, n-1$,

$$U(k,w) = E^{X,\alpha+k,\beta+w}[U(k+1,w_w(\tau_{k+1})) + c_2\tau_{k+1}]$$

where

$$\tau_{k+1} = \inf\{t: k(t)=k+1\}$$

Define

$$V(n,w) = \psi_2(n,w) \qquad w \geq 0.$$

For $k = 0,1,\ldots, n-1$, there exists constants $b_k \geq 0$ increasing in k such that if we set

$$V(k,w) = \begin{cases} U(k,w) & w \geq b_{n-1} \\ (\beta+w)^{\alpha+k}\left[\dfrac{U(k,b_{n-1})}{(\beta+b_{n-1})^{\alpha+k}} + \displaystyle\int_w^{b_{n-1}} dz\, \dfrac{c_2}{(n-k)(\beta+z)^{\alpha+k}}\right. & \\ & b_k \leq w \leq b_{n-1} \\ \qquad \left. + \displaystyle\int_w^{b_{n-1}} dz\, \dfrac{(\alpha+k)V(k+1,z)}{(\beta+z)^{\alpha+k+1}}\right] & \\ \psi_2(k,w) & w \leq b_k \end{cases} \tag{3.13}$$

Then $V(k,w)$ is continuous, $D_w^+ V(k,w)$ exists and is continuous except for at most a finite number of jump discontinuities and moreover satisfies (2.6)-(2.8).

Further,

either b_{n-1} is the unique positive root of
$U(n-1,b_{n-1}) - \psi_2(n-1,b_{n-1})$ or $U(n-1,w) - \psi_2(n-1,w) \leq 0$ (3.14)
for $w \geq 0$ and $b_{n-1} = 0$

and for $k = 0,1,\ldots, n-2$, if

$$f(k,w;b_{n-1}) = (\beta+w)^{\alpha+k}\left(\frac{U(k,b_{n-1})}{(\beta+b_{n-1})^{\alpha+k}} + \int_w^{b_{n-1}} dz \frac{c_2}{(n-k)(\beta+z)^{\alpha+k}}\right.$$
$$\left. + \int_w^{b_{n-1}} dz \frac{(\alpha+k)V(k+1,z)}{(\beta+z)^{\alpha+k+1}}\right) \quad (3.15)$$

then

either b_k is the unique positive root of
$f(k,b_k;b_{n-1}) - \psi_2(k,b_k) = 0$ or for all $w \geq 0$, (3.16)
$f(k,w;b_{n-1}) - \psi_2(k,w) \leq 0$ and $b_k = 0$

Proof: Suppose X satisfies $k(X) = k$. From the fact

$$P^{X,\alpha+k,\beta+w}(\tau_{k+1}\varepsilon dt) = \left(\int_0^\infty d\lambda\ f(\lambda,\alpha+k,\beta+w)\frac{1}{(n-k)\lambda}\ e^{-(n-k)\lambda t}\right)dt$$

it follows by a simple inductive calculation for $k = 0,1,\ldots, n-1$,

$$U(k,w) = c_1 n + \frac{c_2(\beta+w)a_{n-k}}{(n-k)!(\ +k-1)} + \frac{(\beta+w)^2}{(\alpha+n-1)(\alpha+k-1)^2(\alpha+k-2)} \quad (3.17)$$

where

$$a_1 = 1 \qquad a_j = ja_{j-1} + (j-1)! \quad (3.18)$$

One can easily verify for $k = 0,1,\ldots, n-1$

$$AU(k,w) + c_2 = 0 \qquad w \geq 0 \quad (3.19)$$

Let

$$H(k,w) = \frac{\beta+w}{(n-k)(\alpha+k)}\ (A\psi_2(k(w)+c_2) \quad (3.20)$$

From (3.4),

$$H(k,w) = c_1 + \frac{c_2(\beta+w)}{(n-k)(\alpha+k)} - \frac{(\beta+w)^2}{(\alpha+k)^2(\alpha+k-1)^2} \quad (3.21)$$

For large w, $H(k,w) < 0$. Also a quick calculation shows $H(k,w)$ has at most one positive root.

By (3.2) and (3.17) (3.22)

$$U(k,w)-\psi_2(k,w)=(n-k)\left(c_1+\frac{c_2(\beta+w)a_{n-k}}{(n-k)!(n-k)(\alpha+k-1)}-\frac{(\beta+w)^2}{(\alpha+n-1)(\alpha+k-1)^2(\alpha+k-2)}\right)$$

We establish the theorem by backwards induction on k. For $k = n - 1$, by (3.21) and (3.22)

$$U(n-1,w)-\psi_2(n-1,w)=\frac{\alpha+n-1}{\alpha+n-3}H(n-1,w)-\frac{2c_1}{\alpha+n-3}-\frac{c_1(\beta+w)}{(\alpha+n-2)(\alpha+n-3)} \tag{3.23}$$

Thus if b_{n-1} is defined according to (3.14) and $V(n-1,w)$ by (3.13), $V(n-1,w)$ satisfies the regularity conditions, and by (3.19) and (3.23), satisfies the variational inequality (2.6)-(2.8). Also

$$V(n-1,w) - \psi_2(n-1,w) \le V(n,w) - \psi_2(n,w) \qquad w \ge 0.$$

We assume for k+1 that $V(k+1,w)$ is given by (3.13), satisfies the regularity conditions and the variational inequality (2.6)-(2.8). Also

$$0 \le b_{k+1} \le b_{k+2} \le \cdots \le b_{n-1}$$

and for $b_{k+1} \le w \le b_{n-1}$

$$V(k+1,w) - \psi_2(k+1,w) \le V(k+2,w) - \psi_2(k+2,w) \tag{3.24}$$

By (3.22)

$$U(k,w)-\psi_2(k,w)=U(k+1,w)-\psi_2(k+1,w)-\frac{c_2(\beta+w)a_{n-k-1}}{(n-k-1)!(\alpha+k-1)(\alpha+k-2)}$$

(3.25)

$$-\frac{(n-k-1)(\beta+w)^2}{(\alpha+n-1)(\alpha+k)^2(\alpha+k-1)^2(\alpha+k-2)}+c_1+\frac{c_2(\beta+w)}{(n-k)!(\alpha+k-1)}-\frac{(\beta+w)^2}{(\alpha+n-1)(\alpha+k)^2(\alpha+k-1)}$$

Again by (3.22), one can easily verify

$$c_1+\frac{c_2(\beta+w)}{(n-k)(\alpha+k-1)}-\frac{(\beta+w)^2}{(\alpha+n-1)(\alpha+k)^2(\alpha+k-1)}\le U(n-1,w)-\psi_2(n-1,w) \tag{3.26}$$

Hence, if $b_{n-1} = 0$

$$U(k,w) - \psi_2(k,w) \le 0 \qquad w \ge 0$$

and $b_k = 0$. So if $V(k,w)$ is defined by (3.13), it satisfies the regularity condition and the variational inequality (2.6)-(2.8). (3.24) is obviously true.

If $b_{n-1} > 0$, then by (3.25) and (3.26),

$$U(k,b_{n-1}) - \psi_2(k,b_{n-1}) \le U(k+1,b_{n-1}) - \psi_2(k+1,b_{n-1}) \quad (3.27)$$

From (3.13) and (3.20) it follows that for $0 \le w \le b_{n-1}$

$$\frac{f(k,w;b_{n-1})-\psi_2(k,w)}{(\beta+w)^{\alpha+k}} = \frac{U(k,b_{n-1})-\psi_2(k,b_{n-1})}{(\beta+b_{n-1})^{\alpha+k}} + \int_w^{b_{n-1}} dz \frac{\alpha+k}{(\beta+z)^{\alpha+k+1}} H(k,z)$$
$$+\int_w^{b_{n-1}} dz \frac{\alpha+k+1}{(\beta+z)^{\alpha+k+2}} (V(k+2,z)-\psi_2(k+2,z)) \quad (3.28)$$

Also by (3.13) with k+1 in place of k, we have for w, $b_{k+1} \le w \le b_{n-1}$,

$$\frac{V(k+1,w)-\psi_2(k+1,w)}{(\beta+w)^{\alpha+k+1}} = \frac{U(k+1,b_{n-1})-\psi_2(k+1,b_{n-1})}{(\beta+b_{n-1})^{\alpha+k+1}} + \int_w^{b_{n-1}} dz \Big(\frac{\alpha+k+1}{(\beta+z)^{\alpha+k+2}} H(k+1,z)\Big)$$
$$+\int_w^{b_{n-1}} dz \frac{\alpha+k+1}{(\beta+z)^{\alpha+k+2}} (V(k+2,z)-\psi_2(k+2,z)) \quad (3.29)$$

Integrating by parts and using (3.29) yields for w, $b_{k+1} \le w \le b_{n-1}$

$$\int_w^{b_{n-1}} dz \frac{\alpha+k}{(\beta+z)^{\alpha+k+1}} (V(k+1,z)-\psi_2(k+1,z)) = \frac{V(k+1,w)-\psi_2(k+1,w)}{(\beta+w)^{\alpha+k}}$$
$$- \frac{U(k+1,b_{n-1}-\psi_2(k+1,b_{n-1})}{(\beta+b_{n-1})^{\alpha+k}} \quad (3.30)$$
$$-\int_w^{b_{n-1}} dz \frac{\alpha+k+1}{(\beta+z)^{\alpha+k+1}} H(k+1,z) + \int_w^{b_{n-1}} dz \frac{\alpha+k+1}{(\beta+z)^{\alpha+k+1}} [(V(k+1,z)-\psi_2(k+1,z))$$
$$-(V(k+2,z)-\psi_2(k+2,z))]$$

Substituting (3.30) into (3.28) gives for $b_{k+1} \le w \le b_{n-1}$,

$$\frac{f(k,w;b_{n-1})-\psi_2(k,w)}{(\beta+w)^{\alpha+k}} = \frac{U(k,b_{n-1})-\psi_2(k,b_{n-1})}{(\beta+b_{n-1})^{\alpha+k}} - \frac{U(k+1,b_{n-1})-\psi_2(k+1,b_{n-1})}{(\beta+b_{n-1})^{\alpha+k}} \tag{3.31}$$

$$+\frac{V(k+1,w)-\psi_2(k+1,w)}{(\beta+w)^{\alpha+k}} + \int_w^{b_{n-1}} dz \frac{1}{(\beta+z)^{\alpha+k+1}}((\alpha+k)H(k,z)-(\alpha+k+1)H(k+1,z))$$

$$+\int_w^{b_{n-1}} dz \frac{\alpha+k+1}{(\beta+z)^{\alpha+k+1}}(V(k+1,z)-\psi_2(k+1,z))-(V(k+2,z)-\psi_2(k+2,z)))$$

By (3.21)

$$(\alpha+k)H(k,z) - (\alpha+k+1)H(k+1,z) < 0$$

and hence by (3.27), (3.16) with k+1 in place of k, and (3.24),

$$f(k,w;b_{n-1}) - \psi_2(k,w) \le 0 \qquad b_{k+1} \le w \le b_{n-1}.$$

So if b_k is defined according to (3.16), then $b_k \le b_{k+1}$.

Clearly if $V(k,w)$ is defined by (3,13), it satisfies the regularity conditions. Also for $w \ge b_k$

$$AV(k,w) + c_2 = 0$$

and

$$V(k,w) \le \psi_2(k,w).$$

For $0 \le w \le b_k$, (3.28) implies $H(k,w) > 0$ which in turn shows

$$AV(k,w) + c_2 \ge 0$$

Thus $V(k,w)$ satisfies the variational inequality (2.6)-(2.8).

Finally to complete the induction and the proof of the theorem, note that (3.13), (3.27) and (3.31) implies

$$V(k,w) - \psi_2(k,w) \le V(k+1,w) - \psi_2(k+1,w) \qquad b_k \le w \le b_{n-1}.$$

IV. SOLUTION OF THE VARIATIONAL INEQUALITY, GENERAL CASE

We wish to solve for $k = 0,1,\ldots, n-1$

$$AV(k,w) + b(k,w) \ge 0 \tag{4.1}$$

$$V(k,w) \le \psi(k,w) \tag{4.2}$$

and

$$(AV(k,w)+b(k,w))(\psi(k,w)-V(k,w)) = 0 \tag{4.3}$$

A solution must meet the requirements for $k = 0,1,\ldots, n-1$

$$\left.\begin{array}{l} V(k,w) \text{ continuous in } w, \text{ and} \\ D_w^+V(k,w) \text{ exists and is continuous} \\ \text{except for a finite number of} \\ \text{jump discontinuities.} \end{array}\right. \tag{4.4}$$

We assume the following about $b(k,w)$ and $\psi(k,w)$:

(A) 1. For $k = 0,1,\ldots, n$, $b(k,w)$ and $\psi(k,w)$ are continuous in w.

2. For $k = 0,1,\ldots, n$

$b(k;w)$ is bounded

and

$|\psi(k,w)| \leq \mathcal{C}(1+w)^N$, N positive.

3. $D_w^+\psi(k,w)$ exists, is continuous except for a finite number of jump discontinuities, and

$|D_w^+\psi(k,w)| \leq \mathcal{C}(1+w)^N$, N positive.

(B) 1. $b(n,w) > 0$, $w > 0$

2. For all $k = 0,1,\ldots, n-1$ either $A\psi(k,w) + b(k;w) > 0$ for w large or $A\psi(k,w) + b(k,w) < 0$ for w large.

(C) 1. For $k = 0,1,\ldots, n$, $b(k,w)$ and $\psi(k,w)$ are piecewise analytic.

Note that condition (A) imply that if $V(k,w)$ satisfies (4.1)-(4.3) and (4.4) then Theorems 1 and 2 are valid. (See Remark 2.3.)

Before proceeding, we need the following:

Lemma 5. Suppose $U(k+1,w)$ has been define for $w \geq 0$ and is continuous with

$$|U(k+1,w)| \le \mathcal{C}(1+w)^N, \quad N \text{ positive} \tag{4.5}$$

Let

$$\tau_{k+1} = \inf\{t: k(t)=k+1\}$$

and for X satisfying $k(X) = k$ define

$$U(k,w) = E^{X,\alpha+k,\beta+w}[U(k+a,w_w(\tau_{k+1}))+\int_0^{\tau_{k+1}} b(k,w_w(s))ds]$$

Then $U(k,w)$ is continuous, $D_w^+U(k,w)$ exists, and

$$|U(k,w)| \le \mathcal{C}(1+w)^N \tag{4.6}$$

$$|D_w^+U(k,w)| \le \mathcal{C}(1+w)^N, \quad N \text{ positive} \tag{4.7}$$

and

$$AU(k,w) + b(k,w) = 0 \tag{4.8}$$

Proof: Note that

$$P^{X,\alpha+k,\beta+w}(\tau_{k+1}\varepsilon dt) = (\int_0^\infty d\lambda f(\lambda,\alpha+k,\beta+w)\frac{1}{(n-k)\lambda}e^{-(n-k)\lambda t})dt$$

and hence (4.5) implies (4.6) and if (4.8) is valid, (4.5) and (4.6) imply (4.7). Clearly $U(k,w)$ is continuous so we have to prove (4.8).

We show

$$\lim_{h\downarrow 0} \frac{U(k,w+h)-U(k,w)}{h} = -\frac{1}{n-k}b(k,w) + \frac{\alpha+k}{\beta+w}(U(k,w)-U(k+1,w))$$

Define

$$\tau_h = \inf\{t: w(t)=h\}$$

Now on the set $(\tau_h<\tau_{k+1})$

$$U(k,w+h) = U(k,w_w(\tau_h)) \quad \text{a.s.} \quad P^{X,\alpha+k,\beta+w} \tag{4.9}$$

Define $\phi_{\tau_h} : \Omega \to \Omega$ by $\phi_{\tau_h}(\omega)(t) = \omega(\tau_h(\omega)+t)$. By the strong Markov property, on the set $(\tau_h<\tau_{k+1})$,

$$U(k,w_w(\tau_h)) = E^{X,\alpha+k,\beta+w}[U(k+1,w_{w+h}(\tau_{k+1}(\phi_{\tau_h}))(\phi_{\tau_h})) \tag{4.10}$$

$$+\int_0^{\tau_{k+1}(\phi_{\tau_h})} b(k,w_{w+h}(s)(\phi_{\tau_h}))ds \,|\, \mathcal{F}_{\tau_h}]\ \text{a.s. } P^{X,\alpha+k,\beta+w}$$

Also on $(\tau_n < \tau_{k+1})$, a.s. $P^{X,\alpha+k,\beta+w}$

$$\tau_h = \frac{h}{n-k},\quad \tau_{k+1}(\phi_{\tau_h}) = \tau_{k+1} - h/n-k$$

$$w_{w+h}(\tau_{k+1}(\phi_{\tau_h}))(\phi_{\tau_h}) = w + (n-k)\tau_{k+1} \tag{4.11}$$

and $w_{w+h}(s)(\phi_{\tau_h}) = w + (n-k)\tau_{k+1},\quad s \le \tau_{k+1}(\phi_{\tau_h})$

Hence by (4.9)-(4.11)

$$U(k,w+h)P^{X,\alpha+k,\beta+w}(\tau_h<\tau_{k+1}) = E^{X,\alpha+k,\beta+w}[(\tau_h<\tau_{k+1})(U(k+1,w_w(\tau_{k+1})) + \int_{h/n-k}^{\tau_{k+1}} b(k,w_w(s))ds)]$$

Thus

$$U(k,w+h)-U(k,w) = -E^{X,\alpha+k,\beta+w}[(\tau_h<\tau_{k+1})\int_0^{h/n-k} b(k,w_w(s))ds]$$

$$+ E^{X,\alpha+k,\beta+w}[(\tau_h \ge \tau_{k+1})(U(k,w+h)-U(k+1,w_w(\tau_{k+1}))]$$

$$- E^{X,\alpha+k,\beta+w}[(\tau_h \ge \tau_{k+1})\int_0^{\tau_{k+1}} b(k,w_w(s))ds]$$

$$= h\left(-\frac{1}{n-k}\, b(k,w) + \frac{\alpha+k}{\beta+w}(U(k,w)-U(k+1,w))\right) + o(h).$$

Here we have used the facts

$$P^{X,\alpha+k,\beta+w}(\tau_h<\tau_{k+1}) = 1 + O(h)$$

and

$$P^{X,\alpha+k,\beta+w}(\tau_h \ge \tau_{k+1}) = \frac{\alpha+k}{\beta+w}\, h + o(h)$$

Theorem 6. For $\alpha > 1$, under the assumptions (A), (B), and (C) a solution of the variational inequality (4.1)-(4.3) exists satisfying (4.4).

Proof: $V(n,w)$ has been defined as $\psi(n,w)$. For X with $k(X) = k$, let

$$U(k,w)=E^{X,\alpha+k,\beta+w}[U(k+1,w_w(\tau_{k+1}))+\int_0^{\tau_{k+1}} b(k,w_w(s))ds] \tag{4.12}$$

where $U(n,w) = \psi(n,w)$.

Define for $k = 0,1,\ldots, n-1$,

$$V(k,w) = \begin{cases} U(k,w) & w \geq a_k^{-1} \\ g(k,w;a_k^{-1}) & a_k^0 \leq w \leq a_k^{-1} \\ \psi(k,w) & b_k^j \leq w \leq a_k^{j-1} \\ f(k,w;b_k^j) & a_k^j \leq w \leq b_k^j \end{cases} \tag{4.13}$$

where

$$g(k,w;b) = \begin{cases} (\beta+w)^{\alpha+k}\left(\dfrac{U(k,b)}{(\beta+b)^{\alpha+k}} + \displaystyle\int_w^b dz\, \dfrac{b(k,z)}{(n-k)(\beta+z)^{\alpha+k}}\right. & b < \infty \\ \qquad\left. + \displaystyle\int_w^b dz\, \dfrac{\alpha+k}{(\beta+z)^{\alpha+k}}\, V(k+1,w)\right) & \\ \infty & b = \infty \end{cases} \tag{4.14}$$

and

$$f(k,w;b)=(\beta+w)^{\alpha+k}\left(\frac{\psi(k,b)}{(\beta+b)^{\alpha+k}}+\int_w^b dz\, \frac{b(k,z)}{(n-k)(\beta+z)^{\alpha+k}} + \int_w^b dz\, \frac{\alpha+k}{(\beta+z)^{\alpha+k+1}} V(k+1,z)\right) \tag{4.15}$$

Also if

$$R(k,w) = \frac{A\psi(k,w)+b(k,w)}{n-k} + \frac{\alpha+k}{\beta+w}(V(k+1,w)-\psi(k+1,w)) \tag{4.16}$$

and if

$$a_k^{-3} = \max\{a_j^{-1}: j = k+1,\ldots, n\}, \text{ where } a_n^{-1} = 0$$

and

$$a_k^{-2} = \sup\{w: U(k,w)-\psi(k,w)>0\}$$

Define

$$a_k^{-1} = \max\{a_k^{-3}, a_k^{-2}\}$$

and

$$a_k^0 = \begin{cases} \sup\{w\geq 0:\ w\leq a_k^{-1} \text{ and } g(k,w;a_k^{-1})-\psi(k,w)>0\} & a_k^{-1} < \infty \\ \infty & a_k^{-1} = \infty \end{cases}$$

$b_k^1 = \sup\{w\geq 0:\ w\leq a_k^0,\ R(k,w)<0$ and w is a continuity point of $R(k,w)\}$.

$a_k^1 = \sup\{w\geq 0:\ w\leq b_k^1$ and $f(k,w;b_k^1)-\psi(k,w)>0\}$

$\cdots\cdots\cdots\cdots\cdots\cdots\cdots\cdots\cdots\cdots$

$b_k^j = \sup\{w\geq 0:\ w\leq a_k^{j-1},\ R(k,w)<0$ and w is a continuity point of $R(k,w)\}$.

$a_k^j = \sup\{w\geq 0:\ w\leq b_k^j$ and $f(k,w;b_k^j)-\psi(k,w)>0\}$.

Again the proof of the theorem is by backward induction on k. Note that conditions (A) imply by Lemma 5 that $U(k,w)$ is continuous, $D_w^+U(k,w)$ is continuous, and

$$|U(k,w)| \leq \mathscr{C}(1+w)^N, \quad N \text{ positive}$$

and $$AU(k,w) + b(k,w) = 0 \quad w \geq 0 \tag{4.17}$$

Note that for $k(X) = k$, by (4.12) and Dynkin's formula,

$$U(k,w)-\psi(k,w)=E^{X,\alpha+k,\beta+w}[U(k+1,w_w(\tau_{k+1}))-\psi(k+1,w_w(\tau_{k+1})) + \int_0^{\tau_{k+1}}(A\psi(k,w_w(s))+b(k,w_w(s)))ds] \tag{4.18}$$

Now $\psi(n,w) = V(n,w)$. By (4.18), if $A\psi(n-1,w) + b(n-1,w) > 0$ for large w, then $a_{n-1}^{-1} = a_{n-1}^0 = \infty$ and $b_{n-1}^1 < \infty$. If $A\psi(n-1,w) + b(n-1,w) < 0$ for large w, $a_{n-1}^{-1} = a_{n-1}^{-2} < \infty$ and since $U(n-1,w) = g(n-1,w;a_{n-1}^{-1})$ for $w \leq a_{n-1}^{-1}$, $a_{n-1}^{-1} = a_{n-1}^0$. Also a simple exercise shows from (4.17)

$$\frac{U(n-1,w)-\psi(n-1,w)}{(\beta+w)^{\alpha+n-1}} = \int_w^{a_{n-1}^0} dz\, \frac{R(n-1,z)}{(\beta+z)^{\alpha+n-1}} \tag{4.19}$$

which implies $b^1_{n-1} \le a^0_{n-1}$ with equality only if $a^0_{n-1} = 0$.

Condition (C) implies that the zeroes of $R(n-1,w)$ are isolated and again a simple exercise using (4.15) shows

$$\frac{f(n-1,w;b^1_{n-1})-\psi(n-1,w)}{(\beta+w)^{\alpha+n-1}} = \int_w^{b^1_{n-1}} dz \, \frac{R(n-1,z)}{(\beta+z)^{\alpha+n-1}} \tag{4.20}$$

Thus we must have $a^1_{n-1} \le b^1_{n-1}$ with equality only if $b^1_{n-1} = 0$. Iterating the argument yields $a^j_k \le b^j_k \le a^{j-1}_k$ with equality holding only in the case of zero. Moreover condition (C) insures the construction of a^j_{n-1}, b^j_{n-1} will stop at a finite stage.

Clearly $V(n-1,w)$ is continuous, $D^+_w V(n-1,w)$ exists, in continuous except for a finite number of jump discontinuities, and moreover $V(n-1,w)$ must be piecewise analytic. For $w \ge a^{-1}_{n-1} = a^0_{n-1}$, by (4.17)

$$\begin{aligned} &AV(n-1,w) + b(n-1,w) = 0 \\ &\text{and} \quad V(n-1,w) \le \psi(n-1,w) \end{aligned} \tag{4.21}$$

For w, $a^j_{n-1} \le w \le b^j_{n-1}$, by (4.15) we also know (4.21) is true. Since the zeros of $R(n-1,w)$ are isolated, by (4.20) with b^j_{n-1} in place of b^1_{n-1}, we must have for $w < a^j_{n-1}$ but near, $R(n-1,w) > 0$, that is by (4.16),

$$AV(n-1,w) + b(n-1,w) \ge 0 \qquad b^{j-1}_{n-1} \le w \le a^j_{n-1}$$

which show (4.1)-(4.3) for $V(n-1,w)$ defined according to (4.13).

We assume that for $k+1$, $V(k+1,w)$ defined by (4.13) satisfies (4.1)-(4.3) and (4.4). Also we assume that $V(k+1,w)$ is piecewise analytic. Hence (C) implies

(D) the zeroes of $R(k,w)$ are isolated.

By (B.2) and (4.19), either a^{-3}_k and a^{-2}_k are either both finite or both infinite. In the case they are both finite, clearly $a_k^0 \leq a_k^{-1}$. Again by (4.14), if $0 < a_k^0 < \infty$

$$\frac{g(k,w;a_k^0)-\psi(k,w)}{(\beta+w)^{\alpha+k}} = \int_w^{a_k^0} dz \frac{R(k,z)}{(\beta+z)^{\alpha+k}} < 0$$

for $w < a_k^0$ and hence $b_k^1 < a_k^0$. If a_k^{-3} and a_k^{-2} are both infinite, a_k^0 is infinite. This occurs in the case $A\psi(k,w) + b(k,w) > 0$ for large w. Thus b_k^1 must be finite. Arguing as before yields

$$a_k^0 \geq b_k^1 \geq a_k^1 \geq \ldots \geq b_k^j \geq a_k^j$$

with equality holding only in the case of zero. Also the construction of a_k^j, b_k^j stops after a finite number of steps.

By essentially the same argument that was used in the case of n-1, we can show that V(k,w) given by (4.13) satisfies (4.1)-(4.3) and (4.4) and moreover V(k,w) is piecewise analytic. This completes the proof.

Remark 4.1. Condition (C) is very strong, but it may be replaced by any other condition that ensures (D).

References

[1] Anderson, R. F., and Ghosh, J. K., Optimal Stopping in a Reliability Problem, to appear.

[2] Anderson, R. F., and Winkler, W. E., Optimal Stopping in a Reliability Problem, Part II, to appear.

[3] Dynkin, E. B., Markov Processes, Vol. 1, (English Edition), Springer-Verlag, Berlin, 1965.

ON THE HAMILTON-JACOBI APPROACH FOR THE OPTIMAL CONTROL OF DIFFUSION PROCESSES WITH JUMPS

A. Bensoussan
IRIA-Laboria
University Paris IX

INTRODUCTION

In this article, we study the Hamilton-Jacobi Bellman equation corresponding to the stochastic control problem for a diffusion process with jumps, on a bounded domain. The equation is a non linear integer differential equation. We then construct the probability measure of the controlled process and identify the solution of the equation with the infimum of a pay off functional.

We thus develop an approach which is quite analogous to the classical one for diffusions, as described in the book of FLEMING-RISHEL [4]. A different approach, not relying on analytic results, has been developed by BISMUT [3] and LEPELTIER-MARCHAL [8]. At any rate, their approach is confined to the control of the process in the whole space, situation which is somewhat simpler, because we are dealing in this problem with a non local operator.

The martingale problem corresponding to diffusion with jumps has been developed by STROOCK [9] (cf. also KOMATSU [6], LEPELTIER-MARCHAL [7], GIKHMAN-SKOROKHOD [5]). We have chosen here to develop an approach which is well adapted to our problem (in particular for the stopped process and its approximation). It is also self contained, which should make the reading easier, but does not give the most general results on the martingale problem. However, it is sufficient for our purpose.

ISBN 0-12-268380-3

The plan is as follows:

1 - THE HAMILTON-JACOBI EQUATION

1.1. Assumptions and Notation

Let $\mathcal{O}$ be an open subset of R^n such that :

$$\mathcal{O} \text{ bounded, } \mathcal{O} \text{ regular.} \tag{1.1}$$

Let $a_{ij}(x)$, $a_i(x)$ be functions such that :

$$\left| \begin{array}{l} a_{ij} \in W^{1,\infty}(R^n), \ a_i \in L^\infty(R^n) \\ \\ \Sigma\, a_{ij}\ \xi_i\ \xi_j \geq \alpha |\xi|^2, \forall \xi \in R^n, \ \alpha > 0 \end{array} \right. \tag{1.2}$$

we define the differential operator[1]:

$$A = - \sum_{ij} \frac{\partial}{\partial x_i} a_{ij} \frac{\partial}{\partial x_j} + \sum_i a_i \frac{\partial}{\partial x_i} . \tag{1.3}$$

Let now B be a linear operator such that :

---

(1) By virtue of the regularity of the a_{ij}, we may assume $a_{ij}=a_{ji}$ without loss of generality (modifying the a_i's).

$$B \in \mathcal{L}(W_o^{1,p}(\mathcal{O}) \; ; \; W^{-1,p}(\mathcal{O})) \; \forall p \geq 2, \tag{1.4}$$

satisfying :

$$\left| \begin{array}{l} B = B_r^1 + B_r^2 \; ; \\ B_r^1 \in \mathcal{L}(W_o^{1,p}(\mathcal{O}) \; ; \; W^{-1,p}(\mathcal{O})) \\ B_r^2 \in \mathcal{L}(W_o^{1,p}(\mathcal{O}) \; ; \; L^p(\mathcal{O})) \\ ||B_r^1|| = \sigma(r) \to 0, \text{ as } r \to 0, \end{array} \right. \tag{1.5}$$

and :

$$\forall \varphi \in H_o^1(\mathcal{O}), \; \langle B\varphi, \varphi \rangle \leq M ||\varphi|| \; |\varphi| . \tag{1.6}$$

Let now H be a non linear operator such that :

$$H : W_o^{1,p}(\mathcal{O}) \to L^p(\mathcal{O}), \; \forall p \geq 2, \tag{1.7}$$

$$\forall \varphi_1, \varphi_2 \in H_o^1(\mathcal{O}), \text{ then}$$

$$|H(\varphi_1) - H(\varphi_2)|_{L^2(\mathcal{O})} \leq C_o ||\varphi_1 - \varphi_2||_{H_o^1(\mathcal{O})} .$$

The pair B, H satisfies the property :

$$\left| \begin{array}{l} \forall \varphi_1, \varphi_2 \in H_o^1(\mathcal{O}), \; \forall k \in R, \; k \geq 0, \\ \langle B(\varphi_1 - \varphi_2) + H(\varphi_1) - H(\varphi_2), \; (\varphi_1 - \varphi_2 - k)^+ \rangle \\ \qquad \leq C_1 ||(\varphi_1 - \varphi_2 - k)^+|| \; |(\varphi_1 - \varphi_2 - k)^+| \\ \qquad + \gamma_1 k \int_{\mathcal{O}} (\varphi_1 - \varphi_2 - k)^+ \, dx \end{array} \right. \tag{1.8}$$

where C_1, γ_1 are non negative constants independent from φ_1, φ_2, k.[1]

---

(1) We note that $(\varphi_1 - \varphi_2 - k)^+ \in H_o^1(\mathcal{O})$.

Let $a_o(x)$ be a function such that :

$$a_o \in L^\infty(R^n) \ , \ a_o(x) \geq \gamma_o, \ \gamma_o > \gamma_1. \tag{1.9}$$

Let now f be such that :

$$f \in L^p(\mathcal{O}) \ , \quad p > n. \tag{1.10}$$

We consider the problem : to find u such that :

$$Au - Bu - H(u) + a_o u = f, \ u \in W_o^{1,p}(\mathcal{O}). \tag{1.11}$$

We will refer to equation (1.11) as the Hamilton-Jacobi equation.

1.2. Solution of the Hamilton-Jacobi Equation

Our objective is to prove the following :

Theorem 1.1. *Under the assumptions* (1.1), (1.2), (1.4), (1.5), (1.6), (1.7), (1.8), (1.9), (1.10), *there exists one and only one solution of* (1.11).

▨

We will need a series of Lemmas :

Lemma 1.1. *Let* λ *fixed but sufficiently large, then for* $f \in L^2(\mathcal{O})$ *there exists one and only one solution of* :

$$Au_\lambda - Bu_\lambda - H(u_\lambda) + a_o u_\lambda + \lambda u_\lambda = f, \ u_\lambda \in H_o^1(\mathcal{O}). \tag{1.12}$$

Proof. We set $\lambda = \lambda_o + \lambda_1$, and we define on $H_o^1(\mathcal{O})$, a bilinear form :

$$a(v,w) = \langle Av - Bv + a_o v + \lambda_o v, w \rangle$$

which is continuous and satisfies, using (1.6) :

$$\begin{aligned} a(v,v) &\geq \alpha||v||^2 - M'||v|| \ |v| + \lambda_o|v|^2 \\ &\geq \beta||v||^2 \ , \ \forall v \in H_o^1(\mathcal{O}) \end{aligned} \tag{1.13}$$

for a convenient choice of λ_o.

Therefore for w given, $w \in H_o^1(\mathcal{O})$, we can find one and only one $z = S(w)$, such that :

$$Az - Bz + a_o z + \lambda_o z + \lambda_1 z = f + H(w), \; z \in H_o^1(\mathcal{O}). \qquad (1.14)$$

We thus have defined a mapping $S : H_o^1(\mathcal{O}) \to H_o^1(\mathcal{O})$.

It is easy to check that for a convenient choice of λ_1, then one can prove that S is a contraction. Indeed, if w_1, w_2 are two elements of $H_o^1(\mathcal{O})$ and z_1, z_2 the corresponding solutions of (1.14) then we have :

$$a(z_1-z_2, z_1-z_2) + \lambda_1|z_1-z_2|^2 = (H(w_1) - H(w_2), z_1-z_2)$$
$$\leq M\|w_1-w_2\| \, |z_1-z_2|$$

hence :

$$\beta\|z_1-z_2\|^2 + \lambda_1|z_1-z_2|^2 \leq M\frac{h}{2}\|w_1-w_2\|^2 + \frac{M}{2h}|z_1-z_2|^2$$

for any $h > 0$. Taking $h < \frac{2\beta}{M}$, and choosing $\lambda_1 \geq \frac{M}{2h}$, we clearly obtain that S is a contraction. Hence the existence and uniqueness of the solution u_λ of (1.12). ▨

Lemma 1.2. Let r sufficiently small. For $g \in W^{-1,p}(\mathcal{O})$, then there exists one and only one solution of the equation :

$$Az - B_r^1 z + a_o z = g \;, \quad z \in W^{1,p}(\mathcal{O}). \qquad (1.15)$$

Proof. It follows from a fixed point argument. Let $w \in W_o^{1,p}(\mathcal{O})$ given, then $B_r^1 w \in W^{-1,p}(\mathcal{O})$. Therefore there exists one and only one solution ψ of :

$$A\psi + a_o\psi = g + B_r^1 w, \; \psi \in W_o^{1,p}(\mathcal{O}).$$

We thus have defined a mapping $\Sigma_r : W_o^{1,p}(\mathcal{O}) \to W_o^{1,p}(\mathcal{O})$, and clearly:

$$\|\Sigma_r w_1 - \Sigma_r w_2\|_{W_o^{1,p}(\mathcal{O})} \leq C\|B_r^1\| \, \|w_1 - w_2\|$$
$$= C\,\sigma(r)\|w_1 - w_2\|_{W_o^{1,p}(\mathcal{O})} .$$

For r such that $C\,\sigma(r) < 1$, the desired result follows. ▨

Proof of Theorem 1.1. Let us first prove that the solution of (1.12) belongs to $W_o^{1,p}(\mathcal{O})$, if $f \in L^p(\mathcal{O})$, $p \geq 2$ not necessarily stisfying (1.10). Indeed we can write (1.12) as :

$$Au_\lambda - B_r^1 u_\lambda + (a_o + \lambda)u_\lambda = f + B_r^2 u_\lambda + H(u_\lambda).$$

Since $u_\lambda \in H_o^1(\mathcal{O})$, $f + B_r^2 u_\lambda + H(u_\lambda) \in L^2(\mathcal{O})$.

But $L^2(\mathcal{O}) \subset W^{-1,p_o}(\mathcal{O})$, with $\frac{1}{p_o} = \frac{1}{2} - \frac{1}{n}$.

From lemma 1.2, it follows that $u_\lambda \in W_o^{1,p_o}(\mathcal{O})$. Then if $p > p_o$, $f + B_r^2 u_\lambda + H(u_\lambda) \in L^{p_o}(\mathcal{O})$.

But $L^{p_o}(\mathcal{O}) \subset W^{-1,p_1}(\mathcal{O})$, with $\frac{1}{p_1} = \frac{1}{p_o} - \frac{1}{n}$, hence $u_\lambda \in W_o^{1,p_1}(\mathcal{O})$. After a finite number of steps, one can see using a bootstrap argument that $u_\lambda \in W_o^{1,p}(\mathcal{O})$.

In particular, in case (1.10) is satisfied, we obtain that $u_\lambda \in C^o(\bar{\mathcal{O}})$.

Let now $w \in L^\infty(\mathcal{O})$, we can then solve the equation :

$$\begin{cases} Av - Bv - H(v) + a_o v + \lambda v = f + \lambda w, \\ v \in W_o^{1,p}(\mathcal{O}) \end{cases} \tag{1.16}$$

which has one and only one solution. Since $p > n$, $v \in C^o(\bar{\mathcal{O}})$, hence equation (1.16) defines a mapping $T_\lambda : L^\infty(\mathcal{O}) \to L^\infty(\mathcal{O})$.

Let us show that T_λ is a contraction. Let indeed $w_1, w_2 \in L^\infty(\mathcal{O})$ and $v_i = T_\lambda w_i$, $i=1,2$. We have :

$$A(v_1-v_2) - B(v_1-v_2) - H(v_1) + H(v_2) \tag{1.17}$$

$$+ a_o(v_1-v_2) + \lambda(v_1-v_2) = \lambda(w_1-w_2).$$

Let $k > 0$ to be chosen. We multiply (1.17) by $(v_1-v_2-k)^+$ and integrate in $\mathcal{O}$. We get :

$$<A(v_1-v_2-k)^+, (v_1-v_2-k)^+> - \{<B(v_1-v_2)$$

$$+ H(v_1) - H(v_2), (v_1-v_2-k)^+>\} + \int_{\mathcal{O}} (a_o+\lambda)(v_1-v_2-k)^{+2} dx$$

$$+ \int_{\mathcal{O}} (a_o+\lambda)k(v_1-v_2-k)^+ dx = \lambda \int_{\mathcal{O}} (w_1-w_2)(v_1-v_2-k)^+ dx$$

and using (1.8) it follows that :

$$\beta||(v_1-v_2-k)^+||^2 + \lambda_1|(v_1-v_2-k)^+|^2$$

$$\leq C_1||(v_1-v_2-k)^+|| \; |(v_1-v_2-k)^+| \tag{1.18}$$

$$+\int_{\mathcal{O}} (v_1-v_2-k)^+[\gamma_1 k - (\gamma_o+\lambda)k + \lambda||w_1-w_2||_{L^\infty}]dx,$$

where we have assumed that $\lambda = \lambda_o + \lambda_1$ with :

$$<Av, v> + (\gamma_o+\lambda_o)|v|^2 \geq \beta||v||^2 \qquad \forall v \in H_o^1(\mathcal{O}).$$

Now for :

$$k = \frac{\lambda||w_1-w_2||_{L^\infty}}{\gamma_o - \gamma_1 + \lambda}$$

the integral in the right hand side of (1.18) is zero. Clearly for a convenient choice of λ_1, depending only on the values of β and C_1, it follows from (1.18) that :

$$(v_1-v_2-k)^+ = 0$$

or :

$$v_1 - v_2 \leq \frac{\lambda}{\lambda + \gamma_o - \gamma_1} ||w_1 - w_2||_{L^\infty}. \tag{1.19}$$

We now multiply (1.17) by $(v_1-v_2+k)^-$. We obtain :

$$- \langle A(v_1-v_2+k)^-,(v_1-v_2+k)^-\rangle + \{\langle B(v_2-v_1)$$

$$+ H(v_2) - H(v_1),(v_2-v_1-k)^+\rangle\} - \int_{\mathcal{O}} (a_o+\lambda)(v_1-v_2+k)^{-2} \, dx$$

$$- \int_{\mathcal{O}} (a_o+\lambda)k \; (v_1-v_2+k)^- dx = \lambda \int_{\mathcal{O}} (w_1-w_2)(v_1-v_2+k)^- dx$$

where we have taken into account that :

$$(v_1-v_2+k)^- = (v_2-v_1-k)^+.$$

From the choice of k and reasoning as above, it follows that : $(v_1-v_2+k)^- = 0$, hence taking (1.19) into account, we have proven that :

$$||v_1-v_2||_{L^\infty} \leq \frac{\lambda}{\lambda + \gamma_o - \gamma_1} ||w_1-w_2||_{L^\infty} \tag{1.20}$$

and from (1.9) the contraction property is proven.

Since any fixed point of T_λ is a solution of (1.11) and conversely, the desired result is proven.

Remark 1.1. We have also proved that under assumptions (1.1), (1.2), (1.4), (1.5), (1.6), (1.7), (1.9) (but not (1.8)) and for $f \in L^p(\mathcal{O})$, $p \geq 2$ (but not (1.10)), there exists one and only one solution of :

$$Au - Bu - H(u) + (a_o+\lambda)u = f, \quad u \in W^{1,p}(\mathcal{O}) \tag{1.21}$$

provided that λ is sufficiently large. Moreover :

$$||u||_{W_o^{1,p}} \leq C_\lambda |f|_{L^p}. \tag{1.22}$$

1.3. Operators B and H

We now make explicit the operators B and H we are interested in. We need some further notation.

Let $m(dz)$ be a measure on $R^n-\{o\}$, such that :

$$\left| \begin{array}{l} m(dz) \text{ is a positive measure} \\ \int_{\{|z|\le 1\}} |z|^2 m(dz) + \int_{\{|z|>1\}} |z| m(dz) < \infty . \end{array} \right. \tag{1.23}$$

Let next $c_o(x,z)$ be a function defined on $R^n \times R^n$, such that :

$$\left| \begin{array}{l} c_o \ge 0,\ c_o \text{ measurable and bounded} \\ \left|\frac{\partial c_o}{\partial x_k}(x,z)\right| \le K \text{ (constant)}. \end{array} \right. \tag{1.24}$$

Let now $\mathcal{V}$ be a set (called the set of controls) , and functions :

$$\left| \begin{array}{ll} \ell, a_1 : R^n \times \mathcal{V} \to R, & \text{measurable and bounded} \\ h : R^n \times \mathcal{V} \to R^n & \text{measurable and bounded} \end{array} \right. \tag{1.25}$$

$$\left| \begin{array}{l} c_1 : R^n \times \mathcal{V} \times R^n \to R \text{ measurable and bounded} \\ c_1(x,v,z) \ge -1,\ |c_1(x,v,z)| \le C|z| . \end{array} \right. \tag{1.26}$$

Let $\varphi \in \mathcal{D}(R^n)$; we define :

$$B\varphi(x) = \int_{R^n} (\varphi(x+z) - \varphi(x) - z.\nabla\varphi\chi_{|z|\le 1}) c_o(x,z) m(dz) \tag{1.27}$$

$$H\varphi(x) = \operatorname*{Inf}_v \{\ell(x,v) + \nabla\varphi(x).h(x,v) - \varphi(x) a_1(x,v) \tag{1.28}$$

$$+ \int_{R^n} [\varphi(x+z) - \varphi(x) - z.\nabla\varphi\chi_{|z|\le 1}] c_o(x,z) c_1(x,v,z) m(dz)\}.$$

We are going to show the :

Lemma 1.3. The operators B and H satisfy properties (1.4), (1.5), (1.6), (1.7), (1.8).

Proof. Let us prove (1.5), which implies (1.4). We set for $r < 1$:

$$B_r^1 \varphi(x) = \int_{R^n} (\varphi(x+z) - \varphi(x) - z.\nabla\varphi(x))\chi_{|z|\leq r}\, c_o(x,z)m(dz)$$

$$B_r^2 \varphi(x) = \int_{R^n} (\varphi(x+z) - \varphi(x))\chi_{r<|z|}\, c_o\, m(dz)$$

$$- \nabla\varphi(x). \int_{R^n} z\, \chi_{r<|z|\leq 1}\, c_o\, m(dz).$$

We have :

$$B_r^1 \varphi(x) = \int_o^1 d\theta \int_{R^n} (\nabla\varphi(x+\theta z) - \nabla\varphi(x)).z\chi_{|z|\leq r}\, c_o(x,z)m(dz)$$

$$= \int_o^1 d\theta \int_o^\theta d\theta' \int_{R^n} \Sigma \frac{\partial^2\varphi}{\partial x_i \partial x_j}(x+\theta' z)z_i z_j \chi_{|z|\leq r}\, c_o(x,z)m(dz).$$

Let also $\psi \in \mathcal{D}(R^n)$. We obtain :

$$\int_{R^n} B_r^1\varphi(x)\ \psi(x)\ dx = \int_o^1 d\theta \int_o^\theta d\theta' \int_{R^n} dx \int_{R^n} m(dz)$$

$$\left[-\Sigma \frac{\partial\varphi}{\partial x_j}(x+\theta' z)z_i z_j (\frac{\partial c_o}{\partial x_i}(x,z)\psi(x) + \frac{\partial\psi}{\partial x_i}(x)c_o(x,z))\chi_{|z|\leq r} \right] \quad (1.29)$$

hence :

$$\left| \int_{R^n} B_r^1\varphi(x)\psi(x)dx \right| \leq C||\varphi||_{W^{1,p}(R^n)}\, ||\psi||_{W^{1,q}(R^n)} \int |z|^2 \chi_{|z|\leq r} m(dz)$$

where $\frac{1}{p} + \frac{1}{q} = 1$, and C does not depend on φ, ψ or r.

It follows from the estimates (1.29) that $B_r^1 \in \mathcal{L}(W_o^{1,p}(\mathcal{O})\ ;\ W^{-1,p}(\mathcal{O}))$ and :

$$||B_r^1|| \leq C \int |z|^2 \chi_{|z|\leq r}\, m(dz) \to 0 \text{ as } r \to 0.$$

Taking into account that :

$$\int_{\{r<|z|\}} m(dz) \leq \int_{\{|z|>1\}} |z|\, m(dz) + \frac{1}{r^2}\int_{\{|z|\leq 1\}} |z|^2 m(dz) < \infty$$

it is easy to check that $B_r^2 \in \mathcal{L}(W_o^{1,p}(\mathcal{O})\ ;\ L^p(\mathcal{O}))$.

Let us now prove (1.6). It is clearly sufficient to prove the same property with B_r^1 instead of B. But :

$$\varphi(x)(\varphi(x+z) - \varphi(x) - z.\nabla\varphi(x)) \leq \frac{1}{2}[\varphi^2(x+z) - \varphi^2(x) - z.\nabla\varphi^2(x)]$$

hence :

$$<\varphi,\ B_r^1\varphi> \leq \frac{1}{2}\int_{R^n} B_r^1\varphi^2(x)\, dx, \text{ say for } \varphi \in \mathcal{D}(R^n).$$

But formula (1.29) is valid for $\varphi \in \mathcal{D}(R^n)$ and $\psi \doteq 1$, by approximation. Changing φ into φ^2, we obtain :

$$\frac{1}{2}\int_{R^n} B_r^1\varphi^2(x)dx = \int_o^1 d\theta \int_o^\theta d\theta' \int_{R^n} dx \int_{R^n} m(dz)$$
$$\left[-\Sigma\varphi\frac{\partial\varphi}{\partial x_j}(x+\theta' z)z_i z_j \frac{\partial c_o}{\partial x_i}\right]\chi_{|z|\leq r} \qquad (1.30)$$
$$\leq M||\varphi||\ |\varphi|$$

hence (1.6).

Property (1.7) is easy to prove, using the last assumption (1.26). Let us prove (1.8). We first note that :

$$(B\varphi + H\varphi)(x) = \operatorname*{Inf}_v \{\ell(x,v) + \nabla\varphi(x).h(x,v) - \varphi(x)a_1(x,v)$$
$$+ \int_{R^n} [\varphi(x+z) - \varphi(x) - z.\nabla\varphi\chi_{|z|\leq 1}]c(x,v,z)m(dz)\}$$

where $c = c_o(1+c_1) \geq 0$. We note B_ε, H_ε the operators defined in the same way as B, H but with $m_\varepsilon(dz) = \chi_{|z|>\varepsilon} m(dz)$, instead of m(dz). One can check that :

$$\left| \begin{array}{l} H_\varepsilon \varphi \to H\varphi \text{ in } L^2(\mathcal{O}), \forall \varphi \in H_o^1(\mathcal{O}) \\ B_\varepsilon \varphi \to B\varphi \text{ in } H^{-1}(\mathcal{O}) \text{ weakly}, \forall \varphi \in H_o^1(\mathcal{O}) \end{array} \right. \tag{1.31}$$

hence it is sufficient to prove (1.8), with constants C_1, γ_1 not depending on ε.

But :

$$\begin{aligned} B_\varepsilon(\varphi_1-\varphi_2)(x) &+ H_\varepsilon(\varphi_1)(x) - H_\varepsilon(\varphi_2)(x) \leq \operatorname*{Inf}_v\{\ell + \nabla\varphi_2.h - \varphi_2 a_1 \\ &+ \int_{R^n} [\varphi_2(x+z) - \varphi_2(x) - z.\nabla\varphi_2\chi_{|z|\leq 1}]c\, m_\varepsilon(dz) \\ &+ \nabla(\varphi_1-\varphi_2-k).h - (\varphi_1-\varphi_2-k)a_1 + \int_{R^n} [\, (\varphi_1-\varphi_2-k)^+ \\ &\quad (x+z) - (\varphi_1-\varphi_2-k)(x) - z.\nabla(\varphi_1-\varphi_2-k)\chi_{|z|\leq 1}]c\,m_\varepsilon(dz) \\ &- ka_1\} - \operatorname*{Inf}_v\{\ell + \nabla\varphi_2.h - \varphi_2 a_1 + \int_{R^n} [\varphi_2(x+z) \\ &- \varphi_2(x) - z.\nabla\varphi_2\ \chi_{|z|\leq 1}]c\ m_\varepsilon(dz)\} \end{aligned}$$

hence :

$$\begin{aligned} (B_\varepsilon(\varphi_1-\varphi_2)(x) &+ H_\varepsilon(\varphi_1)(x) - H_\varepsilon(\varphi_2)(x))(\varphi_1-\varphi_2-k)^+ \\ &\leq (\varphi_1-\varphi_2-k)^+ \int_{R^n} [\, (\varphi_1-\varphi_2-k)^+(x+z) - (\varphi_1-\varphi_2-k)^+(x) \\ &- z.\nabla(\varphi_1-\varphi_2-k)^+\chi_{|z|\leq 1}]\ c_o m_\varepsilon(dz) + C[|\nabla(\varphi_1-\varphi_2-k)^+(x)| \\ &+ |(\varphi_1-\varphi_2-k)^+(x)|](\varphi_1-\varphi_2-k)^+ + k\gamma_1(\varphi_1-\varphi_2-k)^+(x) \\ &+ C(\varphi_1-\varphi_2-k)^+|\nabla(\varphi_1-\varphi_2-k)^+(x)| \int |z|^2\chi_{|z|\leq 1}\ m_\varepsilon(dz) \end{aligned}$$

$$+ C(\varphi_1-\varphi_2-k)^+(x) \int_{R^n} (\varphi_1-\varphi_2-k)^+(x+z)\chi_{|z|>1} m_\varepsilon(dz)$$

$$+ C \int_0^1 d\theta \int_{R^n} (\varphi_1-\varphi_2-k)^+(x)|\nabla(\varphi_1-\varphi_2-k)^+(x+\theta z)|$$

$$\chi_{|z|\le 1} |z|^2 m_\varepsilon(dz).$$

Integrating in x and using in particular an argument similar to the one used to prove (1.6), we obtain (1.8), which completes the proof of the Lemma. ▨

Remark 1.2. We see that γ_1 is any number such that $a_1^- \le \gamma_1$, $\forall x,v$.

If $a_1 \ge 0$, we can take $\gamma_1 = 0$. ▨

1.4. A Control Problem for a Distributed System

Let v(x) be a measurable mapping from $R^n \to \mathcal{V}$[1]. We introduce the notation :

$$A_v\varphi = A\varphi - \nabla\varphi.h(x,v(x)) + \varphi(x)\ a_1(x,v(x)) \tag{1.32}$$

$$B_v\varphi = \int_{R^n} [\varphi(x+z) - \varphi(x) - z.\nabla\varphi\chi_{|z|\le 1}]c(x,v(x),z)m(dz) \tag{1.33}$$

Clearly A_v and B_v have properties identical to those of A, B respectively. We also set :

$$f_v(x) = f(x) + \ell(x,v(x)) \tag{1.34}$$

We can thus solve the following problem :

$$A_v\phi - B_v\phi + a_0\phi = f_v, \quad \phi \in W_0^{1,p}(\mathcal{O}) \tag{1.35}$$

which has one and only one solution denoted by ϕ_v. We can then state the :

---

(1) We assume that $\mathcal{V}$ is a measurable space.

Theorem 1.2. *Under the assumptions* (1.1), (1.2), (1.9), (1.10), (1.23), (1.24), (1.25), (1.26), *we have the property* :

$$u(x) \leq \phi_v(x) \ , \ \forall v \ , \forall x \in \bar{\mathcal{O}}. \tag{1.36}$$

If in addition we assume that :

$$\mathcal{V} \text{ is compact} \tag{1.37}$$

$$\ell, h, a_1, c_1 \text{ are continuous in } v \tag{1.38}$$

then there exists $\hat{v}(x)$ *such that* :

$$u(x) = \phi_{\hat{v}}(x), \quad \forall x \in \bar{\mathcal{O}}. \tag{1.39}$$

Proof. We have :

$$A_v(\phi-u) - B_v(\phi-u) + a_o(\phi-u) = \ell_v + \nabla u.h_v - ua_{1v}$$

$$+ \int_{R^n} [u(x+z) - u(x) - z.\nabla u\chi_{|z|\leq 1}] \, c_o c_{1v} m(dz) - H(u)$$

$$\in L^p(\mathcal{O}) \geq 0,$$

from which it follows using a reasoning simular to the one used in the proof of theorem 1.1 (Maximum principle type argument) that $\phi-u \geq 0$, hence (1.36).

If (1.37), (1.38) are satisfied, there exists for $u \in W_o^{1,p}(\mathcal{O})$, $\hat{v}(x)$ measurable mapping from $R^n \to \mathcal{V}$, such that the inf is reached in (1.28) (with u instead of φ). Then clearly (1.39) is satisfied.

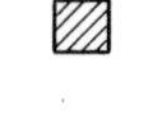

2- PROBABILISTIC INTERPRETATION OF THE HAMILTON-JACOBI EQUATION

2.1. Definitions and Notation

We consider the operators A and B defined by (1.3) and (1.27). We set :

$$g_i(x) = - a_i + \sum_j \frac{\partial a_{ij}}{\partial x_j} \in L^\infty(R^n) \tag{2.1}$$

so that we may write :

$$A = - \sum_{ij} a_{ij} \frac{\partial^2}{\partial x_i \partial x_j} - g_i \frac{\partial}{\partial x_i} . \tag{2.2}$$

Let now $\Omega_o = D([o,\infty[;\ R^n)$ be the space of functions which are right continuous, with left limits endowed with the Skorokhod's metric. We note $x(t;\omega) \equiv \omega(t)$ the canonical process, and define $\mathcal{M}^t = \sigma[x(s),\ o \leq s \leq t]$.

We will say that a probability measure $\hat{P}^x$ on $(\Omega_o, \mathcal{M}^\infty)$ is a <u>solution of the martingale problem for the operator</u> A–B, <u>stopped at the exit of</u> $\mathcal{O}$, <u>starting from</u> $x \in R^n$, <u>if</u> :

$$\hat{P}^x[x(o) = x] = 1 \tag{2.3}$$

$$\forall \varphi \in \mathcal{D}(R^n),\ \varphi(x(t)) + \int_o^t \chi_{\mathcal{O}}(A-B)\varphi(x(s))ds \tag{2.4}$$

is a $\hat{p}^x$ martingale with respect to $\mathcal{M}^t$.

We note $\bar{\mathcal{M}}^t$ the σ-algebra $\mathcal{M}^{t+o}$ completed with the null events of $\mathcal{M}^o$ (with respect to $\hat{p}^x$).

Then by virtue of the right continuity of $x(t)$, the martingale property is also verified with respect to $\bar{\mathcal{M}}^t$, instead of $\mathcal{M}^t$.

<u>Remark 2.1.</u> If $x \notin \mathcal{O}$, then the Dirac measure in $x(t) = x\ \ \forall t$, is a solution of (2.3), (2.4). ▨

Our objective is to construct a <u>strong Markov process</u>, which is a solution of the martingale problem.

2.2. A Brief Overview on Stochastic Calculus

Let $(\Omega, \mathcal{A}, P)$ be a probability space, $\mathcal{F}^t \uparrow \mathcal{F}^t \subset \mathcal{A}$, $\mathcal{F}^t = \mathcal{F}^{t+o}$, $\mathcal{F}^o$ complete. Let $\mathcal{B}$ be the Borel σ-algebra on R^n, and $\mathcal{B}_o$ the class of Borel sets, <u>whose closure does not contain</u> o. We call a <u>measure martingale</u> a random function $\mu(t,A)$, $t \in R^+$, $A \in \mathcal{B}_o$ satisfying

$$\left| \begin{array}{l} \forall A \in \mathcal{B}_o,\ \mu(t,A) \text{ is an } \mathcal{F}^t \text{ martingale, right conti-} \\ \text{nuous with left limits, and :} \\ \mu(t,A\cup B) = \mu(t,A) + \mu(t,B) \text{ if } A,\ B \in \mathcal{B}_o,\ A\cap B = \emptyset \end{array} \right. \tag{2.5}$$

$$\left| \begin{array}{l} \text{if } A,B \in \mathcal{B}_o,\ A \cap B = \emptyset, \text{ then } \mu(t,A)\mu(t,B) \text{ is a} \\ \text{martingale} \end{array} \right. \tag{2.6}$$

$$\left| \begin{array}{l} \mu^2(t,A) - \pi(t,A) \text{ is an } \mathcal{F}^t \text{ martingale, } \forall A \in \mathcal{B}_o, \\ \text{where } \pi(t,A) \text{ is an increasing, continuous process,} \\ \text{and a.s. } \forall t \text{ a measure on } R^n - \{0\}. \end{array} \right. \tag{2.7}$$

One calls $\pi(t,A)$ the increasing process associated with μ, and one sets $\pi(t,\{o\}) = 0$.

One now calls an <u>integer random measure</u> a random function $\nu(t,A)$, $t \in R^+$, $A \in \mathcal{B}$ with values in N or $+\infty$ such that :

$$\nu(t,\{o\}) = 0,\ E\,\nu(t,A) < \infty,\ \forall A \in \mathcal{B}_o \tag{2.8}$$

$$\left| \begin{array}{l} \text{for } A \in \mathcal{B}_o,\ \nu(t,A) \text{ is an increasing process,} \\ \text{adapted with } \mathcal{F}^t, \text{ with jumps of size 1. For } t \\ \text{fixed, } \nu(t,A) \text{ is an additive function of A.} \end{array} \right. \tag{2.9}$$

$$\left| \begin{array}{l} \text{if } \tau_n \uparrow \tau \ (\tau_n,\tau \text{ random times such that } \tau \le T \text{ a.s.}) \\ \text{the } \nu(\tau_n,A) \uparrow \nu(\tau,A) \text{ a.s., } \forall A \in \mathcal{B}_o.\ ^{1} \end{array} \right. \tag{2.10}$$

If $\nu(t,A)$ is an integer random measure, then there exists a unique measure martingale $\mu(t,A)$ such that :

$$\nu(t,A) = \mu(t,A) + \pi(t,A) \tag{2.11}$$

and $\pi(t,A)$ is the increasing process associated with $\mu(t,A)$.

---

(1) We define a measure on $R^+ \times R^m$ by setting :

$$\nu\ ([s,t],A) = \nu(t,A) - \nu(s,A).$$

An integer random measure is called a Poisson measure, if $\pi(t,A)$ is deterministic. It will be called a standard Poisson measure if :

$$\pi(t,A) = t\ m(A) \tag{2.12}$$

and $m(A)$ is called the Levy measure.

We will only work with measure martingales, whose increasing process is absolutely continuous with respect to the Lebesgue measure, and more precisely :

$$\pi(t,A) = \int_o^t \int_A p(s,z)ds\ m(dz) \tag{2.13}$$

where m satisfies (1.23). Let $\mu(t,A)$ be such a measure martingale, and $\varphi(t,z)$ be a random function, measurable with respect to t,z,ω, which is an adapted process for fixed z, and such that :

$$\int_o^t \int_{R^n} |\varphi(s,z)|^2\ p(s,z)ds\ m(dz) < \infty \text{ a.s. } \forall t.$$

Then one can define the stochastic integral

$$\int_o^t \int_{R^n} \varphi(s,z)\ \mu(ds,\ dz).$$

Let $f(x,t)$ be a function which is $C_b^{2,1}$, and $\xi(t)$ be the process (in dimension 1, to simplify) :

$$\xi(t) = \xi(o) + \int_o^t \alpha(s)ds + \int_o^t b(s)dw(s) \tag{2.14}$$
$$+ \int_o^t \int_{R^n} \varphi(s,z)\mu(ds,dz),$$

where $w(s)$ is a standard Wiener process, the following Itô's formula holds true :

$$f(\xi(t),t) = f(\xi(o),o) + \int_o^t \frac{\partial f}{\partial s}(\xi(s),s)ds \tag{2.15}$$

$$+ \int_o^t \frac{\partial f}{\partial x}(\xi(s),s)\alpha(s)ds + \frac{1}{2}\int_o^t \frac{\partial^2 f}{\partial x^2}(\xi(s),s)b^2(s)ds$$

$$+ \int_o^t \int_{R^n} [f(\xi(s) + \varphi(s,z),s) - f(\xi(s),s)$$

$$- \frac{\partial f}{\partial x}(\xi(s),s)\varphi(s,z)]p(s,z)ds\ m(dz)$$

$$+ \int_o^t \frac{\partial f}{\partial x}(\xi(s),s)dw(s) + \int_o^t \int_{R^n} [f(\xi(s) + \varphi(s,z),s)$$

$$- f(\xi(s),s)]\,\mu(dz,\ dz).$$

Formula (2.15) remains valid if $f \in C^{2,1}$ and :

$$\int_o^t \int_{R^n} |f(\xi(s) + \varphi(s,z),s) - f(\xi(s),s) \tag{2.16}$$

$$- \frac{\partial f}{\partial x}(\xi(s),s)\varphi(s,z)|p(s,z)ds\ m(dz) < \infty \quad \text{a.s.}$$

$$\int_o^t \int_{R^n} |f(\xi(s) + \varphi(s,z),s) - \tag{2.17}$$

$$- f(\xi(s),s)|^2 p(s,z)ds\ m(dz) < \infty \quad \text{a.s.}$$

2.3. Solution of the Martingale Problem for $A-B_\varepsilon$

We start to consider the martingale problem for the operator $A-B_\varepsilon$ instead of A-B. We consider an arbitrary probability space $(\Omega,\mathcal{A},P,\mathcal{F}^t)$, on which are given a standard Wiener process on R^n, $w(t)$ and a Poisson measure $\nu_\varepsilon(t,A)$, whose Levy measure is $m_\varepsilon(A)$. One defines $\sigma(x)$ such that :

$$a(x) = \frac{1}{2}\sigma(x)\sigma^*(x) \tag{2.18}$$

where $a(x)$ stands for the matrix $a_{ij}(x)$. Note that σ is Lepschitz.

To the Poisson measure $\nu_\varepsilon(t,A)$ is associated a measure martingale $\mu_\varepsilon(t,A)$. One then considers the equation :

$$\xi_\varepsilon(t) = x + \int_o^t \sigma(\xi_\varepsilon(s))dw(s) + \int_o^t \int_{R^n} z\chi_{|z|\leq 1}\mu_\varepsilon(ds,dz)$$
$$+ \int_o^t \int_{R^n} z\chi_{|z|>1}\nu_\varepsilon(ds,dz). \tag{2.19}$$

<u>Lemma 2.1.</u> <u>There exists one and only one solution of</u> (2.19), <u>which is adapted right continuous with left limits</u>. <u>Moreover</u> :

$$E \sup_{o\leq t\leq T} |\xi_\varepsilon(t)| \leq C(1 + |x|). \tag{2.20}$$

<u>Proof.</u> Since the arguments of the two last integrals in the right hand side of (2.19) do not depend on $\xi_\varepsilon(s)$, one can use an iterative procedure, like in the case of diffusion equations. Details are omitted. ▨

We next consider the process :

$$\alpha_\varepsilon(t) = \sigma^{-1}(\xi_\varepsilon(t))[g(\xi_\varepsilon(t)) \tag{2.21}$$
$$+ \int_{R^n} (1 - c_o(\xi_\varepsilon(t),z))z\chi_{|z|\leq 1} m_\varepsilon(dz)]$$

which for ε fixed is bounded by a deterministic constant.

One then considers the equation :

$$\eta_\varepsilon(t) = 1 + \int_o^t \eta_\varepsilon(s)\alpha_\varepsilon(s)dw(s) + \tag{2.22}$$
$$+ \int_o^t \int_{R^n} \eta_\varepsilon(s)(c_o(\xi_\varepsilon(s),z)-1)\mu_\varepsilon(ds,dz).$$

We have :

Lemma 2.2. The process

$$\eta_\varepsilon(t) = \exp\{ \int_o^t \alpha_\varepsilon(s)dw(s) + \int_o^t \int_{R^n} (c_o(\xi_\varepsilon(s),z)-1)\mu_\varepsilon(ds,dz) \tag{2.23}$$

$$+ \frac{1}{2} \int_o^t |\alpha_\varepsilon(s)|^2 ds - \int_o^t \int_{R^n} (c_o-1-\text{Log } c_o)\nu_\varepsilon(ds,dz)\}$$

is solution of (2.22).

Proof. Note that $c_o - 1 - \text{Log } c_o \geq 0$, so that the right hand side of (2.23) makes sense. Then one applies Itô's formula (2.15) to the function $f(x) = \exp x$, and to the process within brackets. Some caution is necessary for the treatment of the last integral. It must be split into a stochastic integral and an ordinary integral before applying Itô's formula. Since c_o may be 0, the stochastic integral may have no meaning. So one first replaces c_o by $c_o + \delta$, $\delta > 0$, and checks that the Lemma is true in that case. One can then let δ go to 0 in (2.22), (2.23). The desired result follows. ▨

Lemma 2.3. Let $\varphi(x) \in \mathscr{D}(R^n)$. Then the process :

$$\varphi(\xi_\varepsilon(t))\eta_\varepsilon(t) - \varphi(x) + \int_o^t \eta_\varepsilon(s)(A-B_\varepsilon)\varphi(\xi_\varepsilon(s)ds \tag{2.24}$$

is a $P, \mathscr{F}^t$ martingale.

Proof. It follows from applying Itô's formula to the function $f(x,z) = \varphi(x)z$, and to the process $\xi_\varepsilon(t)$, $\eta_\varepsilon(t)$. Again some caution is necessary for the treatment of the last two integrals, in the right hand side of (2.19). The sum must first be approximated by :

$$\int_o^t \int_{R^n} z\chi_{|z|\leq R}\mu_\varepsilon(ds,dz) + \int_o^t \int_{R^n} z\chi_{|z|>1} \, d\lambda \, m_\varepsilon(dz)$$

and then let $R \to +\infty$, after applying Itô's formula. ▨

Lemma 2.4. Let $f(x)$ be a Borel bounded function, then one has :

$$|E \int_0^T f(\xi_\varepsilon(s))ds| \leq C_{T,\varepsilon} |f|_{L^{n+2}(R^n)} . \tag{2.25}$$

Proof. One can assume that f is regular with compact support and ≥ 0. We note from (2.19) that $\xi_\varepsilon(t)$ satisfies the equation :

$$\xi_\varepsilon(t) = x + \int_0^t \sigma(\xi_\varepsilon(s))dw(s) + \tag{2.26}$$

$$+ \int_0^t \int_{R^n} z\, \nu_\varepsilon(ds,dz) - \int_0^t \int_{R^n} z\chi_{|z|\leq 1}\, ds\, m_\varepsilon(dz).$$

One defines the sequence of stopping times :

$$\tau_0 = 0,\ \tau_{j+1} = \inf\{T \geq s \geq \tau_j | \nu_\varepsilon(s,R^n) > \nu_\varepsilon(\tau_j,R^n)\}.$$

Let us define the process :

$$\xi_\varepsilon^j(t) = \xi_\varepsilon(\tau_j) + \int_{\tau_j}^t \sigma(\xi_\varepsilon(s))dw(s) \tag{2.27}$$

$$- \int_{\tau_j}^t \int_{R^n} z\, \chi_{|z|\leq 1}\, ds\, m_\varepsilon(dz)$$

for $t \geq \tau_j$.

Then one has :

$$\xi_\varepsilon(t) = \xi_\varepsilon^j(t),\ \text{for } t \in [\tau_j, \tau_{j+1}],\ \text{if } \tau_j < \tau_{j+1}.$$

But :

$$E \int_0^T f(\xi_\varepsilon(t))dt = \sum_{j=0}^{\infty} E \int_{\tau_j}^{\tau_{j+1}} f(\xi_\varepsilon(t))dt \tag{2.28}$$

$$\leq \sum_{j=0}^{\infty} E\, \chi_{\tau_j < T} \int_{\tau_j}^T f(\xi_\varepsilon^j(t))dt.$$

We consider the solution of :

$$\left|\begin{array}{l} -\dfrac{\partial\phi_\varepsilon}{\partial t} - \sum_{ij} a_{ij}(x)\dfrac{\partial^2\phi}{\partial x_i\partial x_j} + \dfrac{\partial\phi_\varepsilon}{\partial x}.h_\varepsilon = f \\ \phi_\varepsilon(x,T) = 0,\ \phi_\varepsilon \in \mathcal{W}^{2,1,p}(R^n \times (0,T)),\ \ \forall p \geq 2 \end{array}\right. \tag{2.29}$$

where

$$h_\varepsilon = \int_{R^n} z\, \chi_{|z|\leq 1}\, m_\varepsilon(dz).$$

Moreover :

$$||\phi_\varepsilon||_{\mathcal{W}^{2,1,p}(R^n\times(0,T))} \leq C_{\varepsilon,T}|f|_{L^p(R^n)}.$$

Taking $p = n+2$, it follows from Sobolev' injection Theorems that :

$$||\phi_\varepsilon||_{C^o(R^n\times[0,T])} \leq C_{\varepsilon,T}|f|_{L^{n+2}}. \tag{2.30}$$

Using Itô's formula for diffusions, we obtain :

$$E\left[\int_{\tau_j}^T f(\xi_\varepsilon^j(t))dt \,|\mathcal{F}^{\tau_j}\right] = E[\phi_\varepsilon(\xi_\varepsilon^j(\tau_j),\tau_j)\,|\mathcal{F}^{\tau_j}]$$
$$\leq C_{\varepsilon,T}|f|_{L^{n+2}}$$

by virtue of (2.30).

Going back to (2.28), one obtains :

$$E\int_o^T f(\xi_\varepsilon(t))dt \leq C_{\varepsilon,T}|f|_{L^{n+2}} \sum_{j=0}^{\infty} E\,\chi_{\tau_j<T}.$$

But :

$$\sum_{j=0}^{\infty} E\,\chi_{j<T} \leq 1 + E\sum_{j=1}^{\infty}[\nu_\varepsilon(\tau_1,R^n) +\ldots+ \nu_\varepsilon(\tau_j,R^n) - \nu_\varepsilon(\tau_{j-1},R^n) +\ldots]$$
$$= 1 + E\,\nu_\varepsilon(T,R^n) = 1 + T\,m_\varepsilon(R^n)$$

hence (2.25). 

One next defines on $(\Omega,\mathcal{Q})$ a new probability measure P_ε such that :

$$\frac{dP_\varepsilon}{dP}\Big|_{\mathcal{F}^t} = \eta_\varepsilon(t). \qquad (2.31)$$

From the martingale property of $\eta_\varepsilon(t)$ and from Lemma 2.3, it follows that :

$$\varphi(\xi_\varepsilon(t)) - \varphi(x) + \int_0^t (A-B_\varepsilon)\varphi(\xi_\varepsilon(s))ds \text{ is a } P_\varepsilon, \mathcal{F}^t \text{ martingale} \qquad (2.32)$$

We next define

$$\tau_\varepsilon = \inf\{t \geq 0 \mid \xi_\varepsilon(t) \notin \bar{\mathcal{O}}\} \qquad (2.33)$$

and the stopped process

$$\hat{\xi}_\varepsilon(t) = \xi_\varepsilon(t \wedge \tau_\varepsilon). \qquad (2.34)$$

We note by $\hat{P}^x_\varepsilon$ the measure on $(\Omega_0,\mathcal{M}^t)$ image of P_ε by the process $\xi_\varepsilon(.)$. Since :

$$E_\varepsilon \int_0^{T\wedge\tau_\varepsilon} f(\xi_\varepsilon(s))ds \leq E_\varepsilon \int_0^T f(\xi_\varepsilon(s))ds$$

and for $f \geq 0$

$$= E\,\eta_\varepsilon(T) \int_0^T f(\xi_\varepsilon(s))ds$$

$$\leq C_{\varepsilon,T}\left(E \int_0^T f^2(\xi_\varepsilon(s)ds\right)^{\frac{1}{2}}$$

applying Lemma 2.4, we obtain :

$$\left|E_\varepsilon \int_0^{T\wedge\tau_\varepsilon} f(\xi_\varepsilon(s))ds\right| \leq C_{\varepsilon,T}||f||_{L^{2n+4}(R^n)}. \qquad (2.35)$$

<u>Lemma 2.5.</u> <u>The measure</u> $\hat{P}^x_\varepsilon$ <u>is a solution of the martingale problem for the operator</u> $A-B_\varepsilon$, <u>stopped at the exit of</u> $\mathcal{O}$.

Proof. Let η be a functional on Ω_o, which is continuous and $\mathcal{M}^{s_1}$ measurable. We have to prove that, for $\varphi \in \mathcal{D}(R^n)$:

$$\hat{E}^x_\varepsilon\{[\varphi(x(s_2)) - \varphi(x(s_1)) + \int_{s_1}^{s_2} \chi_{\mathcal{O}}(A-B_\varepsilon)\varphi(x(s))ds]\eta\} = 0. \quad (2.36)$$

By definition of $\hat{P}^x_\varepsilon$, (2.36) amounts to :

$$E_\varepsilon\{[\varphi(\xi_\varepsilon(s_2\wedge\tau_\varepsilon)) - \varphi(\xi_\varepsilon(s_1\wedge\tau_\varepsilon)) \quad (2.37)$$
$$+ \int_{s_1}^{s_2} \chi_{\mathcal{O}}(A-B_\varepsilon)\varphi(\xi_\varepsilon(s\wedge\tau_\varepsilon))ds] | \mathcal{F}^{s_1\wedge\tau_\varepsilon}\} = 0.$$

But

$$\int_{s_1}^{s_2} \chi_{\mathcal{O}}(A-B_\varepsilon)\varphi(\xi_\varepsilon(s\wedge\tau_\varepsilon))ds = \int_{s_1\wedge\tau_\varepsilon}^{s_2\wedge\tau_\varepsilon} \chi_{\mathcal{O}}(A-B_\varepsilon)\varphi(\xi_\varepsilon(s))ds.$$

Since

$$\chi_{\mathcal{O}}(A-B_\varepsilon)\varphi = \chi_{\bar{\mathcal{O}}}(A-B_\varepsilon)\varphi \quad \text{a.e.}$$

it follows from (2.35) that

$$E_\varepsilon[\int_{s_1\wedge\tau_\varepsilon}^{s_2\wedge\tau_\varepsilon} \chi_{\mathcal{O}}(A-B_\varepsilon)\varphi(\xi_\varepsilon(s))ds | \mathcal{F}^{s_1\wedge\tau_\varepsilon}]$$
$$= E_\varepsilon[\int_{s_1\wedge\tau_\varepsilon}^{s_2\wedge\tau_\varepsilon} \chi_{\bar{\mathcal{O}}}(A-B_\varepsilon)\varphi(\xi_\varepsilon(s))ds | \mathcal{F}^{s_1\wedge\tau_\varepsilon}] \quad \text{a.s.}$$

hence (2.37) amounts to

$$E_\varepsilon\{[\varphi(\xi_\varepsilon(s_2\wedge\tau_\varepsilon)) - \varphi(\xi_\varepsilon(s_1\wedge\tau_\varepsilon))$$
$$+ \int_{s_1\wedge\tau_\varepsilon}^{s_2\wedge\tau_\varepsilon} \chi_{\bar{\mathcal{O}}}(A-B_\varepsilon)\varphi(\xi_\varepsilon(s))ds] | \mathcal{F}^{s_1\wedge\tau_\varepsilon}\} = 0.$$

By definition of τ_ε

$$\int_{s_1\wedge\tau_\varepsilon}^{s_2\wedge\tau_\varepsilon} \chi_{\bar{\mathcal{O}}}(A-B_\varepsilon)\varphi(\xi_\varepsilon(s))ds = \int_{s_1\wedge\tau_\varepsilon}^{s_2\wedge\tau_\varepsilon} (A-B_\varepsilon)\varphi(\xi_\varepsilon(s))ds$$

and therefore we have to prove that

$$E_\varepsilon\{[\varphi(\xi_\varepsilon(s_2\wedge\tau_\varepsilon)) - \varphi(\xi_\varepsilon(s_1\wedge\tau_\varepsilon))$$
$$+ \int_{s_1\wedge\tau_\varepsilon}^{s_2\wedge\tau_\varepsilon} (A-B_\varepsilon)\varphi(\xi_\varepsilon(s))ds]\,|\mathcal{F}^{s_1\wedge\tau_\varepsilon}\} = 0$$

which is a consequence of (2.32) and Doob's optional sampling theorem. ▨

We can next give the interpretation of some analytic problems. Let now $f \in L^p(\mathcal{O})$, $p > \frac{n}{2}$ and $\lambda > 0$. We consider the problem :

$$A\,u_\varepsilon - B_\varepsilon u_\varepsilon + \lambda\,u_\varepsilon = f, \quad u_\varepsilon \in W^{2,p}(\mathcal{O}) \cap W_o^{1,p}(\mathcal{O}). \tag{2.38}$$

Then there exists one and only one solution of (2.38).Then we have:

<u>Lemma 2.6.</u> <u>The process</u>

$$e^{-\lambda t}\,u_\varepsilon(x(t)) - u_\varepsilon(x) + \int_o^t \chi_{\mathcal{O}}\,f(x(s))e^{-\lambda s}\,ds \tag{2.39}$$

<u>is a</u> $\hat{P}^x_\varepsilon, \mathcal{M}^t$ <u>martingale.</u>

<u>Proof.</u> Let us first assume that $p > 2n+4$.

Let $u^k_\varepsilon \to u_\varepsilon$ in $W^{2,p}(\mathcal{O})$ and $C^o(R^n)$, $u^k_\varepsilon \in \mathcal{D}(R^n)$. Since $B_\varepsilon \in \mathcal{L}(L^p(\mathcal{O})\ ;\ L^p(\mathcal{O}))$, we have :

$$A\, u_\varepsilon^k - B_\varepsilon u_\varepsilon^k + \lambda\, u_\varepsilon^k \to f \text{ in } L^p(\mathcal{O})$$

hence :

$$\chi_{\mathcal{O}}(A\, u_\varepsilon^k - B_\varepsilon u_\varepsilon^k) \to \chi_{\mathcal{O}}(f-\lambda u) \text{ in } L^p(R^n), \text{ as } k \to \infty. \tag{2.40}$$

From Lemma (2.5), we may assert that :

$$u_\varepsilon^k(x(t)) - u_\varepsilon^k(x) + \int_0^t \chi_{\mathcal{O}}(A-B_\varepsilon)u_\varepsilon^k(x(s))ds$$

is a $\hat{P}_\varepsilon^x, \mathcal{M}^t$ martingale. From (2.40) and (2.35), it follows letting k tend to $+\infty$, that :

$$u_\varepsilon(x(t)) - u_\varepsilon(x) + \int_0^t \chi_{\mathcal{O}}(f-\lambda u)(x(s))ds$$

is a $\hat{P}_\varepsilon^x, \mathcal{M}^t$ martingale. By a standard result, it follows that :

$$u_\varepsilon(x(t))e^{-\lambda t} - u_\varepsilon(x) + \int_0^t [\chi_{\mathcal{O}}(f-\lambda u)+\lambda u_\varepsilon](x(s))ds$$

is also a $\hat{P}_\varepsilon^x, \mathcal{M}^t$ martingale. Since $\chi_{\mathcal{O}} u = u$, we obtain the desired result at least when $p > 2n+4$. As a particular case, it follows that :

$$u_\varepsilon(x) = \hat{E}_\varepsilon^x \int_0^\infty e^{-\lambda t}\, \chi_{\mathcal{O}} f(x(t))dt \tag{2.41}$$

hence, by Sobolev's estimates :

$$|\hat{E}_\varepsilon^x \int_0^\infty e^{-\lambda t}\, \chi_{\mathcal{O}} f(x(t))dt| \le C_\varepsilon |f|_{L^p(\mathcal{O})}, \ p > \frac{n}{2}$$

from which, using an approximation scheme, the desired result follows.

▨

Let us next consider the initial value problem :

$$\left| \begin{array}{l} \frac{\partial v_\varepsilon}{\partial t} + (A-B_\varepsilon)v_\varepsilon = 0 \\ \\ v_\varepsilon(x,o) = \varphi(x), \quad v_\varepsilon(x,t) = \varphi(x) \quad \text{for } x \in \partial\mathcal{O} \end{array} \right. \tag{2.42}$$

For $\varphi \in C_b^2(R^n)$, there exists one and only one solution of (2.42), such that $v_\varepsilon \in \mathcal{W}^{2,1,p}(Q)$, with $Q = \mathcal{O} \times]o,T[$, $\forall p \geq 2$. Since (2.36) is valid for $\varphi \in C_b^2(R^n)$, we may write :

$$\varphi(x) = \hat{E}^x \varphi(x(t)) + \hat{E}_\varepsilon^x \int_o^t \chi_{\mathcal{O}}(A-B_\varepsilon)\varphi(x(s))ds. \tag{2.43}$$

Now, by an argument similar to the one used in Lemma 2.6, we have :

$$v_\varepsilon(x,t) - \varphi(x) = - \hat{E}_\varepsilon^x \int_o^t \chi_{\mathcal{O}}(A-B_\varepsilon)\varphi(x(s))ds$$

which, combined with (2.43) yields :

$$v_\varepsilon(x,t) = \hat{E}_\varepsilon^x \; \varphi(x(t)) = \int_{R^n} \hat{P}_\varepsilon(x,t,dy)\varphi(y). \tag{2.44}$$

Formula (2.44) defines a transition probability, corresponding to a Markov process. From the regularity of v_ε , when φ is regular, it follows that the Markov process is Feller.

2.4. Solution of the Martingale Problem for A-B.

We are going to construct a Feller process on the canonical space, which is solution of the martingale problem for A-B, stopped at the exit of $\mathcal{O}$, and which is the weak limit of the family $\hat{P}_\varepsilon^x$, as $\varepsilon \to 0$.

Lemma 2.7. Let $f \in L^p(\mathcal{O})$, $p > n$. For λ large enough (independently from f) one has the estimate :

$$|\hat{E}_\varepsilon^x \int_o^\infty e^{-\lambda t} \chi_{\mathcal{O}} f(x(t))dt| \leq C |f|_{L^p(\mathcal{O})} \tag{2.45}$$

where the constant C does not depend on ε.

Proof. The mathematical expectation on the left side of (2.45) is equal to $u_\varepsilon(x)$. Hence it is sufficient to obtain a priori estimates for the solution of (2.38), which are independent from ε.

This follows from the methods developed in the first part (cf. in particular Remark 1.1). For λ large enough, one has :

$$||u_\varepsilon||_{W_o^{1,p}} \leq C |f|_{L^p} \quad , \quad \forall p \geq 2,$$

and for $p > n$, we obtain (2.45). ▨

Lemma 2.8. The family $\hat{P}_\varepsilon^x$, as $\varepsilon \to o$, is relatively compact in the space of probability measures on Ω_o, endowed with the weak convergence.

Proof. One applies a standard compactness criterion on Ω_o (cf. BILLINGSLEY [2], or LEPELTIER-MARCHAL [7]). One can express the conditions in terms of the original process $\xi_\varepsilon(t)$. Then the result follows from the following estimate : Let τ an $\mathcal{F}^t$ stopping time, such that $\tau \leq T$, then one has :

$$P_\varepsilon[\sup_{o \leq s \leq h} |\xi_\varepsilon(s+\tau) - \xi_\varepsilon(\tau)| > \eta]$$
$$\leq 2n \exp\{-\frac{\lambda}{n}(\eta - k_o h - A) + \frac{\lambda^2}{2} k_1 h(1+e^{|\lambda|})\} + \frac{k_1 h}{A}$$

where k_o, k_1 are constants depending only on the bounds on a_{ij}, a_i, c_o, and the integrals (1.23), λ is any real number, and A is any positive number. The proof of (2.46) follows from martingale estimates and stochastic calculus. For details, cf. A. BENSOUSSAN-J.L. LIONS [1]. ▨

From Lemma 2.8, it follows that we can consider a subsequence of $\hat{P}^x_\varepsilon$, converging weakly towards $\hat{P}^x$.

Lemma 2.9. The measure $\hat{P}^x$ is solution of (2.3), (2.4).

Proof. Since x(t) is continuous in 0, [x(o) = x] is closed for the Skorokhod topology, hence (2.3). To prove (2.4), we notice that it is sufficient to prove that :

$$\varphi(x(t))e^{-\lambda t} + \int_o^t \chi_{\mathcal{O}}(A-B)\,\varphi e^{-\lambda s}ds + \int_o^t \lambda\varphi(x(s))e^{-\lambda s}ds$$

is a $\hat{P}^x, \mathcal{M}^t$ martingale. For $\varphi \in \mathcal{D}(R^n)$, we have :

$$|(B_\varepsilon - B)\varphi|_{L^p(\mathcal{O})} \leq C\,\varphi\int|z|^2\,\chi_{|z|\leq\varepsilon}\,m(dz) \to 0.$$

We also approximate $\chi_{\mathcal{O}}(A-B)\varphi$ by a sequence $H_k \in \mathcal{D}(\mathcal{O})$, converging in $L^p(\mathcal{O})$. By standard properties of the space Ω_o, there exists a countable set of terms $\mathcal{N}$ such that for $t \notin \mathcal{N}$, then the mapping $x(.) \to x(t)$ is continuous on Ω_o, a.s. $\hat{P}^x$. Hence for $s_2 \geq s_1 \notin \mathcal{N}$, and η continuous on Ω_o, bounded and $\mathcal{M}^t$ measurable, one has by the weak convergence:

$$\hat{E}^x_\varepsilon \eta[\varphi(x(s_2))e^{-\lambda s_2} - \varphi(x(s_1))e^{-\lambda s_1} + \int_{s_1}^{s_2} (H_k+\lambda\varphi)(x(s))e^{-\lambda s}ds]$$

$$\to \hat{E}^x \eta[\varphi(x(s_2))e^{-\lambda s_2} - \varphi(x(s_1))e^{-\lambda s_1} + \int_{s_1}^{s_2} (H_k+\lambda\varphi)(x(s))e^{-\lambda s}ds].$$

From this, and the above approximations, it follows, using (2.45), that (2.4) holds true.

▨

Let now $\varphi \in C^2_b(R^n)$, one considers the problem :

$$\frac{dv}{dt} + (A-B)v = 0 \tag{2.47}$$

$$v(x,o) = \varphi(x), \quad v(x,t) = \varphi(x), \quad x \in \partial\mathcal{O}$$

$$v - \varphi \in L^p(o,T;W_o^{1,p}(\mathcal{O})), \ \frac{dv}{dt} \in L^p(o,T;W^{-1,p}(\mathcal{O})), \forall p \geq 2.$$

As in Theorem 1.1., one can obtain an existence and uniqueness result for (2.47). Now $v_\varepsilon \to v$ in $C^o(R^n \times [o,T])$. From (2.44), one obtains :

$$v(x,t) = \hat{E}^x \varphi(x(t)) = \int \varphi(y)\hat{P}(x,t,dy). \tag{2.48}$$

Therefore $\hat{P}^x$ is a Feller process, whose transition probability is $\hat{P}(x,t,dy)$. Hence the limit of $\hat{P}^x_\varepsilon$ is unique.

Finally one considers the problem :

$$Au - Bu + a_o u = f \ ; \quad u \in W_o^{1,p}(\mathcal{O}), \ p > n, \tag{2.49}$$

where $f \in L^p(\mathcal{O})$, $p > n$. One can state the following.

<u>Theorem 2.1.</u> <u>Under the assumptions</u> (1.1), (1.2), (1.10), (1.23), (1.24), <u>then the sequence</u> $\hat{P}^x_\varepsilon$ <u>converges weakly towards</u> $\hat{P}^x$, <u>which is a Feller process</u>, <u>and one has the representation formula</u>:

$$u(x) = \hat{E}^x \int_o^\infty \chi_{\mathcal{O}} f(x(t))(\exp - \int_o^t a_o(x(s))ds)dt. \tag{2.50}$$

▨

To any control function $v(x)$, as in § 1.4, one can now associate a measure $\hat{P}^x_v$. One can then define a functional :

$$J^x(v(.)) = \hat{E}^x_v[\int_o^\infty \chi_{\mathcal{O}} f_v(x(t))(\exp - \int_o^t (a_o + a_{1v})(x(s))ds)\, dt] \tag{2.51}$$

and one can state the :

<u>Theorem 2.2.</u> <u>Under the assumptions of Theorem</u> 1.2. <u>the solution of</u> (1.11) <u>is given explicitly by</u> :

$$u(x) = \underset{v(.)}{\text{Min}} \ J^x(v(.)). \tag{2.52}$$

▨

2.5. Some Remarks

Remark 2.2. In the case of R^n, a different theory can be developed. In particular, one can consider non anticipative controls instead of Markov controls.

Remark 2.3. We have analogous results for the parabolic case.

Remark 2.4. One can also consider variational inequalities and stopping time problems, corresponding to the operator A-B.

BIBLIOGRAPHY

[1] A. BENSOUSSAN ; J. L. LIONS, Contrôle impulsionnel et Inéquations quasi-variationnelles, to be published, DUNOD.

[2] P. BILLINGSLEY, Convergence of Probability Measures, Wiley, 1968.

[3] J. M. BISMUT, Control of Jump processes and applications, Bull. Soc. Math. de France, to be published.

[4] W. H. FLEMING ; R. RISHEL, Optimal Deterministic and Stochastic Control, Springer-Verlag, New York, 197 .

[5] I. M. GIKHMAN ; A. V. SKOROKHOD, Theory of Stochastic Processes, vol. 3 (in Russian).

[6] T. KOMATSU, Markov Processes associated with certain integro-differential operators, Osaka J. Math. 10 (1973), 271-303.

[7] J. P. LEPELTIER ; B. MARCHAL, Problème des martingales et équations différentielles stochastiques associées à un opérateur intégro-différentiel, Ann. Inst. Henri Poincaré, Vol. XII, N°1, 1976, pp. 43-103.

[8] J. P. LEPELTIER ; B. MARCHAL, Sur l'existence de politiques optimales dans le contrôle intégro-différentiel, Ann. Inst. Henri Poincaré, Vol. XII, 1977, pp. 45-97.

[9] D. W. STROOCK, Diffusion Processes Associated with Levy Generators, Z. Wahrscheinlichkeitstheorie Verw. Gebiete 32, 209-244 (1975).

ON THE NUMBER OF DISTINCT SITES VISITED BY A RANDOM WALK

M. D. Donsker* and S. R. S. Varadhan*
Courant Institute, New York University
New York, New York

1. STATEMENT OF THE RESULT

Let Z^d be the lattice in R^d consisting of points with integral coordinates. Let π be a probability distribution on Z^d. That is $\pi = \{\pi_j\}$ where π_j is the probability associated with $j \in Z^d$. Of course $\sum_{j \in Z^d} \pi_j = 1$ and $\pi_j \geq 0$ for each $j \in Z^d$. We consider a sequence $X_1, X_2, \ldots, X_n, \ldots$ of independent identically distributed random variables with values in Z^d. having π for its common distribution. Let $S_n = X_1 + X_2 + \ldots + X_n$ define a random walk in Z^d. We denote by D_n the random set contained in Z^d defined by $D_n = \{j: S_r = j \text{ for some } r, 1 \leq r \leq n\}$ and by $D_n^{\#}$ the cardinality of the set D_n. Our aim to study the asymptotic behavior of $E[\exp[-\nu D_n^{\#}]]$ where expectation is of course with respect to the random variables $X_1, X_2, \ldots$.

Our basic assumption on π is that it is in the domain of normal attraction of a nondegenerate symmetric stable law in R^d. More precisely we have some α in the range $0 < \alpha \leq 2$. If $0 < \alpha < 2$ we have a symmetric measure Ω on the unit sphere $B = \{x: \|x\| = 1\}$ on R^d which is not supported on any lower dimensional subspace. Corresponding to this measure Ω we have a stable process of order α whose infinitesimal generator is given by

*The work of Donsker and Varadhan was supported by the National Science Foundation, Grant No.NSFMCS7702687.

ISBN 0-12-268380-3

$$(\tilde{\Omega}f)(x) = \int_B \Omega(dy) \int_0^\infty \frac{dr}{r^{1+\alpha}} \left| \frac{f(x+ry)+f(x-ry)}{2} - f(x) \right|$$

on smooth functions. The characteristic function of the increment of the process over an interval of length t is given by $\exp[-t\bar{\Omega}(\xi)]$ where

$$\bar{\Omega}(\xi) = \int_B \Omega(dy) \int_0^\infty \frac{dr}{r^{1+\alpha}} [1 - \cos \langle \xi, ry \rangle]$$

If $\alpha = 2$, then of course the stable process is some Brownian motion. Now

$$(\tilde{\Omega}f)(x) = \frac{1}{2} \sum a_{ij} \frac{\partial^2 f}{\partial x_i \partial x_j}$$

where $\{a_{ij}\}$ is a symmetric positive definite matrix. The function $\bar{\Omega}(\xi)$ takes the form

$$\bar{\Omega}(\xi) = \frac{1}{2} \sum a_{ij} \xi_i \xi_j$$

The nondegeneracy can be stated simply as $\bar{\Omega}(\xi) \neq 0$ for $\xi \neq 0$. Our main assumption on π can now be stated explicitly. There is an α in the range $0 < \alpha \leq 2$ and a corresponding $\bar{\Omega}(\xi)$, (coming from some Ω if $0 < \alpha < 2$ and from the matrix $\{a_{ij}\}$ if $\alpha = 2$) such that:

$$\hat{\pi}(\xi) = 1 - \bar{\Omega}(\xi) + o(\|\xi\|^\alpha) \quad \text{as } \|\xi\| \to 0 \tag{1.1}$$

Here $\hat{\pi}(\xi)$ is the characteristic function

$$\hat{\pi}(\xi) = \sum_{j \in Z^d} e^{i\langle \xi, j \rangle} \pi_j \tag{1.2}$$

which is of course a periodic function on R^d of period 2π in each coordinate. We also make the normalization assumption that the random walk does not live on any sub-lattice of Z^d. More precisely

$$\hat{\pi}(\xi) = 1 \quad \text{if and only if} \quad \xi = 2\pi(n_1,\ldots,n_d) \tag{1.3}$$

where $n_1,\ldots,n_d$ are some integers.

Under these assumptions the following theorem holds:

Theorem. For any $\nu > 0$

$$\lim_{n\to\infty} \frac{1}{n^{\frac{d}{d+\alpha}}} \log E\left\{\exp[-\nu D_n^\#]\right\} = -\, k(\nu,\Omega)$$

exists and can be computed as follows:

$$k(\nu,\Omega) = \nu^{\frac{\alpha}{d+\alpha}} \left(\frac{d+\alpha}{\alpha}\right) \left(\frac{\alpha\lambda_\Omega}{d}\right)^{\frac{d}{d+\alpha}}$$

where $\lambda_\Omega = \inf_{G:|G|=1} \lambda(\tilde{\Omega},G)$ and $\lambda(\tilde{\Omega},G)$ is the smallest eigenvalue of $\tilde{\Omega}$ in the region G with zero "boundary" conditions. The infimum is taken over all G such that G is open in R^d and has unit Lebesgue measure.

Remark 1. One can define for any smooth function ϕ on R^d the Dirichlet integral

$$\|\phi\|^2_{\tilde{\Omega}} = -\int_{R^d} (\tilde{\Omega}\phi)(x)\ \phi(x)\ dx$$

Then

$$\lambda(\overline{\Omega},G) = \inf_{\substack{\phi:\int\phi^2(x)dx=1 \\ \phi=0 \text{ outside } G}} \|\phi\|^2_{\tilde{\Omega}}$$

In the spherically symmetric case the infimum is attained when G is a sphere of unit volume.

Remark 2. The formula one obtains is of course identical with the one obtained in [3] for the Wiener sausage or the stable sausage. In fact condition (1.1) ensures the corresponding limit theorem and reduces the problem to the stable case.

We shall now provide a rough outline of the proof. The details will appear elsewhere. The method of proof relies heavily on the techniques developed in [1], [2], [3].

2. LOWER BOUND

It is known [cf. [5]] that $D_n^{\#}$ is typically of the order of magnitude n. However the contribution to $E\{\exp[-\nu D_n^{\#}]\}$ comes mainly from those walks with much smaller value of $D_n^{\#}$ even though their probability is much smaller. The order of magnitude of interest turns out to be $D_n^{\#} \sim kn^{d/(d+\alpha)}$. In order to estimate $P[D_n^{\#} \leq k\, n^{d/(d+\alpha)}]$ from below we use

$$P[D_n^{\#} \leq k\, n^{\frac{d}{d+\alpha}}] \geq P[D_n \subset b_n G] \tag{2.1}$$

where G is some open set containing the origin in R^d and b_n are some constants increasing to ∞ like $n^{1/(d+\alpha)}$. By using the limit theorem one can estimate $P[D_n \subset b_n G]$ in terms of the probability of the corresponding stable process staying in G for times up to $n^{d/(d+\alpha)}$. The number k in (2.1) is of course related to the volume of G. When one carries out the computation one gets for the lower bound

$$\varliminf_{n\to\infty} \frac{1}{n^{d/(d+\alpha)}} \log E[\exp\{-\nu D_n^{\#}\}] \geq -[\nu|G| + \lambda(\tilde{\Omega}, G)]$$

Optimizing over all G leads to the lower bound contained in the theorem.

3. UPPER BOUND

The upper bound is much more complicated to derive. One has to combine a version of the local limit theorem with the results on large deviations. The basic idea is to convolute the occupation distribution with the uniform distribution over a single cube in R^d. In this process one obtains a density on R^d defined by

$$\phi_n(x) = \frac{1}{n} \sum_{r=1}^{n} \chi(x - S_r)$$

where $\chi(x)$ is the indicator function of the cube $\{x: |x_i| \leq \frac{1}{2} \text{ for } 1 \leq i \leq d\}$. One can then think of $D_n^{\#}$ as $|\{x: \phi_n(x) > 0\}|$. We then prove for the random density $\phi(x)$ large deviation results of the type proved in [2] in the strong or L_1-topology. For technical reasons one has to reduce everything to a torus of large size and then let the size increase to infinity. The techniques are very similar to those used in [2] and [3]. But the transition to $\phi_n(x)$ is made possible here because of some estimates coming from local limit theorems.

4. REFERENCES

[1] Donsker, M. D., and Varadhan, S. R. S., Asymptotic evaluation of certain Markov process expectations for large times I, Comm. Pure Appl. Math., Vol. 28, 1975, pp. 1-47.

[2] Donsker, M. D., and Varadhan, S. R. S., Asymptotic evaluation of certain Markov process expectations for large times II, Comm. Pure Appl. Math., Vol. 28, 1975, pp. 279-301.

[3] Donsker, M. D., and Varadhan, S. R. S.,

Asymptotics for the Wiener Sausage, Comm. Pure Appl. Math., Vol. 28, 1975, pp. 525-565.

[4] Kac, M., Probabilistic methods in some problems of scattering theory, Rocky Mount. Jour. Math., Vol. 4, No. 3, 1974, pp. 511-537.

[5] Spitzer, F., Principles of random walk, Van Nostrand, 1964.

ON DUALITY FOR MARKOV PROCESSES

E. B. Dynkin[1]

Department of Mathematics
Cornell University
Ithaca, New York

1. The idea of duality plays an important role in applications of stochastic analysis to potential theory and partial differential equations.

The following definition prevails in the literature of the last decade: two Markov processes (x_t, P) and (x_t^*, P^*) are in duality relative to a given σ-finite measure m if their transition functions are connected by the relation

$$m(dx)p_t(x,dy) = m(dy)p_t^*(y,dx). \tag{1}$$

(In this definition, only stationary transition functions are considered which became rather common in 1950's and 1960's.) No relation between paths of x_t and x_t^* is requested: the processes can even be defined on different spaces Ω and Ω^*.

However the real source of duality is the fact that each Markov process can be considered in two directions of time (for example, this is a way Kolmogorov forward and backward differential equations are deduced).

In the present paper, a theory of duality is outlined in which functions p and p^* connected by formula (1) are interpreted as a forward and a backward transition function of a single Markov

[1]Supported by NSF grant MCS 77-03543 A01.

ISBN 0-12-268380-3

process with random birth and death times. To simplify the comparison with the traditional approach, we start from the stationary case, but then the theory is developed in its natural frames without any homogeneity assumptions.

First this approach was suggested in [1] (see also [2,3]). Our aim is to present basic ideas as well as some results obtained recently.

2. Suppose that two stationary transition functions p and p* in a standard Borel space $(E,\mathcal{B})$ are connected by formula (1). Suppose, in addition, that $m(E) = 1$ and $p_t(x,E) = p^*_t(x,E) = 1$ for all $x \in E$. Then

$$\int_B m(dx)p_t(x,B) = m(B).$$

Applying Kolmogorov Theorem, we construct a stochastic process (x_t,P) on the time interval $R = (-\infty,+\infty)$ such that

$$P\{x_{t_1} \in dy_1,\ldots,x_{t_n} \in dy_n\}$$

$$= m(dy_1)p_{t_2-t_1}(y_1,dy_2)\cdots p_{t_n-t_{n-1}}(y_{n-1},dy_n).$$

For this process, m is the probability distribution at time t, p is a forward transition function and p* is a backward transition function:

$$P\{x_t \in B\} = m(B);$$

$$P\{x_t \in B|x_u,u \leq s\} = p_{t-s}(x_s,B) \quad \text{a.s. } P;$$

$$P\{x_s \in B|x_u,u \geq t\} = p_{t-s}(x_t,B) \quad \text{a.s. } P.$$

Instead of two unrelated processes on the time interval $[0,+\infty)$ we have a single process on $(-\infty,+\infty)$.

3. Now we consider the general case. Let a measurable space $(E_t,\mathcal{B}_t)$ be associated with each $t \in R$. A function $p(s,x;t,B)$, $s < t \in R$, $x \in E_s$, $B \in \mathcal{B}_t$ is a <u>forward transition function</u> if it is $\mathcal{B}_s$-measurable in x, is a measure with respect to B and if the following conditions are satisfied:

$$p(s,x;\ t,E_t) \leq 1, \tag{2}$$

$$\int_{E_t} p(s,x;\ t,dy)p(t,y;\ u,B) = p(s,x;u,B) \tag{3}$$

for $s < t < u \in R$ (Chapman-Kolmogorov equation).

A <u>backward transition function</u> $p^*(s,B;\ t,y)$, $s < t \in R$, $B \in \mathcal{B}_s$, $y \in E_t$ is a $\mathcal{B}_t$-measurable function in y and a measure with respect to B such that

$$p^*(s,E_s;\ t,x) \leq 1, \tag{4}$$

$$\int_{E_t} p^*(s,B;\ t,y)p^*(t,dy;\ u,z) = p^*(s,B;u,z) \tag{5}$$

for $s < t < u \in R$.

Let a σ-finite measure $m_t(B)$ on E_t be given for each $t \in R$. We say that <u>transition functions</u> p <u>and</u> p* <u>are in duality relative to</u> m <u>if</u>

$$m_s(dx)p(s,x;\ t,dy) = p^*(s,dx;\ t,y)m_t(dy). \tag{6}$$

We assume that all $(E_t,\mathcal{B}_t)$ are standard Borel spaces and that the transition functions p and p* are <u>normal</u>, which means that

$$p(s,x;\ t,E_t) \uparrow 1 \quad \text{as } t \downarrow s, \tag{7}$$

$$p^*(s,E_s;\ t,y) \uparrow 1 \quad \text{as } s \uparrow t. \tag{8}$$

Then there exists a stochastic process (x_t,P) with the birth time α and the death time β such that, for all $t_1 < t_2 < \cdots < t_n$,

$$\begin{aligned} &P\{\alpha < t_1,\ x_{t_1} \in dy_1,\ldots,x_{t_n} \in dy_n,\ t_n < \beta\} \\ &= m_{t_1}(dy_1)p(t_1,y_1;\ t_2,dy_2)\cdots p(t_{n-1},y_{n-1};\ t_n,dy_n) \\ &= p^*(t_1,dy_1;\ t_2,y_2)\cdots p^*(t_{n-1},dy_{n-1};\ t_n,y_n)m_{t_n}(dy_n). \end{aligned} \tag{9}$$

It follows from here that

$$P\{\alpha < t < \beta,\ x_t \in B\} = m_t(B), \tag{10}$$

$$P\{\alpha < t < \beta,\ x_t \in B \mid x_u, u \le s\} = p(s, x_s;\ t, B) \tag{11}$$
$$\text{a.s. } P,\ \alpha < s < \beta,$$

$$P\{\alpha < s < \beta,\ x_s \in B \mid x_u, u \ge t\} = p^*(s, B;\ t, x_t) \tag{12}$$
$$\text{a.s. } P,\ \alpha < t < \beta.$$

The relations (11) and (12) mean that (x_t, P) is a Markov process with a forward transition function p and a backward transition function p*.

4. Constructing a process subject to conditions (9) is a particular case of a general problem: to construct a stochastic process on a random time interval starting from a given system of finite-dimensional distributions. We start with precise definitions.

A stochastic process on a random time interval is determined by the following elements:

(i) a measure space $(\Omega, \mathcal{F}, \mathbf{P})$;

(ii) two measurable functions $\alpha(\omega) < \beta(\omega)$ on Ω with values in the extended real line $[-\infty, +\infty]$;

(iii) for each $t \in R$, a measurable mapping $x_t(\omega)$ of the set $\{\omega; \alpha(\omega) < t < \beta(\omega)\}$ into a measurable space $(E_t, \mathcal{B}_t)$.

Let Λ be a finite subset of R. We denote by $(E_\Lambda, \mathcal{B}_\Lambda)$ the product space of $(E_t, \mathcal{B}_t)$, $t \in \Lambda$ and we denote by x_Λ a collection of x_t, $t \in \Lambda$. We write $s < \Lambda$ if $s < t$ for all $t \in \Lambda$, and $u > \Lambda$ if $u > t$ for all $t \in \Lambda$. The formula

$$m_\Lambda(B_\Lambda) = P\{\alpha < \Lambda < \beta,\ x_\Lambda \in B_\Lambda\} \tag{13}$$

defines a measure m_Λ on $(E_\Lambda, \mathcal{B}_\Lambda)$. The collection of measures m_Λ for all finite $\Lambda \subset R$ is called a system of finite-dimensional distributions of $(x_t, \mathbf{P})$. The question is: under what conditions on $\{m_\Lambda\}$ does there exist a process $(x_t, \mathbf{P})$ satisfying relations (9)?

It follows from (13) that

$$m_{\Lambda_1 t \Lambda_2}(B_{\Lambda_1} \times E_t \times B_{\Lambda_2}) = m_{\Lambda_1 \Lambda_2}(B_{\Lambda_1} \times B_{\Lambda_2}) \quad \text{if } \Lambda_1 < t < \Lambda_2. \tag{14}$$

Further (13) implies that

$$m_\Lambda(B_\Lambda) = m_{s\Lambda}(E_s \times B_\Lambda) + P\{s \leq \alpha < \Lambda,\ x_\Lambda \in B_\Lambda\}, \text{ if } s < \Lambda; \tag{15}$$

$$m_\Lambda(B_\Lambda) = m_{\Lambda u}(B_\Lambda \times E_u) + P\{x_\Lambda \in B_\Lambda,\ \Lambda < \beta \leq u\}, \text{ if } u > \Lambda; \tag{16}$$

$$m_\Lambda(B_\Lambda) + m_{\Lambda u}(E_s \times B_\Lambda \times E_u) = m_{s\Lambda}(E_s \times B_\Lambda) + m_{\Lambda u}(B_\Lambda \times E_u)$$

$$+ P\{s \leq \alpha < \Lambda < \beta \leq u,\ x_\Lambda \in B_\Lambda\} \quad \text{if } s < \Lambda < u. \tag{17}$$

Hence[1]

$$m_{s\Lambda}(E_s \times B_\Lambda) \uparrow m_\Lambda(B_\Lambda) \quad \text{as } s \uparrow \Lambda, \text{ if } m_\Lambda(B_\Lambda) < \infty; \tag{18}$$

$$m_{\Lambda u}(B_\Lambda \times E_u) \uparrow m_\Lambda(B_\Lambda) \quad \text{as } u \downarrow \Lambda, \text{ if } m_\Lambda(B_\Lambda) < \infty; \tag{19}$$

$$m_{s\Lambda}(E_s \times B_\Lambda) + m_{\Lambda u}(B_\Lambda \times E_u) \leq m_\Lambda(B_\Lambda) + m_{s\Lambda u}(E_s \times B_\Lambda \times E_u) \tag{20}$$

$$\text{for } s < \Lambda < u.$$

Assume that all $(E_t, \mathcal{B}_t)$ are standard Borel spaces and all m_Λ are σ-finite measures. Then conditions (14),(18),(19),(20) are not only necessary but also sufficient for existence of $(x_t, \mathbf{P})$. The proof will be published in another place (it is based on the same ideas as the proof of a particular case given by Kuznecov in [4]).

The system of measures m_Λ defined by the right side of formula (9) satisfies the conditions (14),(18),(19) and (20) (this follows from (2),(3),(7) and (8)). Therefore there exists a process (x_t, P) for which (9) and hence (10),(11) and (12) are valid.

It is useful to have an expression of $P(\Omega)$ through $\{m_\Lambda\}$. Put

$$c_t = m_t(E_t), \quad c_{st} = m_{st}(E_s \times E_t). \tag{21}$$

---

[1] $s \to \Lambda$ means $s \uparrow t$, where t is the smallest element of Λ.

If $c_s = \infty$ for some s, then $\mathbf{P}(\Omega) = \infty$. If all c_s are finite, we associate with each finite set $\Lambda = \{s_0 < s_1 < \cdots < s_n\}$ a real number

$$c(\Lambda) = \sum_{k=0}^{n} c_{s_k} - \sum_{k=1}^{n} c_{s_{k-1}s_k}. \tag{22}$$

It follows from (15) that $\mathbf{P}(\Omega)$ is equal to the supremum of $c(\Lambda)$ over all finite sets Λ (also $\mathbf{P}(\Omega) = \lim c(\Lambda_n)$ for every increasing sequence Λ_n with the union everywhere dense in R).

5. In no. **4** infinite measures **P** were admitted. Of course, if P is infinite, $\mathbf{P}(A)$ cannot be interpreted as a probability of A. However, a little modification of the definitions is sufficient to get a rather realistic model of certain phenomena.

Let a point x_t of a measurable space $(E_t, \mathcal{B}_t)$ be fixed for each t of an open interval (α, β). Then we say that x_t, $\alpha < t < \beta$ is a trajectory in $(E_t, \mathcal{B}_t)$. Now suppose that $(\Omega^0, \mathcal{F}^0, \mathbf{P}^0)$ is a probability space and that, for every $\omega_0 \in \Omega^0$, a finite (maybe empty) or countable set of trajectories $x_t^k(\omega_0)$, $\alpha_k(\omega_0) < t < \beta_k(\omega_0)$ in $(E_t, \mathcal{B}_t)$ are given. We say that $(x_t^k, \mathbf{P}^0)$ is a <u>multi-valued stochastic process</u> if, for each $t \in R$ and $B_t \in \mathcal{B}_t$, the number of trajectories satisfying the condition $\alpha_k(\omega_0) < t < \beta_k(\omega_0)$, $x_t^k(\omega_0) \in B_t$ is $\mathcal{F}^0$-measurable.

The principal result of no. **4** can be proved in the following stronger form. Assume that $(E_t, \mathcal{B}_t)$ are standard Borelian and m_Λ are σ-finite. Then (14),(18),(19) and (20) are necessary and sufficient for the existence of a multi-valued stochastic process $(x_t^k, \mathbf{P}^0)$ such that $m_\Lambda(B_t)$ is equal to an expected number of trajectories subject to the condition $\alpha_k(\omega_0) < \Lambda < \beta_k(\omega_0)$, $x_\Lambda(\omega_0) \in B_\Lambda$.

If $(E_t, \mathcal{B}_t)$ does not depend on t and if m_Λ are invariant under translation, we have a stationary multi-valued process on a random time interval. In this case the expected number s of births and deaths in time interval Δ depend only on the length of Δ. Hence either a.s. all particles live from $-\infty$ to $+\infty$ or the expected number of particles is infinite. A typical situation in the second case is: there exist a finite number of particles at each time t; each particle has a finite life time but the total number of particles is infinite.

6. Now we apply the results of no. **4** to construct a Markov process corresponding to a transition function p, an excessive measure m and an excessive function h (the process defined by formula (9) is a particular case).

Operators T_t^s are associated with a transition function p which act on measures by formula

$$\nu T_t^s(B) = \int_{E_s} \nu(dx)p(s,x;\ t,B)$$

and on positive measurable functions by formula

$$T_t^s f(x) = \int_{E_t} p(s,x;\ t,dy)f(y).$$

Let m_t be a σ-finite measure on $(E_t,\mathcal{B}_t)$ and let

$$m_s T_t^s \uparrow m_t \quad \text{as } s \uparrow t.$$

Then we say that m is an <u>excessive measure</u> (relative to p). Analogously, if, for each t, h^t is a measurable function on $(E_t,\mathcal{B}_t)$ with values in $[0,+\infty]$ and if

$$T_t^s h^t \uparrow h^s \quad \text{as } t \downarrow s,$$

then we say that h is an <u>excessive function</u>. Put

$$m_\Lambda(dx_1,dx_2,\ldots,dx_n) = m_{t_1}(dx_1)p(t_1,x_1;t_2,dx_2) \cdots p(t_{n-1},x_{n-1};t_n,dx_n)h^{t_n}(x_n) \tag{23}$$

for $\Lambda = \{t_1 < t_2 < \cdots < t_n\}$. If $h^t < \infty$ a.s. m_t for all t, then all measures m_Λ are σ-finite. It is easy to see that conditions (14),(18),(19) and (20) are satisfied. Hence there exists a stochastic process $(x_t, \mathbf{P})$ on a random time interval with finite-dimensional distributions (23). To mention explicitly the dependence of $\mathbf{P}$ on m and h, we shall write $\mathbf{P}_m^h$.

If the transition function p is normal, then $h = 1$ is an excessive function; the process P_m^1 coincides with the process described in no. **3**. If $h^s(x)$ is an arbitrary strictly positive finite excessive function, then $(x_t, \mathbf{P}_m^h)$ is a Markov process with the transition function

$$p^h(s,x;\ t,dy) = \frac{1}{h^s(x)}\, p(s,x;\ t,dy)h^t(y).$$ [1]

Quantities

$$c_t = m_t(h^t) = \int_{E_t} h^t(x)m_t(dx),$$

$$c_{st} = m_s T_t^s h^t = \int_{E_s}\int_{E_t} m_s(dx)p(s,x;\ t,dy)h^t(y)$$

introduced in (21) are bilinear functions of m, h. So are $c(\Lambda)$ defined by (22). Since $P_m^h(\Omega) = \lim c(\Lambda_n)$ for some sequence Λ_n,

$$P_m^h(\Omega) = \langle m,h \rangle$$

is also a bilinear function of m, h.

Suppose that $\ell^t(y)$ is a positive function on E_t and λ is a measure on R. If an integral

$$h^s(x) = \int_s^\infty \int_{E_t} p(s,x;\ t,dy)\ell^t(y)\lambda(dt) = \int_s^\infty T_t^s \ell^t(x)\lambda(dt) \tag{24}$$

has a meaning, it defines an excessive function h and

$$\langle m,h \rangle = \int_{-\infty}^{+\infty} m_t(\ell^t)\lambda(dt) \tag{25}$$

---

[1]In the general case, p^h is a normal transition function on $E_t' = \{0 < h^t(x) < \infty\}$ and, if $h^t < \infty$ a.s. m_t for all t, the process $(x_t, \mathbf{P}_m^h)$ can be considered on the reduced state space E_t'.

for each excessive measure m. Analogously, if an integral

$$m_t(\Gamma) = \int_{-\infty}^{t} \int_{E_t} \gamma(ds,dx)p(s,x;\ t,\Gamma) \tag{26}$$

has a meaning for a measure γ, then it defines an excessive measure m and

$$<m,h> = \int_{-\infty}^{+\infty} \gamma(ds,dx)h^s(x) \tag{27}$$

for every excessive function h.

We say that $m_t(\Gamma)$ is an <u>entrance law</u> at time s_0, $-\infty \leq s_0 < +\infty$ if

$$m_t = 0 \qquad \text{for } t \leq s_0,$$

$$m_s T_t^s = m_t \qquad \text{for } t > s > s_0.$$

Analogously $h^s(x) \geq 0$ is an <u>exit law</u> at time u_0, $-\infty < u_0 \leq +\infty$ if

$$h^s = 0 \qquad \text{for } t \geq u_0,$$

$$T_t^s h^t = h^s \qquad \text{for } s < t < u_0.$$

Obviously all entrance laws are excessive measures and all exit laws are excessive functions. If m is an entrance law, at time s_0, then, for an arbitrary excessive function h, $\alpha = s_0$ a.s. $\mathbf{P}_m^h$ and $m_t(h^t) \uparrow <m,h>$ as $t \downarrow s_0$. If h is an exit law at time u_0, then, for an arbitrary excessive measure m, $\beta = u_0$ a.s. P_m^h and $m_t(h^t) \uparrow <m,h>$ as $t \uparrow u_0$. If m is an entrance law at s_0 and h is an exit law at time $u_0 > s_0$, then the process $(x_t,\mathbf{P}_m^h)$ has non-random birth time s_0 and death time u_0 and $\mathbf{P}_m^h(\Omega) = m_t(h^t)$ for all $s_0 < t < u_0$.

7. Up to now we were interested in constructing Markov processes with given transition functions. Now we discuss the converse problem — constructing transition functions for a given Markov process.

A stochastic process $(x_t,\mathbf{P})$ on a random time interval is called a Markov process, if

$$\mathbf{P}(AB|x_t) = \mathbf{P}(A|x_t)\ \mathbf{P}(B|x_t) \quad \text{a.s. } \mathbf{P},\ \alpha < t < \beta$$

for each $t \in R$, $A \in \mathcal{F}_{<t}$ and $B \in \mathcal{F}_{>t}$ (here $\mathcal{F}_{<t}$ is the minimal σ-algebra in Ω which contains all sets $\{\alpha<s<\beta,\ x_s \in B\}$ for $s < t$, $B \in \mathcal{B}_s$; the definition $\mathcal{F}_{>t}$ is similar). We say that p is a forward transition function for $(x_t,\mathbf{P})$ if (2),(3) and (11) hold. Analogously p* is a backward transition function for (x_t,P) if (4),(5) and (12) are satisfied.

Not all Markov processes have transition functions. However, both forward and backward transition functions do exist if the following condition is satisfied:

(AC) For every $s < t \in R$, the two dimensional distribution

$$m_{st}(B) = P\{(x_s,x_t) \in B\}, \quad B \in \mathcal{B}_s \times \mathcal{B}_t$$

is absolutely continuous with respect to the product $m_s \times m_t$ of the corresponding one-dimensional distributions.

Under condition (AC) a Radon-Nykodim derivative $\rho(s,x;\ t,y)$ of m_{st} with respect to $m_s \times m_t$ can be chosen in such a way that

$$\int_{E_t} \rho(s,x;\ t,y)m_t(dy)\rho(t,y;\ u,z) = \rho(s,x;\ u,z)$$

$$\text{for all } s < t < u \in R,\ x \in E_s,\ z \in E_u,$$

$$\int_{E_t} \rho(s,x;\ t,y)m_t(dy) \leq 1 \quad \text{for all } s < t \in R,\ x \in E_s,$$

$$\int_{E_s} m_s(dx)\rho(s,x;\ t,y) \leq 1 \quad \text{for all } s < t \in R,\ y \in E_t.$$

The forward and the backward transition functions for $(x_t,\mathbf{P})$ can be defined by formulas

$$p(s,x;\ t,dy) = \rho(s,x;\ t,y)m_t(dy) \tag{28}$$

and

$$p^*(s,dx;\ t,y) = m_s(dx)\rho(s,x;\ t,y). \tag{29}$$

The function $\rho(s,x;\ t,y)$ is excessive in s, x relative to the forward transition function (28) and is excessive in t, y relative to the backward transition function (29). It plays the same role in general theory as the fundamental solution of two adjoint parabolic differential equations does in the theory of diffusion. We call ρ a fundamental density for the process $(x_t,\mathbf{P})$.

If the process $(x_t,\mathbf{P})$ is stationary, then there exists a stationary fundamental density $\rho(s,x;\ t,y) = \rho_{t-s}(x,y)$. Formulas (28) and (29) define in this case stationary transition functions.

8. An advanced theory of Markov processes is based on a concept of a Markov process as a family of measures on the space $(\Omega,\mathcal{F})$ rather than a single measure $\mathbf{P}$. Suppose that a sub-σ-algebra $\mathcal{F}(I)$ of σ-algebra $\mathcal{F}$ is given for each open interval I and that $\{\alpha < t < \beta,\ x_t \in B\} \in \mathcal{F}(I)$ for $t \in I$, $B \in \mathcal{B}_t$. Let probability measures $\mathbf{P}_{t,x}$ on $\mathcal{F}_{>t} = \mathcal{F}(t,+\infty)$ and $\mathbf{P}^{t,x}$ on $\mathcal{F}_{<t} = \mathcal{F}(-\infty,t)$ be defined for each $t \in R$, $x \in E_t$ such that:

A. $\alpha = t$ a.s. $\mathbf{P}_{t,x}$ (this means that, if $\alpha \neq t$ everywhere on $A \in F_{>t}$, then $P_{t,x}(A) = 0$).

A*. $\beta = t$ a.s. $\mathbf{P}^{t,x}$.

B. For each $A \in \mathcal{F}_{>t}$,

$$\mathbf{P}\{A|\mathcal{F}_{<t},x_t\} = \mathbf{P}_{t,x_t}(A) \quad \text{a.s. } P,\ \alpha < t < \beta.$$

B*. For each $A \in \mathcal{F}_{<t}$,

$$\mathbf{P}\{A|\mathcal{F}_{>t},x_t\} = \mathbf{P}^{t,x_t}(A) \quad \text{a.s. } P,\ \alpha < t < \beta.$$

C. For every $s < t$, $A \in \mathcal{F}_{>t}$,

$$\mathbf{P}_{s,x}\{A|\mathcal{F}(s,t),x_t\} = \mathbf{P}_{t,x_t}(A) \quad \text{a.s. } P_{s,x},\ t < \beta.$$

C*. For every $u > t$, $A \in \mathcal{F}_{<u}$,

$$\mathbf{P}^{u,x}\{A|\mathcal{F}(t,u),x_t\} = \mathbf{P}^{t,x_t}(A) \quad \text{a.s. } P^{u,x},\ \alpha < t.$$

Then we say that $X = (x_t, \mathbf{P}, \mathbf{P}_{t,x}, \mathbf{P}^{t,x})$ is _a Markov process relative to_ $\mathcal{F}(I)$. Measures $\mathbf{P}_{t,x}$ and $\mathbf{P}^{t,x}$ are called _forward and backward transition probabilities_.

Denote by $\mathcal{E}$ the union of all E_t. To each $u \in R$, $A \in \mathcal{F}_{<u}$ and $B \in \mathcal{F}_{>u}$, there correspond real-valued functions on $\mathcal{E}$ defined by formulas

$$f(t,x) = \begin{cases} \mathbf{P}_{t,x}(A) & \text{for } t \leq u, \\ 0 & \text{for } t > u, \end{cases}$$

$$g(t,x) = \begin{cases} \mathbf{P}^{t,x}(B) & \text{for } t \geq u, \\ 0 & \text{for } t < u. \end{cases}$$

These functions generate the natural measurable structure in $\mathcal{E}$.

If $(x_t, \mathbf{P})$ satisfies condition (AC) of no. **7**, the transition probabilities can be defined by the formulas:

$$\mathbf{P}_{t,x}(A) = \int_A \rho(t,x;\, u,x_u) P(d\omega) \qquad \text{for } t < u,\ A \in \mathcal{F}_{>u}, \tag{30}$$

$$\mathbf{P}^{t,x}(A) = \int_A \rho(s,x_s;\, t,x) P(d\omega) \qquad \text{for } s < t,\ A \in \mathcal{F}_{<s}, \tag{31}$$

where $\rho(s,x;\, u,y)$ is a fundamental density.[1]

Conditions A and A*, B and B*, C and C* are examples of dual statements which can be obtained from each other by time reversal.

We consider here processes on an open time interval (α,β). Processes on half-open intervals $[\alpha,\beta)$, $(\alpha,\beta]$ and on closed interval $[\alpha,\beta]$ can be treated in a similar way.

9. Most applications of stochastic processes to classical analysis depend on certain regularity properties of paths. We introduce two dual forms of regularity conditions: R-regularity and L-regularity.

---

[1] Formula (30) defines the measure $\mathbf{P}_{t,x}$ on each σ-algebra $\mathcal{F}_{>u}$ with $u > t$. To get a measure on $\mathcal{F}_{>t}$, we need some natural restrictions on the family $\mathcal{F}(I)$ (see [2,3]).

A Markov process $X = (x_t, \mathbf{P}, \mathbf{P}_{t,x}, \mathbf{P}^{t,x})$ on the time interval $[\alpha, \beta)$ is called R-<u>regular</u> if:

(a) functions[2]

$$\mathbf{P}_{t,x_t} Y^u, \quad Y^u \in \mathcal{F}_{>u}$$

are right-continuous in t on $\Delta_u = [\alpha,\beta) \cap (-\infty, u)$ a.s. P' where P' is any of measures $\mathbf{P}, \mathbf{P}_{s,x}$, $s < u$, $x \in E_s$;

(b) functions

$$\mathbf{P}^{t,x_t} Y_s, \quad Y_s \in \mathcal{F}_{<s}$$

are right-continuous in t on $\Delta^s = [\alpha,\beta) \cap [s, +\infty)$ a.s. $\mathbf{P}'$ if $\mathbf{P}'Y < \infty$ where P' is any of measures $\mathbf{P}, \mathbf{P}^{u,x}, u > s$, $x \in E_u$.

(c) for each $u > t$, x_t is (t,u)-measurable, and $\mathbf{P}_{t,x}\{x_t = x\} = 1$;

(d) the union $\mathcal{E} = \cup E_t$ with the natural measurable structure is a standard Borel space; if $f(t,x)$ is a measurable function on $\mathcal{E}$, then the restriction of $f(t, x_t(\omega))$ to the set $\alpha < s < t < u < \beta$ is $\mathcal{B}(s,u) \times \mathcal{F}(s,u)$-measurable (here $\mathcal{B}(s,u)$ is the Borel σ-algebra on the interval (s,u)).

To get the definition of L-<u>regularity</u>, one needs to replace the word "right" by "left" and to do other obvious changes.

Now let $X = (x_t, \mathbf{P}, \mathbf{P}_{t,x}, \mathbf{P}^{t,x})$ be an arbitrary Markov process with standard Borel state spaces $(E_t, \mathcal{B}_t)$. Then there exists an R-<u>regularization</u> of X, i.e., an R-regular Markov process $X^+ = (x_{t+}, \mathbf{P}, \mathbf{P}_{t+,x}, \mathbf{P}^{t+,x})$ with the same space $(\Omega, \mathcal{F})$ and the same σ-algebras $\mathcal{F}(I)$ such that, for an arbitrary everywhere dense subset Λ of R, the following relations hold:

(i) for each $u \in R$, $Y^u \in \mathcal{F}_{>u}$,

---

[2]If $\mathcal{F}$ is a σ-algebra and P is a measure on $\mathcal{F}$, then writing $Y \in \mathcal{F}$ means that Y is a positive $\mathcal{F}$-measurable function, and PY means an integral of Y with respect to **P**.

$$\mathbf{P}_{t+,x_{t+}} Y^u = \lim_{\substack{r \downarrow t \\ r \in \Lambda}} \mathbf{P}_{r,x_r} Y^u \quad \text{for all } t \in \Delta_u \text{ a.s. } \mathbf{P}'$$

where $\mathbf{P}'$ is any of measures $\mathbf{P}, \mathbf{P}_{s,x}, \mathbf{P}_{s+,x}$;

(ii) for each $s \in R$, $Y_s \in \mathcal{F}_{<s}$,

$$\mathbf{P}^{t+,x_{t+}} Y_s = \lim_{\substack{r \downarrow t \\ r \in \Lambda}} \mathbf{P}^{r,x_r} Y_s \quad \text{for all } t \in \Delta^s \text{ a.s. } \mathbf{P}'$$

where $\mathbf{P}'$ is any of measures $\mathbf{P}, \mathbf{P}^{u,x}, \mathbf{P}^{u+,x}$.

There exists, of course, also an L-regularization $X^- = (x_{t-}, \mathbf{P}, \mathbf{P}_{t-,x}, \mathbf{P}^{t-,x})$ of the process X related to X by conditions dual to (i) and (ii). The processes X^+ and X^- form together one Markov process $(x_v, \mathbf{P}, \mathbf{P}_{v,x}, \mathbf{P}^{v,x})$ with the time parameter v taking values in the split real line V (the latter contains two points t+ and t- corresponding to each real number t and also points $+\infty$ and $-\infty$).

A regular split process corresponding to a given Markov process X was constructed first in [2] under additional assumption (AC). The restriction was recently removed by Kuznecov [5]. Only processes with fixed birth and death times were treated in [5]. The general case can be investigated using the fact that a measure P corresponding to an arbitrary Markov process can be represented as an integral of measures corresponding to processes on fixed time intervals. We shall discuss these topics in another publication.

REFERENCES

1. Dynkin, E. B., Integral representation of excessive measures and excessive functions, Russian Math. Surveys, 27, 1 (1972), 43-84.

2. Dynkin, E. B., Markov representations of stochastic systems, Russian Math. Surveys, 30, 1 (1975), 64-104.

3. Dynkin, E. B., On a new approach to Markov processes, Proceedings of the Third Japan - USSR Symposium on Probability Theory, Springer, Berlin, 1976 (Lecture Notes in Mathematics 550, 42-62).

4. Kuznecov, S. E., Construction of Markov processes with random birth and death times, Teoriya Veroyatnostei i ee Primeneniya, 18, 3 (1973), 596-601 (English translation in Theory of Probability and Its Applications).

5. Kuznecov, S. E., Construction of a regular split process, Teoriya Veroyatnostei i ee Primeneniya, 22, 4 (1977), 791-812 (English translation in Theory of Probability and Its Applications).

STOCHASTIC DYNAMICAL SYSTEMS AND THEIR FLOWS

K. D. ELWORTHY
University of Warwick, Coventry, England

SUMMARY

The basic theory of stochastic differential equations on manifolds is reviewed together with some of the properties of their solutions and some geometric examples. The approach used works equally well for certain infinite dimensional manifolds, in particular for separable Hilbert manifolds. Using this infinite dimensional theory, under suitable conditions a stochastic differential equation on a compact manifold M is shown to lift to a diffeomorphism group of M and to have full solutions there. These solutions are the 'flow' of the original equation.

This technique was used by Ebin and Marsden [7] for ordinary differential equations. It gives straightforward proofs of some of the results of Baxendale and of Malliavin concerning the almost everywhere smoothness with respect to the initial points of a suitable version of the flow of a stochastic differential equation: results which seem to have been new for equations on $\mathbb{R}^n$ as well as on manifolds. Other applications are given: the uniform convergence, with respect to initial point, of solutions of ordinary differential equations to solutions of stochastic ones is discussed in §6; and a version of Itô's lemma for functionals of Sobolev class $W^{2,2}$ is in §10.

1. STOCHASTIC DIFFERENTIAL EQUATIONS ON HILBERT SPACE

Let w denote a standard Brownian motion on $\mathbb{R}^n$ with probability space $\{\Omega, F, \mu\}$. For a real Hilbert space H the basic theory of stochastic integral equations for H-valued processes driven by w works in essentially the same way as for finite dimensional state

ISBN 0-12-268380-3

spaces: it is only necessary to cast the finite dimensional theory into a basis free notation. In particular if $X:H \to L(\mathbb{R}^n;H)$ and $A:H \to H$ are globally Lipshitz, where $L(\mathbb{R}^n;H)$ denotes the space of linear maps, the equation

$$x(t) = x_o + \int_0^t X(x(s))dw(s) + \int_0^t A(x(s))ds$$

with a given initial value x_o, has (up to equivalence) a unique non-anticipating solution

$$x:[0,\infty) \times \Omega \to H$$

with continuous sample paths. As usual the proof can be reduced to an application of the contraction mapping principle to the mapping induced on the space of continuous non-anticipating functions $x:[0,t] \to L^2(\Omega, F,\mu;H)$.

This can be easily extended to the case where w is itself an infinite dimensional Brownian motion, e.g. see [2], and various authors have extensions to wider classes of Banach spaces. In particular A. Neidhardt [20] shows that the basic estimates and results also go through easily when H belongs to a more general class of Banach spaces, including the L_p spaces for $2 \leq p < \infty$. Consequently the results of the next section apply equally well to manifolds modelled on such spaces.

2. STOCHASTIC DIFFERENTIAL EQUATIONS ON MANIFOLDS

The approach which we will use seems to have first appeared in an article by J.M.C. Clark [5]. As Clark says a lot of the work was already done in Itô's 1950 paper [13]. Clark uses Fisk-Stratonovich integrals, but we will use Itô integrals throughout (though not Itô differentials!).

Let M be a separable metrizable C^2 manifold modelled on the Hilbert space H. Let T_mM denote the tangent space at m to M. Suppose V is a vector field on M and $\{X(m)\}_{m\in M}$ is a family of linear maps $X(m):\mathbb{R}^n \to T_mM$. A *localization* of the stochastic dynamical system (X,V) at the point m of M will be a triple

$\Lambda = ((U,\phi),U_o,\lambda)$ where:

(i) (U,ϕ) is a C^2 chart about m, with ϕ mapping the open set U of M onto a bounded open set W of H.

(ii) U_o is an open neighbourhood of m in U with $\overline{W}_o \subset W$, where $W_o = \phi(U_o)$.

(iii) $\lambda:H \to [0,1]$ is C^1 with supp $\lambda \subset W$ and $\lambda|W_o \equiv 1$.

(iv) If $X_\Lambda:H \to L(E;H)$ is defined by $X_\Lambda(h) = \lambda(h)\phi_*(X)(h)$, both X_Λ and $V_\Lambda + \frac{1}{2}$ tr $DX_\Lambda.X_\Lambda$ are globally Lipshitz.

Here $\phi_*(X)$ is the local representative of X in the chart (U,ϕ): $\phi_*(X)(h) = T_{\phi^{-1}(h)}\phi\circ X(\phi^{-1}(h))$. Also tr $DX_\Lambda.X_\Lambda:H \to H$ is the map

$$h \mapsto \sum_{i=1}^{n} DX_\Lambda(h)(X_\Lambda(h)e_i,e_i)$$

for the orthonormal base $e_1,\ldots,e_n$ of $\mathbb{R}^n$.

We will assume that a localization exists at each point of M; this will certainly be true if X and V are of class C^2 and C^1 respectively. Given a localization Λ and times $t \geq r \geq 0$ we have the equation

$$y(t)=y(r)+\int_r^t X_\Lambda(y(s))dw(s)+\int_r^t V_\Lambda(y(s))ds+\frac{1}{2}\int_r^t trDX_\Lambda.X_\Lambda(y(s))ds \quad (1)$$

For $a \geq 0$, a process $x:[a,\xi) \times \Omega \to M$, defined up to a stopping time ξ, will be called a *solution of the stochastic differential equation* $dx = Xdw + Vdt$ if it is non-anticipating, has continuous sample paths, and if there is a cover of M by localizations Λ such that if $\tilde{\phi}:M \to H$ is a measurable extension of ϕ and $r \geq a$ then $\tilde{\phi} \circ x$ agrees with the solution of (1), for $y(r) = \tilde{\phi}\circ x(r)$, almost surely on $\{x(r) \in U_o\}$ from time r up to the first exit time after r of that solution from W_o.

Given two such solutions $x_i:[a,\xi_i) \times \Omega \to M$, $i = 1,2$ with $x_1(a) = x_2(a)$ almost surely, it is straightforward to check that their restrictions to $[a,\xi_1 \wedge \xi_2) \times \Omega$ agree a.s., and furthermore they determine a solution defined on $[a,\xi_1 \vee \xi_2) \times \Omega$. This is then easily extended to the case of any countable number of such solutions. Letting ξ denote the least upper bound (in the sense of [21]) of the stopping times corresponding to all solutions with given initial value $x(0)$ we obtain a *maximal solution* $x:[0,\xi) \times \Omega \to M$, defined up to the *explosion time* ξ, and this solution is essentially unique. Finally we can check that $\xi > 0$ almost surely.

3. PROPERTIES OF SOLUTIONS

A cover C of a subset A of M by localizations is *uniform* if there exist $r > 0$, $K < \infty$ such that for $\Lambda \in C$

$$\{h \mid \|h\| < 2r\} \subset W_o$$

$$\|X_\Lambda(h)\|_{L(\mathbb{R}^n,H)} \leq K \text{ and } \|V_\Lambda(h) + \frac{1}{2} \operatorname{tr}DX_\Lambda . X_\Lambda(h)\|_H \leq K \text{ all } h \in H$$

and such that $\{\phi^{-1}(\|h\|<r): \Lambda \in C\}$ covers A.

The following is due to Itô [13], see also [5]; as these authors point out, it includes the usual linear growth criterion for the case $M=H$. Slight modifications are needed in the proof for the infinite dimensional case:

NON-EXPLOSION CRITERION. If there is a uniform cover of M then $\xi = \infty$ a.s.

Slightly more can be extracted: if a subset A of M has a uniform cover then, with probability one, if explosion occurs it occurs after leaving A; that is $\lim_{t\to\xi(w)} \chi_A(x(t,w)) = 0$ a.s. for $\xi(w) < \infty$.

Our definitions are designed to be diffeomorphism invariant. From this, c.f. [5]:

RESTRICTION TO SUBMANIFOLDS. Suppose N is a closed submanifold of M and $V(m) \in T_mN$, $X(m)e \in T_mN$, all $m \in N$ and $e \in \mathbb{R}^n$; thereby inducing $(X|N, V|N)$ on N. If $x(0) \in N$ then $x(t) \in N$ almost surely for $t \in [0,\xi)$ and x is a maximal solution of $dx = X|N\, dw + V|N\, dt$.

Consequently given (X,V) on M we can take a closed embedding of M in some (finite or infinite dimensional) Hilbert space $\overline{H}$

and extend X, V to $\overline{X}$, $\overline{V}$ on $\overline{H}$. Our solution x on M can then be considered as a solution of $dy = \overline{X}\, dw + \overline{V}\, dt$ on $\overline{H}$. This enables us to read off properties of solutions of equations on manifolds from the corresponding properties of solutions of S.D.E. on linear spaces. This is the way the next three properties can be proved. For simplicity they are stated only for the case $\xi \equiv \infty$.

PEICEWISE LINEAR APPROXIMATION. Let $\pi = (t_1,\dots,t_k)$ be a partition of a fixed time interval $[0,T]$, $0 = t_1 \le \dots \le t_k = T$. For a Brownian path w let w_π be the piecewise linear approximation.

$$w_\pi(t) = (t_{j+1}-t_j)^{-1}[(t_{j+1}-t)w(t_j)+(t-t_j)w(t_{j+1})]$$
$$t_j \le t \le t_{j+1}$$

Suppose the family of ordinary differential equations on M

$$\frac{dx_\pi}{dt} = X\frac{dw_\pi}{dt} + V, \quad x_\pi(0) = x(0)$$

have solutions $\quad x_\pi\colon [0,T] \times \Omega \to M$

then if $\xi \equiv \infty$ the family $\{x_\pi\}_\pi$ converges to x uniformly on $[0,T]$ in probability as mesh π tends to 0.

This follows as described above from the corresponding result of McShane [19], on the linear space $\overline{H}$. Another method of approximation has been used extensively by Malliavin [16], [17], see also §6 below.

DIFFERENTIABILITY WITH RESPECT TO INITIAL VALUE. When $\xi \equiv \infty$ there is a 'flow map' $F_t\colon M \to L^0(\Omega,F;M)$, $\quad 0 \le t < \infty$ into the space of M-valued random variables, such that $\{F_t(m)\,|\,0 \le t \le \infty\}$ is a solution with initial value $m \in M$. When X is C^2 and V is C^1 we can form a stochastic dynamical system $(\delta X, \delta V)$ on the tangent bundle TM of M by taking

$$\delta X(\ )e_i = S\cdot TX^i \; : TM \to T^2M = TTM \quad i = 1,\dots,n$$

where $\quad S\colon T^2M \to T^2M$

is the symmetry map, given over an open set U of H as

$$S:U \times H \times H \times H \to U \times H \times H \times H$$
$$(x,u,v,w) \to (x,v,u,w)$$

and where $TX^i : TM \to T^2M$

is the tangent map of the vector field $X^i = X(\ )e_i$ on M. The vector field δV on TM is defined similarly, as $S \bullet TV$.

The results in Gikman and Skorohod [12] can be applied in an ambient space $\bar{H}$ and we can deduce that $(\delta X, \delta V)$ has full solutions on TM, with a corresponding flow $\delta F_t : TM \to L^0(\Omega, \mathcal{F}; TM)$. Moreover this is the derivative in probability of F_t in the sense that if $f:M \to E$ is a C^1 map into a Banach space E then $df \circ \delta F_t$ is the derivative in probability of $f \circ F_t$. This will be examined in more detail below; see also [16], [18].

ITÔ FORMULAE. We will give three versions of the Itô transformation formula. For the third one we are assuming that M has had a linear connection imposed on it. For the first, if $e \in \mathbb{R}^n$, $S(t,m)e$ denotes the solution of the ordinary differential equation:

$$\frac{dy}{dt} = X(y(t))e \ , \quad y(0) = m,$$

while $X^i = X(\ )e_i$, $S^i(t,m) = S(t,m)e_i$, $i = 1,\dots,n$.

Let $f:M \to G$ be a C^2 map into the Hilbert space G. Then

$$f(x(t) = f(x(0)) + \int_0^t \frac{d}{dr} f \circ S(r,x(s)) \Big|_{r=0} dw(s)$$
$$+ \frac{1}{2} \sum_{i=1}^n \int_0^t \frac{d^2}{dr^2} f \circ S^i(r,x(s)) \Big|_{r=0} ds$$
$$+ \int_0^t df \bullet V(x(s))ds$$
$$= f(x(0)) + \int_0^t df \circ X(x(s))dw(s) + \frac{1}{2} \sum_i \int X^i(X^i f)(x(s))ds$$
$$+ \int_0^t df \circ V(x(s))ds$$

$$= f(x(0)) + \int_0^t df\bullet X(x(s))dw(s) + \frac{1}{2}\sum_i \int_0^t df\bullet \nabla X^i(X^i)(x(s))ds$$
$$+ \frac{1}{2}\sum_i \int_0^t \nabla df(X^i(x(s)),X^i(x(s))ds + \int_0^t df\circ V(x(s))ds$$

VARIANT: ITÔ DIFFERENTIAL EQUATIONS. For a manifold M with a given linear connection we could take '$dx = X\,dw + V\,dt$' to mean what we have defined to be $dx = Xdw - \frac{1}{2}\ \Sigma\nabla X^i(X^i)dt + V\,dt$. In other words we subtract off the 'Stratonovich terms'. This definition has the following properties:

(1) It agrees with the usual Itô definition of $dx = Xdw + Vdt$ on $\mathbb{R}^n$.

(2) Existence and uniqueness will be true for X, V only locally Lipshitz.

(3) It is not diffeomorphism invariant: it depends on the connection.

This is essentially the approach used by Baxendale in [1], [2].

4. EXAMPLE

Let N be a Riemannian manifold, $\dim N = n < \infty$. We will give a construction of Brownian motion on N, as described for example in [9], [10].

We will take $M = O(N)$, the orthonormal frame bundle of N, [15]. Let $\pi : O(N) \to N$ be the projection, so if $u \in O(N)$ then u is an isometry

$$u : \mathbb{R}^n \to T_{\pi(u)}N.$$

The Levi-Civita connection of N determines

$$X : O(N) \times \mathbb{R}^n \to TO(N),\ X(u,-) \in L(\mathbb{R}^n, T_u TO(N))$$

such that the solution curve $S(t,u)e$ of $X(\ ,e)$ is the horizontal lift of the geodesic γ on N with $\gamma(0) = \pi(u)$ and $\dot\gamma(0) = u(e)$. Alternatively $X(u,e)$ is the unique horizontal vector at u such that $T\pi \circ X(u,e) = u(e)$, [15 page 119]. For a maximal solution

$u: [0,\xi) \times \Omega \to O(N)$

of $\quad du = X\, dw\,, \quad u(0) = u_o$

set $\quad x = \pi \circ u: [0,\xi) \times \Omega \to N.$

Then, x is a *Brownian motion* on N, starting at $x_o = \pi(u_o)$. Because of the invariance properties of X, it depends on the choice of $u_o \in \pi^{-1}(x_o)$ only up to an orthogonal transformation of the original Brownian motion w on $\mathbb{R}^n$. In order to check that the infinitesimal generator of the projected process is the Laplace-Beltrami operator of N apply the first Itô formula above to the composition

$$O(N) \xrightarrow{\pi} N \xrightarrow{f} \mathbb{R}.$$

In fact remembering that $u(t): \mathbb{R}^n \to T_{x(t)}N$, and using also the second definition of X we have, for f of class C^2

$$f(x(t)) = f(x_o) + \int_0^t df \circ u(x(s))dw(s) + \frac{1}{2}\int_0^t \Delta f(x(s))ds.$$

The process u can be considered as the 'horizontal lift' of the Brownian motion, and for any $v \in Tx_oN$, $u(t)u_o^{-1}v$ represents 'stochastic parallel transport along x', c.f. [6], [8], [11], [13]. Apart from the role this plays in the 'diffusion of tensors' it has other important uses: for example the Radon-Nikodym derivative which occurs in the Girsanov-Cameron-Martin formula comparing the measures induced on the space of paths on M starting at x_o by the Brownian motion and by the minimal process with infinitesimal generator $A(f) = \frac{1}{2}\Delta f + V(f)$, V a vector field on N, has the form

$$\exp\left\{\int_0^t \langle V(x(s)), u(s)dw(s)\rangle_{x(s)} - \frac{1}{2}\int_0^t |V(x(s))|^2_{x(s)}\, ds\right\}.$$

Corresponding to X on O(N) we have δX on TO(N) and so a flow δF_t on TO(N). For a horizontal lift $\tilde{v}_o \in T_{u_o}O(N)$ of a vector $v_o \in T_{x_o}N$ we can project the process $\delta F_t(\tilde{v}_o)$ to TN by $T\pi$. The resulting process v on TN, which is not in general a Markov process, is in some sense a derivative of the Brownian motion in the direction v. Use of our first version of the Itô formula together with the structure equations of the connection shows that

$$u(t)^{-1} v(t) = u_o^{-1} v_o + \int_0^t A(s)dw(s) - \frac{1}{2}\int_0^t u(s)^{-1} K(v(s),-)^{\#} ds$$

where A is a process on the space of real skew-symmetric $n \times n$-matrices given by

$$A(t) = \int_0^t u(s)^{-1} R(u(s)dw(s),v(s))u(s)$$
$$+ \frac{1}{2}\int_0^t [u(s)^{-1} \sum_i R(u(s)e_i,u(s)A(s)e_i)u(s)]ds$$
$$+ \frac{1}{2}\int_0^t [u(s)^{-1} \sum_i \nabla R(u(s)e_i)(u(s)e_i,v(s))u(s)]ds.$$

Here R denotes the curvature tensor of N and K the Ricci tensor with $K(\alpha,-)^{\#}$ defined by $\langle K(\alpha,-)^{\#},\beta\rangle = K(\alpha,\beta)$. The sign conventions of Kobayashi and Nomizu [15] are used. By the Bianchi identities the term involving ∇R can be expressed in terms of ∇K. In [18], Malliavin gives a product integral formula for v.

Some computations are simplified by using our first version of the Itô formula together with the fact that the flows $\delta S(-,-)e$ of $\delta X(\ )e$ project under $T\pi$ to Jacobi fields along geodesics of N. In fact I. Caldwell [3] showed that δF_t is associated to an indefinite Riemannian metric on the manifold TN whose geodesics are just such Jacobi fields, [25], in much the same way as F_t is associated to the Riemannian metric of N; c.f. [18].

5. INDUCED DIFFUSION ON THE DIFFEOMORPHISM GROUPS

Suppose now that M is compact and C^∞ with dim M = m. For $s > \frac{m}{2}$ we can consider the space $H^s(M;M)$ of maps $h:M \to M$ of Sobolev class $H^s = W^{2,s}$. This is a C^∞ Hilbert manifold. For $s > \frac{m}{2} + 1$ let $\mathcal{D}^s \equiv \mathcal{D}^s(M)$ denote the open subset of $H^s(M,M)$ consisting of diffeomorphisms. The following facts are taken from [7], see also [22]: assuming henceforth that $s > \frac{m}{2} + 1$

(i) $\mathcal{D}^s$ is a C^∞ manifold and a topological group under composition.

(ii) For $h \in \mathcal{D}^s$, $R_h : \mathcal{D}^s \to \mathcal{D}^s$, $R_h(\alpha) = \alpha \circ h$, is C^∞.

(iii) For a C^∞ manifold N the map $\phi_r : H^{s+r}(M;N) \times \mathcal{D}^s(M) \to H^s(M;N)$, $\phi_r(f,h) = f \circ h$, is C^r, $r = 0,1,\ldots$

(iv) The tangent space $T_h \mathcal{D}^s$ to $\mathcal{D}^s$ at a point h can be identified with the Hilbertable space of H^s maps $f : M \to TM$ with $\tau \circ f = h$, where $\tau : TM \to M$ is the projection. In particular the tangent bundle $T\mathcal{D}^s$ can be identified with $H^s(M;TM)$.

Assume that the vector fields $X(\ )e$, $e \in \mathbb{R}^n$, on M are of class H^{s+2}. By (iv) this means they lie in the tangent space at the identity $T_{id}\mathcal{D}^{s+2}$. Define

$$\tilde{X} : \mathcal{D}^s \times \mathbb{R}^n \to T\mathcal{D}^s$$

by
$$\tilde{X}(h,e)(y) = X(h(y))e \qquad y \in M$$

or equivalently $\tilde{X}^i(h) = (R_h)_* X^i = \phi_2(X^i,h) \qquad i = 1,\ldots,n$

taking N = TM in (iii).

Then $\tilde{X}$ is C^2 since ϕ_2 is C^2, and it is right invariant. For V of class H^{s+1} define a right invariant vector field $\tilde{V}$ on $\mathcal{D}^s$ in the same way; $\tilde{V}$ will be C^1. By (ii), the existence of inverses in $\mathcal{D}^s$, and the right invariance, $\mathcal{D}^s$ has a uniform cover for $(\tilde{X},\tilde{V})$. Consequently there is a solution

$$\tilde{F} : [0,\infty) \times \Omega \to \mathcal{D}^s$$

of
$$d\tilde{F} = \tilde{X}dw + \tilde{V}dt, \quad \tilde{F}(0) = id : M \to M.$$

For $x_o \in M$ the evaluation map $ev_{x_o} : \mathcal{D}^s \to M$, $h \to h(x_o)$, is C^∞.

We can therefore apply Itô's formula to $\phi \circ ev_{x_o}(\tilde{F}(t))$ where $\phi : M \to \mathbb{R}^m$ is C^2 and extends the chart map of a localization $\Lambda = ((U,\phi),U_o,\lambda)$ of (X,V). This shows that for each $x_o \in M$, $\tilde{F}(t)(x_o)$ is a solution of $dx = Xdw + Vdt$, $x(0) = x_o$. In other words $\tilde{F}(t)$ is a version of the flow map F_t of (X,V). We have proved the following version of theorems of Malliavin [16], [17] and of Baxendale [1], [2]. The method we have used comes from the proof by Ebin and Marsden of the corresponding theorem for O.D.E, [7].

THEOREM For M compact and C^∞, if X, V are of class H^{s+2} and H^{s+1} respectively where $s > \frac{\dim M}{2} + 1$ then the equation $dx = Xdw + Vdt$ has a solution flow consisting of diffeomorphisms of class H^s, with the time evolution being represented almost surely by continuous paths into the H^s topology.

In particular if X, V are C^∞ the diffeomorphisms can be chosen to be C^∞ also.

Use of the derivative flow δF_t as in [7] would enable us to show that $\tilde{F}(t)$ lies in $\mathcal{D}^{s+1}$. Also if we are willing to move away from Hilbert manifolds, for example using Neidhardt [20], no doubt the same method will work for diffeomorphisms of class $W^{p,s}$, $s > \frac{m}{p} + 1$, $p \geq 2$ assuming X,V to be $W^{p,s+2}$ and $W^{p,s+1}$ respectively. For O.D.E. this has been ascribed to H. Brezis. However, in any case the methods of Malliavin and of Baxendale require rather less differentiability.

6. UNIFORMITY OF PIECEWISE LINEAR APPROXIMATION

Applying the approximation result of §3 to $d\tilde{F} = \tilde{X}dw + \tilde{V}dt$ on $\mathcal{D}^s$ when M is compact and $X \in H^{s+2}$, $V \in H^{s+1}$, we see, in the notation of §3, that the convergence of x_π to x is in H^s with respect to the initial point, in particular it is uniform in x(0). As an important special case of this:

THEOREM For X and V both C^∞ and M compact, let $\tilde{F}_\pi : [0,\infty) \times \Omega \to C^\infty(M,M)$ represent the flow of the family of O.D.E. on M, indexed by Ω, $\frac{dx_\pi}{dt} = X \frac{dw_\pi}{dt} + V$. Give $C^\infty(M;M)$ the topology of uniform convergence of all derivatives. Then, as mesh $\pi \to 0$, $\tilde{F}_\pi$ converges uniformly on each $[0,T]$ in probability to a map $\tilde{F} : [0,\infty) \times \Omega \to \mathcal{D}^\infty(M)$ which is a version of the flow of the S.D.E., $dx = Xdw + Vdt$.

This theorem was proved by Malliavin [16], [17] with his regularizations w_ε replacing w_π. These are defined by

$$w_\varepsilon(t) = \int_0^\varepsilon \frac{1}{\varepsilon} \ w(t+s)u(\frac{s}{\varepsilon})ds$$

for a suitable bump function u. His proof was direct and he used this result to *deduce* that the S.D.E. has a flow consisting of diffeomorphisms.

7. SUBGROUPS OF THE DIFFEOMORPHISM GROUP

It may well happen that X^i, V lie in tangent spaces at the identity to certain chains of subgroups of diffeomorphisms satisfying conditions (i),...,(iv) of §5, for example volume preserving diffeomorphisms, or symplectic diffeomorphisms, [7], [22]. The analysis of §5 shows that the flow $\tilde{F}_t$ can then be taken to lie in this class of diffeomorphisms. Thus if M is Riemannian and X^i,V are all divergence free it will follows that $\tilde{F}$ is a process on the space of volume preserving diffeomorphisms.

If M = O(N) as in §4, then M has a natural O(n)-invariant Riemannian metric such that π is a Riemannian submersion, [23]. Using [23] together with [15] Propositions III 5.4 and IV 2.3 we see that div $X^i = 0$, $i = 1,\ldots,n$. Thus we have Malliavin's result [18], that the flow $\tilde{F}_t$ on O(N) is volume preserving, at least for X,V sufficiently regular and N compact.

8. METHOD OF BAXENDALE

The existence of continuous flow is proved in [1], [2] under rather more general conditions than will be described here: in particular the driving Brownian motion is allowed to be infinite dimensional, and special attention is given to sample continuity with respect to parameters when X,V are parametrized. The basic tool is the following, [24]:

THEOREM OF TOTOKI. For a compact subset M of $\mathbb{R}^q$ and a metric space (K,d), if a process

$$z:M \to L^o(\Omega, F,\mu;K)$$

satisfies $\mu\{\omega:d(z(m_1,\omega), z(m_2,\omega)) > \delta\} \le \beta\, \delta^{-\alpha} \, \| m_1-m_2 \|^{q+\gamma}$

each $\delta > 0$, some positive α,β,γ, then it has a sample continuous version.

Returning to our manifold M, assumed compact, we can embed it in some Euclidean space $\mathbb{R}^q$ and extend X,V to $\overline{X}$ and $\overline{V}$ as in §3. If X,V are C^2 and C^1 respectively we can take $\overline{X}$, $\overline{V}$ to be C^2 and C^1 and both of compact support. Consequently (e.g. [12] Section 8) the flow map

$$\overline{F}_t:M \to L^p(\Omega, F,\mu;\mathbb{R}^q)$$

is Lipshitz for $1 < p < \infty$. Applying Totoki's theorem we see that the above conditions on X,V ensure that the flow map F_t of $dx = Xdw + Vdt$ has a sample continuous version $F_t:M \times \Omega \to M$. Further regularity can then be deduced as in [16] using δF_t.

Although both this method and that of Malliavin require less differentiability of X,V than the diffeomorphism approach it is nevertheless often useful to know that the flow $\tilde{F}$ is a solution of a stochastic differential equation: e.g. in §6 and in §10 below.

9. NON-COMPACT MANIFOLDS

Compactness is not assumed in either [2], or [16], though some conditions are required in its stead. Non-explosion is not enough: the equation $dx = dw$ on $\mathbb{R}^2-\{0\}$ is non-explosive for each initial point but the flow would have to be $\tilde{F}_t(x) = x+w(t)$, $x \in \mathbb{R}^2-\{0\}$, and for any non-void open set U, $\tilde{F}_t(U)$ is not defined for all time. Perhaps this is made to look more impressive by taking a diffeomorphism of $\mathbb{R}^2-\{0\}$ with the cylinder $S^1 \times (-\infty,\infty)$ and considering the S.D.E. induced on the cylinder from $dx = dw$.

When $M = \mathbb{R}^m$ and X,V satisfy suitable asymptotic conditions it should be possible to apply the theory in [4] to carry through the analysis of §5 on some groups of diffeomorphisms of $\mathbb{R}^m$, but not enough seems to be known about diffeomorphism groups of more general non-compact manifolds to get much further. However local results for non-compact manifolds can be deduced from the compact case. In this connection it is useful to note that if $M, \partial M$ is a compact C^∞ manifold with boundary and X,V vanish on ∂M then by [7] the analysis of §5 works equally well on the groups of H^s diffeomorphisms of M which restrict to the identity on ∂M.

10. ITÔ FORMULA FOR FUNCTIONALS OF CLASS H^2.

Let M be compact and C^∞ with $C^1(M)$ denoting the Banach manifold of C^1 diffeomorphisms $h:M \to M$. If $f \in H^2(M:\mathbb{R}^p)$ the induced map $f_* : C^1(M) \to L^2(M;\mathbb{R}^p)$, $f_*(h) = f \circ h$, is C^2. To see this note that by the analogue of (ii) §5 for C^1 it is enough to check it on a neighbourhood of the identity map of M. Then give M a Riemannian metric and take a sequence $f^i:M \to \mathbb{R}^p$, $i = 1,2,\ldots$, of C^∞ maps such that f^i, ∇f^i, and $\nabla^2 f^i$ converge to f, ∇f and $\nabla^2 f$ in L^2. By (iii) §5, each f^i_* is C^2, so it suffices to check that f^i_* together with its first two derivatives converge uniformly in the domain of some chart about the identity for $C^1(M)$. This is easily done using the fact that we are dealing with diffeomorphisms and that their Jacobian determinants are bounded away from 0 in a neighbourhood of the identity. From this, the fact that $\mathcal{D}^s(M) \to C^1(M)$ is C^∞ if $s > \frac{m}{2} + 1$, the results of §5, and the infinite dimensional Itô formula, we obtain:

THEOREM. For M compact and C^∞, with X and V of class H^{s+2} and H^{s+1} respectively, $s > \frac{\dim M}{2} + 1$, if $f \in H^2(M;\mathbb{R}^p)$ then, w.p.1, almost surely in $y \in M$,

$$f(\tilde{F}_t(y)) = f(y) + \int_0^t df \circ X(\tilde{F}_s(y))dw(s) + \frac{1}{2}\sum_i \int_0^t X^i(X^i f)\tilde{F}_s(y)ds$$
$$+ \int_0^t df \circ V(\tilde{F}_s(y))ds$$

where df and $X^i(X^i f)$ are taken in L^2 and the integrals on the right hand side are integrals with values in L^2.

The other versions in §3 of the Itô formula could have been used equally well. No doubt the conditions on X and V are unnecessarily stringent.

ACKNOWLEDGMENTS

Sections 1 to 4 are extracted from duplicated notes of seminars given by J. Eells and the author at the University of Warwick, c.f. [8], [9], [10]: and are based on joint work with Eells. The remaining sections were inspired by the work of Malliavin and of Baxendale. I would particularly like to thank P. Baxendale for several conversations and for showing me a preliminary version of [2].

REFERENCES

(1) P. Baxendale, Measures and Markov processes on function spaces, *Bull. Soc. math. France, Mémoire* 46, 1976, 131-141.

(2) P. Baxendale, Wiener processes on manifolds of maps, *J. Diff. Geom.* to appear.

(3) I.C. Caldwell, A relation between a lift of connections and a lift of stochastic dynamical systems, M.Sc. dissertation, University of Warwick, 1976.

(4) M. Cantor, Perfect fluid flows over $\mathbb{R}^n$ with asymptotic conditions, *J. Funct. Anal. 18*, (1975), 73-84.

(5) J.M.C. Clark, An Introduction to stochastic differential equations on manifolds, Geometric Methods in Systems Theory (D.Q. Mayne, R.W. Brockett, Eds.) Reidel, Holland, 1973.

(6) E.B. Dynkin, Diffusion of tensors, *Soviet Math.Dokl. 9* (1968) 532-534, = *Dokl. Akad.Nauk SSSR, 179,* (1968), No.6.

(7) D.G.Ebin and J.Marsden, Groups of diffeomorphisms and the motion of an incompressible fluid, *Ann. of Math. 92* (1970), 102-163.

(8) J. Eells and K.D. Elworthy, Wiener integration on certain manifolds; Some Problems in Non-linear Analysis, *C.I.M.E. IV* (1971), 67-94.

(9) J. Eells and K.D. Elworthy, Stochastic dynamical systems, Control Theory and Topics in Functional Analysis, Vol. III, International Atomic Energy Agency, Vienna (1976), 179-185.

(10) K.D. Elworthy, Measures on infinite dimensional manifolds, Functional Integration and its applications, (A.M. Arthurs, Ed.) Oxford University Press, 1975.

(11) J.B. Ferebee, Parallel translation and diffusions on the tangent bundle, Ph.D. dissertation, Princeton University,1972.

(12) I.I. Gihman and A.V. Skorohod, Stochastic Differential Equations, Ergebn. der Math. 72. Springer 1972.

(13) K. Itô, On stochastic differential equations on a differentiable manifold 1, *Nagoya Math.J.,1,* 35-47 (1950).

(14) K. Itô, Stochastic parallel displacement, Probabilistic Methods in Differential Equations, Lecture Notes in Mathematics Vol. 451, Springer (1975), 1-7.

(15) S. Kobayashi and K. Nomizu, Foundations of Differential Geometry, Vol. 1, Interscience (1963).

(16) P.Malliavin, Stochastic calculus of variation and hypoelliptic operators, Report No. 13, (1976) Institut Mittag-Leffler.

(17) P.Malliavin, Un principe de transfert et son application au Calcul des Variations, *C.R.Acad.Sc. Paris, 284, série A,* (1977), 187-189.

(18) P.Malliavin, Champs de Jacobi stochastiques, *C.R.Acad.Sc. Paris, 285, série A,* (1977), 789-792.

(19) E.J.McShane, Stochastic Calculus and Stochastic Models, Academic Press, 1974.

(20) A. Neidhardt, Personal Communication.

(21) E.Nelson, An existence theorem for second order parabolic equations, *Trans. Amer. Math. Soc., 88* (1958), 414-429.

(22) H. Omori, Infinite Dimensional Lie Transformation Groups, Lecture Notes in Mathematics, Vol. 427, Springer 1974.

(23) B. O'Neill, The fundamental equations of a submersion, *Michigan Math. J. 13* (1966), 459-469.

(24) H. Totoki, A method of construction of measures on function spaces and its applications to stochastic processes, *Mem. Fac. Sci. Kyushu Univ. Ser A, 15,* (1961), 178-190.

(25) J. Vilms, Connections on tangent bundles, *J. Diff. Geom. 1.* (1967), 235-243.

SOME MEASURE-VALUED POPULATION PROCESSES

W. H. Fleming[1]

Division of Applied Mathematics
Brown University
Providence, Rhode Island

and

M. Viot

Institut de Recherche d'Informatique et
d'Automatique
Rocquencourt, France

This paper is concerned with stochastic models in population genetics, with a continuum of possible genetic types. The results are formulated in terms of measure-valued Markov processes. The state space for such a process consists of all possible distributions of types in a population. Conditions for existence and uniqueness of solutions to corresponding martingale problems are obtained, as well as the Feller property of these measure-valued processes. The paper consists of a statement of main results. Detailed proofs will appear elsewhere.

1. Finite Dimensional Case

The technique of diffusion approximation is well known in population genetics theory [1, Chap. 8]. With this technique certain Markov chains describing gene frequency processes are replaced by corresponding Markov diffusions. Suppose that there are a finite number of gene types (alleles); and let p_j denote the frequency of type $j = 1,\dots,J$. The vector $p = (p_1,\dots,p_J)$ represents the distribution of types in a population. We have $p \in \Sigma^J$, where Σ^J is the standard $(J-1)$-dimensional simplex. The diffusion approximation is a process with Σ^J as state

[1]This work was partly supported by the National Science Foundation under MCS76-07247.

ISBN 0-12-268380-3

and with generator $\mathfrak{A}^J$ a second order partial differential operator of the form

$$\mathfrak{A}^J f(p) = \sum_{j=1}^{J} g_j(p) f_{p_j} + \sum_{i,j=1}^{J} (p_i \delta_{ij} - p_i p_j) f_{p_i p_j}, \tag{1.1}$$

for any $f \in C^2(\Sigma^J)$. See [3], [4], [7], and [1] when $J = 2$. Here δ_{ij} is the Kronecker symbol. In (1.1) the second order terms correspond to chance fluctuations in gene frequencies (called "random genetic drift"). The coefficients $g_j(p)$ in first order terms represent such genetic forces as mutation and natural selection.

2. The Generator $\mathscr{G}$

We wish to describe a diffusion model corresponding to the one in §1, when a continuum of genetic types are possible. Our work was motivated in part by interest in the "ladder model" of Ohta and Kimura [6], [5] for the distribution of electrophoretic states. See §6.

Let x denote a genetic type; $x \in S$, where S denotes the space of possible types. We assume in this paper that S is a compact metric space. Let

$$\mathscr{B}(S) = \{\text{Borel subsets of } S\}$$

$$C(S) = \{\text{continuous real-valued functions on } S\}$$

$$\mathscr{M}(S) = \{\text{measures } \mu \text{ on } \mathscr{B}(S): \mu \geq 0, \mu(S) = 1\}.$$

Each $\mu \in \mathscr{M}(S)$ is a possible distribution of types in a population. We are interested in $\mathscr{M}(S)$-valued processes which model the same kinds of genetic forces (including random genetic drift, mutation, selection) as the finite dimensional diffusion with with generator (1.1).

For $\beta \in C(S)$, $\mu \in \mathscr{M}(S)$ let

$$\langle\beta,\mu\rangle = \int_S \beta(x) d\mu(x).$$

We give $\mathcal{M}(S)$ a metric equivalent to weak* convergence of sequences in $\mathcal{M}(S)$.

Let $\mathcal{L}: \mathcal{S} \to C(S)$ be a linear operator representing mutational forces, where $\mathcal{S}$ is a dense subset of $C(S)$. For an example corresponding to the ladder model see §6. Let $H(\mu,\beta)$ be a function representing such genetic forces as natural selection. Certain assumptions on $\mathcal{L}, H$ will be made in §'s 4, 5. The generator $\mathcal{G}$ for the $\mathcal{M}(S)$-valued process is as follows. For brevity write $\mathcal{M} = \mathcal{M}(S)$. Let

$$\mathcal{D}_0 = \{\phi: \phi(\mu) = F(<\beta_1,\mu>,\ldots,<\beta_K,\mu>), \quad \beta_1,\ldots,\beta_K \in \mathcal{S},\ F \in C^2(R^K),\ K = 1,2,\ldots\} \tag{2.1}$$

and $\mathcal{D}$ the linear subspace of $C(\mathcal{M})$ spanned by $\mathcal{D}_0$. It can be shown that $\mathcal{D}$ is dense in $C(\mathcal{M})$. We define $\mathcal{G}\phi$ for $\phi \in \mathcal{D}$ by requiring that $\mathcal{G}\phi$ be linear, and that for $\phi \in \mathcal{D}_0$:

$$\mathcal{G}\phi(\mu) = \sum_{k=1}^{K} (<\mathcal{L}\beta_k,\mu> + H(\mu,\beta_k))F_{z_k}(\ldots) \tag{2.2}$$

$$+ \sum_{k,\ell=1}^{K} (<\beta_k\beta_\ell,\mu> - <\beta_k,\mu><\beta_\ell,\mu>)F_{z_k z_\ell}(\ldots).$$

Here ... stands for the vector $z = (z_1,\ldots,z_K)$ with $z_k = <\beta_k,\mu>$.

3. Martingale Problem

We consider the following "canonical" sample space

$$\Omega = C([0,\infty);\mathcal{M})$$

$$\mathcal{F} = \mathcal{B}(\Omega),$$

where Ω has the topology of uniform convergence on each finite subinterval $[0,T] \subset [0,\infty)$.

We consider measure-valued processes $Y(t)$ on the canonical sample space:

$$Y(t)(\omega) = \omega(t), \quad \text{for all} \quad \omega \in \Omega.$$

The probability measure P on $(\Omega, \mathscr{F})$ is to be chosen to solve the following martingale problem corresponding to the generator $\mathscr{G}$. Let $\mathscr{F}_t$ be generated by $\{\omega(s)\colon\ 0 \leq s \leq t\}$. For each $\phi \in \mathscr{D}$ let

$$M_\phi(t) = \phi[Y(t)] - \phi[Y(0)] - \int_0^t \mathscr{G}\phi[Y(v)]dv.$$

Martingale Problem. Given $\mu^o \in \mathscr{M}$, find a probability measure P such that

(a) $P(Y(0) = \mu^o) = 1.$

(b) $M_\phi(t)$ is a $(\{\mathscr{F}_t\}, P)$-martingale for each $\phi \in \mathscr{D}$.

(c) The increasing process $\langle\langle M_\phi(t)\rangle\rangle$ satisfies, for each $\phi \in \mathscr{D}_0$,

$$\langle\langle M_\phi(t)\rangle\rangle = \int_0^t \psi_\phi[Y(v)]dv,$$

where

$$\psi_\phi(\mu) = \sum_{k,\ell=1}^{K} (\langle\beta_k\beta_\ell,\mu\rangle - \langle\beta_k,\mu\rangle\langle\beta_\ell,\mu\rangle)F_{z_k}(\ldots)F_{z_\ell}(\ldots).$$

4. Existence

Our method is as follows. We first approximate the martingale problem by corresponding problems for finite dimensional diffusions (§1). Then we verify a Prokhorov-type compactness condition and pass to the limit. The finite dimensional problems are obtained by discretizing the linear operator $\mathscr{L}$.

Definition. $\mathscr{L}$ is positively discretizable if there exist finite subsets $S^1, S^2, \ldots$ of S,

$$S^r = \{x_1^r,\ldots,x_{J^r}^r\}, \quad r = 1,2,\ldots,$$

and $\mathscr{L}^1,\mathscr{L}^2,\ldots$ such that:

$$\text{(i)} \quad \mathscr{L}^r\beta(x_i^r) = \sum_{j=1}^{J^r} \theta_{ij}^r\beta(x_j^r) \quad \text{for all} \quad \beta \in C(S),$$

$$\theta_{ij}^r \geq 0 \quad \text{for} \quad i \neq j, \; \theta_{ii}^r = -\sum_{j\neq i} \theta_{ij}^r, \quad r = 1,2,\ldots;$$

$$\text{(ii)} \quad \max_{x\in S^r} |\mathscr{L}^r\beta(x) - \mathscr{L}\beta(x)| \to 0 \quad \text{as} \quad r \to \infty, \text{ for all } \beta \in \mathscr{S}.$$

We assume that $H \in C(\mathscr{M}(S) \times C(S))$, and that H has the following property. Given any finite set $\tilde{S} \subset S$, $\tilde{S} = \{x_1,\ldots,x_J\}$ there exist $h_1,\ldots,h_J \in C^\infty(\Sigma^J)$ such that

$$H(\mu,\beta) = \sum_{j=1}^{J} h_j(p)\beta(x_j) \tag{4.1}$$

for each $\mu = \sum_{i=1}^{J} p_i\delta_{x_i}$ in $\mathscr{M}(\tilde{S})$ and $\beta \in C(S)$.

Let $\mathscr{G}^r$ denote the generator when $\mathscr{L}$ is replaced by $\mathscr{L}^r$ in (2.2). Let $h_j^r(p)$ be as in (4.1) when $\tilde{S} = S^r$, and

$$g_j^r(p) = \sum_{i=1}^{J^r} p_i\theta_{ij}^r + h_j^r(p), \; j = 1,\ldots,J^r.$$

Let $\mathfrak{A}^{J^r}$ be the corresponding generator in (1.1). Then, for $\phi \in \mathscr{D}_0$,

$$\mathscr{G}^r\phi(\mu) = \mathfrak{A}^{J^r} f^r(p), \quad \mu \in \mathscr{M}(S^r),$$

$$\mu = \sum_{i=1}^{J^r} p_i\delta_{x_i^r}, \quad f^r(p) = \phi(\mu).$$

For $r = 1,2,\ldots$ let $\mu^{or} \in \mathscr{M}(S^r)$, $\mu^{or} \to \mu^o$ as $r \to \infty$. For each r there exists a measure P^r on $(\Omega,\mathscr{F})$ solving the martingale problem for $\mu^{or}, \mathcal{G}^r$; moreover, $P^r(Y(t) \in \mathcal{M}(S^r)) = 1$. By rather standard estimates $\{P^1, P^2, \ldots\}$ satisfies the Prokhorov compactness condition. Moreover, for each $\phi \in \mathcal{D}_0$

$$\lim_{r\to\infty} \max_{\mu\in\mathcal{M}(S^r)} |\mathcal{G}^r\phi(\mu) - \mathcal{G}\phi(\mu)| = 0.$$

A solution P to the martingale problem for μ^o, $\mathcal{G}$ is obtained as the limit of P^r for a subsequence of r.

5. Uniqueness; Markov and Feller Properties

We now impose additional assumptions. First of all we take $H = 0$ (thus, the genetic forces acting are mutation and random genetic drift.) We also make a hypothesis (condition U below) which will guarantee the uniqueness of moments of certain finite dimensional distributions. These moments obey a system of differential equations, which we write in a weak form as (5.1), (5.2) below.

Let $u(\beta_1, \ldots, \beta_K, t)$ denote a continuous real-valued function on $C(S) \times \ldots \times C(S) \times [0,\infty)$, such that u is multilinear and symmetric in $\beta_1, \ldots, \beta_K$. Let $(x_1, \ldots, x_K) \in S \times \ldots \times S$. The notation $\mathcal{L}_{x_k}$ indicates that $\mathcal{L}$ acts in the variable x_k. Let $\mathcal{L}^*_{x_k}$ denote the (weak sense) adjoint of $\mathcal{L}_{x_k}$:

$$\mathcal{L}^*_{x_k} u(\beta_1, \ldots, \beta_K, t) = u(\beta_1, \ldots, \beta_{k-1}, \mathcal{L}\beta_k, \beta_{k+1}, \ldots, \beta_K, t),$$

$$\text{for all } \beta_1, \ldots, \beta_K \in \mathcal{S}.$$

For $K = 1$ the equation for $u(\beta,t)$ is

$$\frac{du}{dt} = \mathcal{L}^* u, \text{ if } \beta \in \mathcal{S},\ t \geq 0. \tag{5.1}$$

For $K > 1$, let

$$v = 2 \sum_{k<\ell} u(\beta_1, \ldots, \beta_{k-1}, \beta_k \beta_\ell, \beta_{k+1}, \ldots, \beta_{\ell-1}, 1, \beta_{\ell+1}, \ldots, \beta_K, t)$$

The equation for $u(\beta_1, \ldots, \beta_K, t)$ is

$$\frac{du}{dt} = \sum_{k=1}^{K} \mathcal{L}^*_{x_k} u - K(K-1)u + v, \text{ if } \beta_1, \ldots, \beta_K \in \mathcal{S}, \ t \geq 0. \tag{5.2}$$

Equations (5.1), (5.2) are considered with the following initial data. Given $\mu^o \in \mathcal{M}$

$$u(\beta_1, \ldots, \beta_K, 0) = \langle \beta_1, \mu^o \rangle \ldots \langle \beta_K, \mu^o \rangle \tag{5.3}$$

If P solves the martingale problem in §3, then a solution to (5.1)-(5.3) is

$$u(\beta_1, \ldots, \beta_K, t) = E^P(\langle \beta_1, Y(t) \rangle \ldots \langle \beta_K, Y(t) \rangle). \tag{5.4}$$

This corresponds to taking in (2.1) $F(z) = z_1 \ldots z_K$, a monomial in $z_1, \ldots, z_K$.

Let us now assume that the operator $\mathcal{L}$ has:

Uniqueness Property (U). Equations (5.1), (5.2) and the initial data (5.3) uniquely determine $u(\beta_1, \ldots, \beta_K, t)$ for all $K = 1, 2, \ldots,\ \beta_1, \ldots, \beta_K \in \mathcal{S},\ t \geq 0$.

By property (U) and (5.4), $E^P \phi[Y(t)]$ is unique for $\phi(\mu) = F(\langle \beta_1, \mu \rangle, \ldots, \langle \beta_K, \mu \rangle)$ if $F(z_1, \ldots, z_K)$ is any polynomial. Then $E^P \phi[Y(t)]$ is also unique for all $\phi \in C(\mathcal{M})$, given μ^o. We may then write here E_{μ^o} instead of E^P. Let

$$T_t \phi(\mu) = E_\mu \phi[Y(t)]. \tag{5.5}$$

Theorem. $\{T_t\}$ is a semigroup on $C(\mathcal{M})$.

This theorem is proved by combining the technique used to prove existence outlined in §4 and appropriate use of property (U).

By using properties of regular conditional distributions it can be shown that the conditional expectations satisfy, for all

$\phi \in C(\mathcal{M})$, $s < t$

$$E^P\{\phi[Y(t)] | \mathcal{F}_s\} = T_{t-s}\phi[Y(s)], \quad \text{P-almost surely.}$$

Then

$$\tilde{u}(\beta_1,\ldots,\beta_K,t,\omega) = T_{s+t}\phi[\omega(s)]$$

with $\phi(\mu) = \langle\beta_1,\mu\rangle \ldots \langle\beta_K,\mu\rangle$, $\omega \in \Omega$, satisfies (5.1), (5.2), and (5.3) with μ^o replaced by $\omega(s)$. From this follows uniqueness of conditional moments, from which one can get:

Theorem. The solution P to the martingale problem in §3 is unique, given μ^o. This collection of measures $P(= P(\mu^o))$ defines a Markov process with the Feller property. The semigroup associated with this Markov process is $\{T_t\}$.

A sufficient condition for property (U) is that $\mathcal{L}$ be diagonalizable, in the following sense: there exist eigenfunctions $\beta_1,\beta_2,\ldots$ for $\mathcal{L}$ (each $\beta_k \in \mathcal{S}$, $\beta_1(x) \equiv 1$) such that finite linear combinations of $\beta_1,\beta_2,\ldots$ are dense in $C(S)$. More generally, property (U) holds if $\mathcal{L}$ is triangulizable, in the following sense. Let $C_c(S)$ denote the space of complex-valued, continuous functions on S, and $\mathcal{S}_c$ the dense subspace of functions with real and imaginary parts in $\mathcal{S}$. Triangulizable means that $\beta_1,\beta_2,\ldots$ in $\mathcal{S}_c$ exist, $\beta_1(x) \equiv 1$, such that

$$\mathcal{L}\beta_\ell = \sum_{\ell=1}^{k} c_{k\ell}\beta_\ell$$

and finite linear combinations of $\beta_1,\beta_2,\ldots$ are dense in $C_c(S)$.

6. An Example

Let $S = [-a,a]$ and $\mathcal{L} = d^2/dx^2$, with reflecting boundary conditions. We take $\mathcal{S} = \{\beta \in C^2[-a,a]: \beta'(\pm a) = 0\}$. For $\mathcal{L}^r$

we take the usual discretization of d^2/dx^2, with second differences replacing second derivatives. Since $\mathscr{L}$ is diagonalizable, property (U) holds.

Equation (5.1) is now simply the heat equation written in a weak form, with reflecting boundary conditions. The solution $u(\beta,t) = E^P(\langle\beta,Y(t)\rangle)$ corresponds for $t > 0$ to a C^∞ function $m(x,t)$ such that

$$\frac{\partial m}{\partial t} = \frac{\partial^2 m}{\partial x^2}, \quad \frac{\partial m}{\partial x}(\pm a,t) = 0,$$

$$E^P(\langle\beta,Y(t)\rangle) = \int_{-a}^{a} \beta(x)m(x,t)dx.$$

We call $m(x,t)$ the <u>mean density</u> of type x at time t. For $K = 2$ there exists $m_2(x_1,x_2,t)$ for $x_1 \neq x_2$, $t > 0$ such that

$$\frac{\partial m_2}{\partial t} = \frac{\partial^2 m_2}{\partial x_1^2} + \frac{\partial^2 m_2}{\partial x_2^2} - 2m_2 + 2m(x_1)\delta(x_1-x_2),$$

$$E^P(\langle\beta_1,Y(t)\rangle\langle\beta_2,Y(t)\rangle) = \int_{-a}^{a}\int_{-a}^{a} \beta_1(x_1)\beta_2(x_2)m_2(x_1,x_2,t)dx_1dx_2,$$

with normal derivative $\partial m_2/\partial\nu = 0$ on the boundary of the square $S \times S$. Here δ is the Dirac function. We call $m_2(x_1,x_2,t)$ the mean joint density of (x_1,x_2). Similar equations are obtained for the mean joint density $m_K(x_1,\ldots,x_K,t)$ when $K > 2$, using (5.2)

In the ladder model of Ohta-Kimura [6] the type space consists of the integers, and mutation produces transitions to nearest neighbor types with equal frequency left and right. After a rescaling of variables one gets a continuous analogue of the ladder model, with $S = (-\infty,\infty)$, $\mathscr{L} = d^2/dx^2$. While the real line S is not compact, there is no difficulty in extending the existence and uniqueness results in §4 to include this case. In the ladder model the distribution of electrophoretic types tends to move erratically with time along the line S. For this reason Moran [5] used the term "wandering distribution". Instead of the type distribution itself, Ohta-Kimura and Moran considered the distribution of the difference $\xi = x_2 - x_1$ in the types x_1, x_2 of the two randomly selected individuals in the population. In the continuous-type version of the ladder model let us denote by $Z(t)$ the measure-valued process representing the distribution of differences in types. Then

$$\langle \beta, Z(t) \rangle = \langle \beta * Y(t), Y(t) \rangle$$

where $*$ denotes convolution, and β is any continuous function on $(-\infty,\infty)$ with compact support. Let

$$I(\xi,t) = \int_{-\infty}^{\infty} m_2(x, x+\xi, t)\,dx.$$

We calll $I(\xi,t)$ the mean density of types differing by ξ. Then

$$\frac{\partial I}{\partial t} = \frac{\partial^2 I}{\partial \xi^2} - 2I + 2\delta(\xi).$$

As $t \to \infty$, $I(t,\xi)$ tends to an equilibrium value

$$\lim_{t\to\infty} I(t,\xi) = 2^{-\frac{1}{2}} \exp(-2^{\frac{1}{2}}|\xi|).$$

See [6, p. 203] for the corresponding formula in the discrete state ladder model.

We also introduce

$$I_2(\xi_1,\xi_2,t) = \int m_3(x, x+\xi_1, x+\xi_2)\,dx$$

$$J(\xi_1,\xi_2,t) = \int\int m_4(x_1, x_1+\xi_1, x_2, x_2+\xi_2)\,dx_1\,dx_2,$$

and call $J(\xi_1,\xi_2,t)$ the mean joint density of types differing by ξ_1,ξ_2. From (5.2) with $K = 2,3,4$ one can get a system of three partial differential equations for I, I_2, J. These can be studied with the aid of Fourier transforms in ξ,ξ_1,ξ_2, furnishing an alternative approach to the technique of Moran [5, Part II] for studying numerically mean joint densities.

ACKNOWLEDGMENTS

The authors wish to acknowledge helpful suggestions from D.A. Dawson and G. Papanicolaou. The idea of using measure-valued processes to model population genetics phenomena was suggested by Dawson's paper [2].

REFERENCES

[1] J.F. Crow and M. Kimura, An Introduction to Population Genetics Theory, Harper and Row, New York, 1970.

[2] D.A. Dawson, Stochastic evolution equations and related measure processes, J. Multivariate Anal. 5(1975), 1-52.

[3] S.N. Ethier, A class of degenerate diffusion processes occurring in population genetics, Comm. Pure Appl. Math. 29(1976), 483-493.

[4] S. Karlin and J. McGregor, On some stochastic models in genetics, in Stochastic Models in Medicine and Biology (ed. J. Gurland), 247-279. University of Wisconsin Press, Madison, 1964.

[5] P.A.P. Moran, Wandering distributions and the electrophoretic profile, I, II Theor. Popn. Biol. 8(1975), 318-330; 10(1976), 145-149.

[6] T. Ohta and M. Kimura, A model of mutation appropriate to estimate the number of electrophoretically detectable alleles in a finite population, Genet. Res. Camb. 22(1973), 201-204.

[7] K-I. Sato, Diffusion processes and a class of Markov chains related to population genetics, Osaka Math. J. 13(1976), 631-659.

OPTIMAL STOPPING FOR RANDOM EVOLUTION OF MULTI-DIMENSIONAL POISSON PROCESSES WITH PARTIAL OBSERVATIONS

Avner Friedman[1]

Department of Mathematics
Northwestern University
Evanston, Illinois

1. THE MODEL

Let ω be a variable point in the space $\Omega = D([0,\infty);R^m)$ of right continuous left limited functions $x(t)$ from $t \geq 0$ into R^m. Denote by $\mathcal{F}$ the σ-field of subsets of Ω generated by $(x(t) \in A)$, A any Borel set in R^m and $x(\cdot) = \omega \in \Omega$ defined by $x(t,\omega) = \omega(t)$. Let $\lambda^\nu = (\lambda_1^\nu,\ldots,\lambda_m^\nu)$ $(\nu = 1,\ldots,n)$ be distinct vectors with positive components, and denote by $P_x^{\lambda^\nu}$ the m-dimensional Poisson process in $(\Omega,\mathcal{F})$ with parameter λ^ν. Thus

$$P_x^{\lambda^\nu}(x_1(t) = x_1 + k_1,\ldots,x_m(t) = x_m + k_m)$$

$$= \frac{(\lambda_1^\nu t)^{k_1}\cdots(\lambda_m^\nu t)^{k_m}}{k_1!\cdots k_m!} e^{-(\lambda_1^\nu +\cdots+ \lambda_m^\nu)t}$$

where the k_i are nonnegative integers.

Let $\theta(t)$ be a right continuous left limited Markov process with states $1,\ldots,n$ and infinitesimal matrix $(q_{i,j})$, defined in a probability space $(\Omega',\mathcal{F}')$. We introduce the random evolution

[1]This work is partially supported by National Science Foundation Grant MC 575-21416 A01.

ISBN 0-12-268380-3

of the $P_x^{\lambda\nu}$ dictated by the Markov process $\theta(t)$. This can be done, for instance, by defining the family of probabilities $P_{i,x}$ ($i = 1,\dots,n; x \in R^m$) in the product space $(\Omega \times \Omega', \mathcal{F} \times \mathcal{F}')$ and verifying the strong Markov property; this is entirely analogous to the diffusion case treated by Anderson and Friedman (2). We now set

$$\tilde{\mathcal{F}}_t = \sigma(x(u),\ 0 \le u \le t),\quad \mathcal{F}_t = \tilde{\mathcal{F}}_{t+0},$$

$$\tilde{\mathcal{M}}_t = \sigma((\theta(u),x(u));\ 0 \le u \le t),\quad \mathcal{M}_t = \tilde{\mathcal{M}}_{t+0}.$$

Suppose a machine kicks out the product $x(t)$ with a statistics $P_{x(t)}^{\lambda\nu}$ where ν can take any value from 1 to n. We observe $x(t)$, but we do not know the attached statistics. We do know, however, that $P_{x(t)}^{\lambda\nu}$ is produced when the machine is at the level of efficiency ν, and that this is the case precisely when $\theta(t) = \nu$. When at level ν, the rate of cost of production is γ_ν, whereas the product always sells at a rate N. Thus, if $f(j) = c_j$, where $c_j = \gamma_j - N$, then the running cost of the machine is $f(\theta(t))$.

At a sequence of times $\tau = (\tau_1, \tau_2, \dots, \tau_\ell, \dots)$ we are going to inspect the interior of the machine in order to discover the level of efficiency. Each inspection incurs the cost K, and the following procedure is adopted:

If at time $t = \tau_\ell$ the level of efficiency is j, then we execute one of the following three options:

(a) move the machine from level j to a lower level i at the given cost b_{ji} (the expense of repair);

(b) if $j < n$, leave the machine in the same level with no cost (this is a special case of (a) if we set $b_{jj} = 0$);

(c) if $j = n$, shut off the machine with cost b.

Denoting the initial state of the machine by i_0, the total cost for production, inspections and shut-off, when discounted by a factor α, $\alpha > 0$, is given by

$$J_{i_0,x}(\tau) = E_{i_0,x}\Big[Ke^{-\alpha\tau_1} + \int_0^{\tau_1} e^{-\alpha s} f(\theta(s))$$

$$+ \sum_{j=1}^{n-1} I_{\theta(\tau_1)=j} \min_{1 \le i \le j} \{b_{ji} + E_{i,x(\tau_1)}[\cdots]\} e^{-\alpha\tau_1}$$

$$+ I_{\theta(\tau_1)=n} \min\{b, \min_{1 \le i \le n-1} \{b_{ni} + E_{i,x(\tau_1)}[\cdots]\} e^{-\alpha\tau_1}\Big].$$

The problem we consider in this work is to find $\bar{\tau}$ such that

$$J_{i_0,x}(\bar{\tau}) \le J_{i_0,x}(\tau) \quad \text{for any } \tau.$$

It is called a problem of quality control.

To make the definition of τ more precise, introduce the shift operator

$$\phi_s: x(t) \to x(t + s)$$

and take

$$\tau_1 = \sigma_1, \quad \tau_{m+1} = \tau_m + \sum_{\ell=1}^{n} I_{\theta(\tau_m)=\ell}\, \sigma_{m+1,\ell}(\phi_{\tau_m})$$

where σ_1, $\sigma_{j,\ell}$ are $\mathcal{F}_t$ stopping times. The corresponding sequence $\tau = (\tau_1,\ldots,\tau_\ell,\ldots)$ is then called an inspection sequence. Notice that

$$\tau_{m+1} \in \sigma(\mathcal{F}_t, \theta(\tau_1),\ldots,\theta(\tau_m)) \subset \mathcal{M}_t.$$

Thus, τ_{m+1} is a stopping time with respect to σ-subfields of the fields $\mathcal{M}_t$ of the Markov process $(\theta(t),x(t))$. This makes the problem of optimizing $J_{i_0,x}(\tau)$ a non-Markovian problem (with partial observations).

2. REFORMULATION OF THE PROBLEM

We introduce the notation

$$e_j = (\delta_{j1},\dots,\delta_{jn}),$$

$$\Pi = \{(p_1,\dots,p_n);\ p_i > 0,\ \Sigma p_i = 1\},\quad \bar{\Pi} = \text{closure of } \Pi,$$

$$\Pi_0 = \bar{\Pi} \setminus \{e_n\},$$

$$P^{\mu,x} = \sum_{i=1}^{n} p_i P_{i,x} \quad \text{where } \mu = (p_1,\dots,p_n) \in \bar{\Pi},$$

$$P^{\mu} = P^{\mu,0},\quad E^{\mu,x} = E^{P^{\mu,x}},\quad E^{\mu} = E^{\mu,0},$$

$$p(t) = p^{\mu}(t) = (p_1(t),\dots,p_n(t))$$

where

$$p_j(t) = E^{\mu,x}[\theta(t) = j \mid \mathcal{F}_t].$$

THEOREM 1. The process $p^{\mu}(t)$, $P^{\mu}, \mathcal{F}_t$ is a strong right continuous Markov process. It is given in terms of $x(t)$ and $q_{i,j}$ by

$$p_j(t) = \sum_{i=1}^{n} \frac{p_i \bar{p}_{i,j}(t)}{\sum_{\ell=1}^{n} p_\ell z_{i,\ell}(0,t) \sum_{k=1}^{n} \bar{p}_{\ell,k}(t)}$$

where

$$\bar{p}_{i,j}(t) = \sum_{\rho=0}^{\infty}{}^{*} \sum_{(i,\gamma_1,\dots,\gamma_\rho,j)}{}' \int_0^t du_{i,\gamma_1} e^{-q_i u_{i,\gamma_1}} q_{i,\gamma_1} \int_0^{t-u_{i,\gamma_1}} du_{\gamma_1,\gamma_2}$$

$$e^{-q_{\gamma_1} u_{\gamma_1,\gamma_2}} q_{\gamma_1,\gamma_2} \cdots \int_0^{t-u_{i,\gamma_1}-u_{\gamma_1,\gamma_2}-\dots-u_{\gamma_{\rho-1},\gamma_\rho}} du_{\gamma_\rho,j}\, e^{-q_{\gamma_\rho} u_{\gamma_\rho,j}}$$

$$q_{\gamma_\rho,j}\, e^{-q_j(t-u_{i,\gamma_1}-\dots-u_{\gamma_{\rho-1},\gamma_\rho}-u_{\gamma_\rho,j})}\, z_{i,\gamma_1}(u_{i,\gamma_1}, u_{i,\gamma_1}+u_{\gamma_1,\gamma_2})$$

$$z_{i,\gamma_2}(u_{i,\gamma_1}+u_{\gamma_1,\gamma_2}, u_{i,\gamma_1}+u_{\gamma_1,\gamma_2}+u_{\gamma_2,\gamma_3})\cdots$$

$$z_{i,j}(u_{i,\gamma_1}+u_{\gamma_1,\gamma_2}+\dots+u_{\gamma_{\rho-1},\gamma_\rho}+u_{\gamma_\rho,j}, t) \quad (1 \le i,j \le n), \tag{2.1}$$

$$z_{i,j}(s,t) = \exp\Big\{\sum_{k=1}^{m} \log \frac{\lambda_k^j}{\lambda_k^i}\cdot(x_k(t) - x_k(s)) - \langle\lambda^j - \lambda^i\rangle(t-s)\Big\},$$

$$\langle\lambda^j - \lambda^i\rangle = \sum_{\ell=1}^{m}(\lambda_\ell^j - \lambda_\ell^i), \qquad q_\ell = -q_{\ell,\ell}.$$

Here we have used the following notation:

$$\sum_{(i,\gamma_1,\ldots,\gamma_\rho,j)}{}'$$

indicates the sum taken over all $\gamma_1,\ldots,\gamma_\rho$ with each γ_ℓ varying over $1,2,\ldots,n$ in such a way that

$$i \neq \gamma_1 \neq \gamma_2 \neq \cdots \neq \gamma_{\rho-1} \neq \gamma_\rho \neq j,$$

i.e., $\gamma_\ell \neq \gamma_{\ell+1}$ for $\ell = 1,\ldots,\rho$ and $\gamma_1 \neq i$, $\gamma_\rho \neq j$. The notation

$$\sum_{\rho=0}^{\infty}{}^{*}$$

indicates that if $i \neq j$ then the summation is taken over $\rho = 0,1,2,3,\ldots$, whereas if $i = j$ then the summation is over $\rho = -1,1,2,3,\ldots$. When $\rho = 0$ there is only one integration on the right hand side of (2.1) (generally for any ρ the number of integrations is $\rho+1$); the term in (2.1) corresponding to $\rho = -1$ (when $i = j$) is understood to be $\exp(-q_i t)$.

The proof is the same as in Anderson and Friedman (2) where the diffusion case was treated, except that the $z_{i,j}$ are now different.

THEOREM 2. If $u \in C^1(\bar{\Pi})$ and $\mu = (p_1,\ldots,p_n) = p$, then

$$\frac{1}{t} E^{\mu}(u(p(t) - u(p)) \to Au(p)$$

where

$$Au(p) = \sum_{j=1}^{n}\sum_{i=1}^{n}[p_i q_{ij} + p_i p_j\langle\lambda^i - \lambda^j\rangle]\frac{\partial u}{\partial p_j}$$

$$+ \sum_{j=1}^{m}\sum_{i=1}^{n} p_i\lambda_j^i[u(\Lambda_j p) - u(p)],$$

and

$$\Lambda_j p = \left(\frac{\lambda_j^k \, p_k}{\sum_{\ell=1}^{n} \lambda_j^\ell p_\ell} \right)_{k=1,\ldots,n} .$$

The proof is obtained by direct computation, using Theorem 1. We evaluate the expectation separately on the sets where $x(t) = x$ and where $x_j(t) = x_j + 1$ for one j; the remaining event has probability $O(t^2)$.

We shall now assume for simplicity that in the model described in §1, we do not have the option (a) and that in (c) shut-off is compulsory (i.e., $\tau_{m+1} = \infty$ on the set $\theta(\tau_m) = n$) with $b = 0$. However, all the results of this work extend to the general model; see §7.

THEOREM 3. (a) If $p = (0,\ldots,0,i_0,0,\ldots,0)$, then $J_{i_0,x} = \tilde{J}_{p,x}$ where

$$\tilde{J}_{p,x}(\tau) = E^{p,x}\Big[Ke^{-\alpha\tau_1} + \int_0^{\tau_1} e^{-\alpha s} \sum_{j=1}^{n} c_j p_j(s)ds$$

$$+ \sum_{\ell=1}^{\infty} \sum_{i=1}^{n-1} p_i(\tau_\ell)e^{-\alpha\tau_\ell} E^{i,x(\tau_\ell)}\Big[K + \int_0^{\sigma_{\ell,i}} e^{-\alpha s} \sum_{j=1}^{n} c_j p_j(s)ds\Big]\Big];$$

(b) Defining $\tilde{J}_{p,x}$ by (a), for any $p \in \Pi_0$, the function

$$V(p) = \inf \tilde{J}_{p,x}(\tau)$$

does not depend on x and satisfies

$$V(p) = \inf_\theta E^p\Big[\int_0^\theta e^{-\alpha s} \sum_{j=1}^{n} c_j p_j(s)ds + e^{-\alpha\theta}\big(K + \sum_{j=1}^{n-1} p_j(\theta)V(e_j)\big)\Big] \tag{2.2}$$

where θ varies over the set $\mathcal{A}$ of all $\mathcal{F}_t$ stopping times.

The proof is the same as for the diffusion case treated by Anderson and Friedman (2); it is based on Lemmas 3.1, 3.2, 3.6 of (2) which remain true for right continuous processes.

The important feature of (2.2) is that the cost function depends only on the process $p(t)$ whereas, at the same time, the decision variable θ is a stopping time with respect to the $\mathcal{F}_t$. This makes the problem of finding $V(p)$ a standard Markovian problem of optimal stopping:

$$V(p) = \inf_{\theta \in \mathcal{A}} E^p[e^{-\alpha\theta} \psi(p(\theta)) + \int_0^\theta e^{-\alpha s} g(p(s))ds]$$

where

$$g(p) = \sum_{i=1}^{n} c_i p_i, \quad \psi(p) = K + \sum_{i=1}^{n-1} p_i V(e_i). \tag{2.3}$$

Proceeding formally one can derive for V the system of inequalities

$$\begin{aligned} &Au - \alpha u + g \geq 0, \\ &u \leq \psi, \\ &(Au - \alpha u + g)(u - \psi) = 0. \end{aligned} \tag{2.4}$$

Such a system is called a <u>variational inequality</u>; ψ is called the <u>obstacle</u>. Since ψ depends on the unknown solution, we actually refer to (2.4) as a <u>quasi variational inequality</u> (q.v.i.) (cf. Bensoussan and Lions (3)). For literature on variational inequalities, see Bensoussan and Lions (5) and Friedman (7).

In the diffusion case, Anderson and Friedman (2) proved that the corresponding q.v.i. has a unique regular solution. However the analytic part of the methods they used cannot be carried over for the present problem. Only in the case $n = 1$, $m = 2$ (2.3) was solved analytically by Anderson and Friedman (1).

In the following sections we shall show that the minimal cost V is a regular solution of the q.v.i. (2.3) and we shall study the nature of the <u>continuation set</u>

$$C = \{p;\ u < \psi\},$$

the <u>stopping set</u>

$$S = \{p;\ u = \psi\},$$

and the free boundary Γ separating them.

The methods we use are almost entirely probabilistic, following the thesis of Robin (9) and a recent paper by Friedman and Robin (8).

3. THE EQUATION $Au - \alpha u = f$

It will be convenient to work with the new variables

$$y_j = \frac{p_j}{p_1} \qquad (2 \le j \le n).$$

Then Π_0 is mapped onto R^+_{n-1}. Setting $v(y) = u(p)$, $Au(p)$ becomes

$$\tilde{A}v(y) = \sum_{j=2}^{n} M_j \frac{\partial v}{\partial y_j} + \sum_{k=1}^{m} \sum_{j=1}^{n} \frac{\lambda_k^j y_j}{1+Y} [v(\tilde{\Lambda}_k y) - v(y)]$$

where $y_1 \equiv 1$, $Y = y_2 + \cdots + y_n$,

$$M_j = \sum_{i=1}^{n} (q_{i,j} - q_{i,1} y_j) y_i + \langle \lambda^1 - \lambda^j \rangle,$$

$$\tilde{\Lambda}_k y = (\mu_k^2 y_2, \ldots, \mu_k^j y_j, \ldots, \mu_k^n y_n), \qquad \mu_k^j = \frac{\lambda_k^j}{\lambda_k^1}.$$

From now on we assume that

$$\begin{aligned} &q_{i,1} = 0 \quad \text{if } i > 1, \quad q_{1,j_0} > 0 \quad \text{for some } j_0 > 1, \\ &\hat{q}_j \equiv q_1 - q_j + \langle \lambda^1 - \lambda^j \rangle > 0 \qquad \text{for } 2 \le j \le n. \end{aligned} \tag{3.1}$$

It follows that

$$M_j = M_j(y) = \sum_{\substack{i=1 \\ i \ne j}}^{n} q_{i,j} y_i + \hat{q}_j y_j$$

where $\hat{q}_j > 0$, $q_{i,j} \ge 0$ $(j \ne i)$ and $q_{1,j_0} > 0$ for some $j_0 > 1$.

Consider the system of differential equations

$$dz_j/dt = M_j(z) \qquad (2 \le j \le n) \tag{3.2}$$

with initial condition $z_j(0) = z_j^0$ where $z^0 = (z_2^0,\ldots,z_n^0)$ varies in the domain

$$\{z^0 \in R_{n-1}^+ ;\ \prod_{j=2}^{n} z_j^0 = 1\}.$$

The solution $z = z(t;z^0)$ is a curve $\gamma(z^0)$ in R_{n-1}^+, which tends to ∞ (in fact, each component tends to ∞) if $t \to \infty$ and exits R_{n-1}^+ as t gets to be sufficiently small and negative. Furthermore, the curves $\gamma(z^0)$ cover in a 1-1 way the entire quadrant R_{n-1}^+.

Consider the partial differential operator

$$\sum_{j=2}^{n} M_j(y) \frac{\partial u}{\partial y} - \alpha u = f \qquad \text{in } R_{n-1}^+ \tag{3.3}$$

where f is a bounded continuous function. If a bounded and C^1 solution u exists then

$$\frac{d}{dt} u(z(t;z^0)) - \alpha u(z(t;z^0)) = f(z(t;z^0)) \tag{3.4}$$

and therefore, necessarily,

$$u(z(t;z^0)) = \int_t^\infty e^{-\alpha(s-t)} f(z(s;z^0))ds. \tag{3.5}$$

If f is in C^1, this is, in fact, a bounded and C^1 solution.

For any bounded and continuous f, we now _define_ the solution u of (3.3) by (3.5). Then u is bounded and continuous, $du(z(t,y^0)/dt$ exists for each z^0, (3.4) holds, and

$$\|u\| \leq \frac{\|f\|}{\alpha}, \tag{3.6}$$

$$\left\|\frac{du}{dt}\right\| \leq 2\|f\| \tag{3.7}$$

where $\|\ \ \|$ is the L^∞ norm in R_{n-1}^+.

Next we introduce the semigroup $\Phi(t)$ of the process $p(t)$:

$$\Phi(t)u = E^p[u(p(t))] \qquad (u \in C^0(\bar{\Pi})),$$

or, in terms of the y(t) process $(y(t) = (p_2(t)/p_1(t), \ldots, p_n(t)/p_1(t)))$

$$\tilde{\Phi}(t)v(y) = (\Phi(t)u)(p).$$

LEMMA 1. (a) $\Phi(t)g$ is continuous if g is continuous; (b) $(\Phi(t)g)(p) \to g(p)$ if $t \to 0$ and g is continuous; (c) if g is continuously differentiable then

$$Ag = \lim_{t \to 0} \frac{\Phi(t)g - g}{h}$$

converges pointwise and boundedly, and thus (by Dynkin (6; p. 50)) g belongs to the domain D_A of the weak infinitesimal generator A.

The proof of (a), (b) is directly from the explicit form of $\Phi(t)g$. From (a), (b) it follows that Φ is a Feller semigroup when Π (or R^+_{n-1}) are endowed with the euclidean topology.

We denote by $C^0_b(R^+_{n-1})$ the class of bounded and continuous functions in R^+_{n-1}. When we write

$$-Aw + \alpha w = f, \qquad w \in D_A \cap C^0_b(R^+_{n-1}) \tag{3.8}$$

we understand by Aw the application of the infinitesimal generator A to w. On the other hand, when we write

$$-\tilde{A}w + \alpha w = f \tag{3.9}$$

we mean that

$$-\frac{dw}{dt} - \sum_{k=1}^{m} \sum_{j=1}^{n} \frac{\lambda^j_k y_j}{1+Y} [w(\tilde{\Lambda}_k y) - w(y)] + \alpha w = f$$

where $w = w(z(t,z^0))$. The interpretation of this latter equation was discussed above.

LEMMA 2. For any $f \in C^0_b(R^+_{n-1})$ there exists a unique solution w of (3.8), and w also satisfies (3.9).

PROOF. Since, by Lemma 1, $\Phi(t)$ is a Feller process, we can use Dynkin (6; p. 24) to deduce that (3.8) has a unique solution v^0, and

$$v^0(y) = E^y\left[\int_0^\infty e^{-\alpha s} f(y(s))ds\right].$$

We now follow Friedman and Robin (8) and solve successively, for a fixed $\mu > 0$,

$$-\tilde{A}v^k + (\alpha + \mu)v^k = g + Qv^{k-1} \qquad (g = f + \mu v^0)$$

where

$$Qw = \sum_{k=1}^{m} \sum_{j=1}^{n} \frac{\lambda_k^j y_j}{1+Y}(w(\tilde{\Lambda}_k y) - w(y)).$$

Using (3.6),(3.7) we get

$$\|v^k - v^{k-1}\| \leq \frac{C}{\alpha+\mu}\|v^{k-1} - v^{k-2}\|,$$

$$\left\|\frac{d}{dt} v_k\right\| \leq C,$$

where C is a constant independent of k, μ. It follows that $v = \lim v^k$ is a solution of (3.9). If we show that $v = v^0$, then the proof is complete.

Now, if f is smooth then, by direct calculation, v^0 is continuously differentiable and therefore, by Lemma 1,

$$-\tilde{A}v^0 + \alpha v^0 = f.$$

Hence

$$-\tilde{A}(v - v^0) + (\alpha + \mu)(v - v^0) = 0,$$

implying

$$\|v - v^0\| \leq \frac{C}{\alpha+\mu}\|v - v^0\|,$$

that is, $v = v^0$ (provided we choose $\mu > C - \alpha$).

For general f, take f_j smooth, $\|f_j - f\| \to 0$. The corresponding solution $w = w_j$ of Lemma 2 then satisfies:

$$\begin{aligned} &-Aw_j + \alpha w_j = f_j, \qquad w_j \in D_A \cap C_b^0(R_{n-1}^+), \\ &\|dw_j/dt\| \leq C, \qquad \|w_j - w\| \to 0 \end{aligned} \tag{3.10}$$

for some w, and

$$w^j(y) = E^y\Big[\int_0^\infty e^{-\alpha s} f_j(y(s))ds\Big].$$

Taking $j \to \infty$ in the last relation we obtain

$$w(y) = E^y\Big[\int_0^\infty e^{-\alpha s} f(y(s))ds\Big],$$

so that $w \in D_A \cap C_b^0(R_{n-1}^+)$. From (3.10) we also obtain that w satisfies (3.9).

4. SOLUTION OF THE QUASI VARIATIONAL INEQUALITY

Consider the problem

$$-Au_\varepsilon + \alpha u_\varepsilon + \frac{1}{\varepsilon}(u_\varepsilon - \psi)^+ = g. \tag{4.1}$$

THEOREM 4. For any g, ψ in $C_b^0(R_{n-1}^+)$ there exists a unique solution u_ε of (4.1) such that

$$u_\varepsilon \in D_A, \quad u_\varepsilon \text{ and } \frac{du_\varepsilon}{dt} \text{ belong to } C_b^0(R_{n-1}^+).$$

The proof is based on Lemma 2 and is similar to the corresponding proof of Theorem 2.1 in Friedman and Robin (8). We note that (4.1) is equivalent to

$$u_\varepsilon(y) = \int_0^\infty e^{-\alpha t} \Phi(t)[g - \frac{1}{\varepsilon}(u_\varepsilon - \psi)^+]dt.$$

We now take $\varepsilon \to 0$. As in Friedman and Robin (8) we can estimate

$$\frac{1}{\varepsilon}(u_\varepsilon - \psi)^+ \leq C$$

so that

$$\Big\|\frac{du_\varepsilon}{dt}\Big\| \leq C.$$

We can also give u_ε a probabilistic interpretation which shows that $u_\varepsilon(y) \to u(y)$ uniformly in y, as $\varepsilon \to 0$, where

$$u(y) = \inf_\theta E^p[e^{-\alpha\theta} \psi(p(\theta)) + \int_0^\theta e^{-\alpha s} g(p(s))ds], \tag{4.2}$$

where the relation $y \longleftrightarrow p$ is as before. We conclude that

$$\begin{aligned} &-\tilde{A}u + \alpha u \leq g \quad \text{a.e.}, \qquad u \leq \psi, \\ &(-\tilde{A}u + \alpha u - g)(u - \psi) = 0 \quad \text{a.e.} \end{aligned} \tag{4.3}$$

We can thus state:

THEOREM 5. The infimum with u(y) defined by (4.2) is in $C_b^0(R_{n-1}^+)$, and it satisfies the variational inequality (4.3) a.e.
More precisely, for any z^0, it satisfies (4.3) a.e. in t, and $\| du(z(t;z^0))/dt \| < \infty$.

The operator $-\tilde{A}+\alpha$ is not coercive. Therefore the standard uniqueness proof for variational inequalities (or the proof given in Friedman and Robin (8)) does not work.

We note however that if u is continuously differentiable then by applying Dynkin's formula we can deduce from (4.3) that u must be given by (4.2). Applying this observation to a C^1 solution u* of

$$\begin{aligned} &-\tilde{A}u^* + \alpha u^* \leq g^*, \qquad u^* \leq \psi^*, \\ &(-Au^* + \alpha u^* - g^*)(u^* - \psi^*) = 0, \end{aligned} \tag{4.4}$$

we obtain the following comparison result:

THEOREM 6. If u* is a C^1 solution of (4.4) and if $g^* \geq g$, $\psi^* \geq \psi$, then $u^* \geq u$, where u is defined in Theorem 5.

We shall prove later on that, under some conditions, the continuation set $\tilde{C} = \{y;\ u(y) < \psi(y)\}$ is bounded. The uniqueness proof in Friedman and Robin (8) can be extended, in this case, to show that there are no other solutions of the variational inequality with a bounded continuation set. (We use here the fact that in the class of functions of bounded support, $-\tilde{A} + \alpha + \lambda$ is coercive for some λ large.)

We now specialize the variational inequality to $g = \Sigma c_i p_i$ and

$$\psi(p) = K + \sum_{i=1}^{n-1} p_i V^{k-1}(e_i)$$

and find a solution $u = V^k(p)$.

Proceeding with suitable V^1, we can show (as in the diffusion case in Anderson and Friedman (2), see also Bensoussan and Lions (4)) that

$$V^k(p) \downarrow V(p)$$

and $u(y) = V(p)$ is a solution of the q.v.i. (4.3), (2.3).

As in Anderson and Friedman (2) one can immediately characterize an optimal sequence of inspections: an inspection should be made as soon as $p(t)$ hits S, given that $p(0)$ is at the state of the previous inspection.

In the following sections we study the location of the continuation set and the shape of the free boundary.

5. BOUNDEDNESS OF THE CONTINUATION SET

We shall denote by $\tilde{C}$, $\tilde{S}$, $\tilde{\Gamma}$ the image in R^+_{n-1} of the continuation set C, the stopping set S, and the free boundary Γ.

Let D be a region in R^+_{n-1}. Denote by θ_D the exit time from D. Suppose $E^y\theta_D < \infty$ for any $y \in D$ and let $\tilde{D}$ be a set which contains all the points $y(\theta_D)$, where $y(0)$ is in D. We can now state a sharper form of Theorem 6 (whose proof, however, is the same):

THEOREM 6'. If u^* is a C^1 solution of (4.3) and if $g^* \geq g$, $\psi^* \geq \psi$ in $D \cup \tilde{D}$, then $u^* \geq u$ in D.

LEMMA 3. The following formulas hold:

$$\tilde{A}\,\frac{1}{1+Y} = -\,\frac{q_1}{1+Y}\,, \qquad \tilde{A}\,\frac{y_\ell}{1+Y} = \frac{\sum_{i=1}^{n} q_{i,\ell}\,y_i}{1+Y}\,.$$

The proof is by direct computation.

Setting

$$w = K + \sum_{\ell=1}^{n-1} \frac{y_\ell V(e_\ell)}{1 + Y} - u$$

we find that

$$\tilde{A}w - \alpha w \geq h, \quad w \leq 0, \quad (\tilde{A}w - \alpha w - h)w = 0 \tag{5.1}$$

where

$$h(y) = \frac{1}{1 + Y}\left(B_1 + \sum_{j=2}^{n} B_j y_j\right) \tag{5.2}$$

and

$$B_j = c_j + \sum_{k=1}^{n-1} q_{j,k} V(e_k) - \alpha V(e_j) - \alpha K \quad (V(e_n) = 0). \tag{5.3}$$

THEOREM 7. (a) If $B_j > 0$ for $2 \leq j \leq n$ then $\tilde{C}$ is a compact set. (b) If $B_{j_0} < 0$ for some $j_0 \geq 2$, then $\tilde{C}$ is unbounded.

PROOF. For any $y_0 \in R^+_{n-1}$ with $|y_0|$ sufficiently large, introduce the set

$$D = \{y \in R^+_{n-1};\ |y - y_0| \leq C_1 |y_0|\}, \quad C_1 > 1.$$

Applying Dynkin's formula to $1/(1+Y)$ we find that $E^y(\theta_D) < \infty$. By examining the formula for the process $y(t)$ (which is obtainable from Theorem 1) we find that if $y(0) \in D$ then $y(\theta_D) \in \tilde{D}$ where

$$\tilde{D} \subset \{y \in R^+_{n-1}; |y - y'| \leq \lambda_* |y'| \text{ for some } y',\ |y' - y_0| = C_1 |y_0|\}.$$

Here λ_* depends on the λ_i^j, but not on C_1. We choose $C_1 > 2\lambda_*$. We wish to apply Theorem 6' to w, w^* with

$$w^* = N \frac{|y - y_0|^2}{1 + |y|^2}, \quad \psi^* = \psi = 0, \quad N > 0.$$

For this we must show that

$$N \frac{|y - y_0|^2}{1 + |y|^2} \geq M \text{ in } \tilde{D}, \quad h^* \equiv \tilde{A}w^* - \alpha w^* \leq h \text{ in } D,$$

where $M = \| w \|$. Choosing $N > 8M$ the first inequality holds; the second inequality then follows if $|y_0|$ is sufficiently large, since the left hand side is bounded whereas $h \to \infty$ if $|y| \to \infty$ (by (5.2) and the assumption that $B_j > 0$ if $j \geq 2$).

Applying Theorem 6' we get $w \leq w^*$ in D. In particular, $w(y_0) \leq w^*(y_0) = 0$. This completes the proof of (a).

To prove (b) notice that if $B_{j_0} < 0$ then $h(y) \to -\infty$ if $|y| \to \infty$, y in some cone V. Now, if $\tilde{C}$ is bounded then $w \equiv 0$ in $V \cap \{|y| > R\}$ for some $R > 0$. But in this set we then have

$$\tilde{A}w - \alpha w \text{ is bounded}, \quad h \to -\infty \quad \text{if } |y| \to \infty,$$

which contradicts the first inequality in (5.1).

6. THE SHAPE OF THE FREE BOUNDARY

In this section we specialize by assuming that

$$\lambda_k^j \geq \lambda_k^1 \qquad (2 \leq j \leq m,\ 1 \leq k \leq n). \tag{6.1}$$

From the formula for the process $y(t)$ we then deduce that

$$y_j(t) \uparrow \quad \text{if } t \uparrow. \tag{6.2}$$

Introduce the planar region

$$\Sigma = \{y \in R^+_{n-1};\ B_1 + \sum_{j=2}^{n} B_j y_j = 0\}.$$

THEOREM 8. If $B_j > 0$ for $j \geq 2$ and if (6.1) holds then the free boundary $\tilde{\Gamma}$ coincides with Σ.

The proof follows from the stochastic interpretation of w, noting that

$$h \gtrless 0 \quad \text{if } B_1 + \sum_{j=2}^{n} B_j y_j \gtrless 0$$

and recalling (6.2). One can give an alternate proof using comparison functions.

It is easily seen that

$$\frac{du}{dt} = \frac{d}{dt} u(z(t,z^0))$$

is continuous across Σ, that is, $dw/dt \to 0$ if $z(t,z^0)$ approaches Σ from the set where $h(y) < 0$.

Theorem 8 and the last remark were proved by Anderson and Friedman (1) in the special case $n = 2$, $m = 1$; the proof given in that paper does not extend to general n, m.

7. THE GENERAL MODEL

All the results of this work extend to the general model of §1. The only difference is that in the q.v.i. for V, the constraint ψ is now given by

$$\psi(p) = K + \sum_{j=1}^{n-1} p_j \min_{1 \le i \le j} [b_{ji} + V(e_i)] + p_n \min\Big\{b, \min_{1 \le i \le n-1} [b_{ni} + V(e_i)]\Big\}.$$

REFERENCES

1. Anderson, R. F. and Friedman, A., A quality control problem and quasi variational inequalities, J. Rat. Mech. Analys., 63 (1977), 205-252.

2. Anderson, R. F. and Friedman, A., Multi-dimensional quality control problems and quasi variational inequalities, Trans. Amer. Math. Soc., to appear.

3. Bensoussan, A. and Lions, J. L., Nouvelles Methodes en Contrôle Impulsionnel, Appl. Math. and Optimization, 1 (1975), 289-312.

4. Bensoussan, A. and Lions, J. L., Contrôle impulsionnel et temps d'arrêt inéquations variationnelles et quasi-variationnelles d'évolution, Cahiers de mathématiques de la décision, #7523, Univ. of Paris 9, Dauphine, 1975.

5. Bensoussan, A. and Lions, J. L., Temps d'arrêt et contrôle impulsionnel, Dunod, Paris, 1978.

6. Dynkin, E. B., Markov Processes, vol. 1, Springer-Verlag, 1965, Berlin.

7. Friedman,,A., Stochastic Differential Equations and Applications, vol. 2, Academic Press, New York, 1976.

8. Friedman, A. and Robin, M., The free boundary for variational inequalities with non-local operators, SIAM J. Control and Optimization, 16 (1978), 347-372.

9. Robin, M., Contrôle impulsionnel de processus de markov, Thesis, IRIA, Rocquencourt, March, 1977.

THE TAIL σ-FIELD OF ONE DIMENSIONAL DIFFUSIONS

Bert Fristedt[1]
Steven Orey[1]

Department of Mathematics
University of Minnesota
Minneapolis, Minnesota

Let $X = (X_t)$ be a diffusion with stationary transition probabilities on a one dimensional interval J. It is assumed that all interior points of J are regular and there is no killing. Then X is determined by its scale function $p(x)$ and speed measure $m(dx)$, and it has a differential generator

$$L = \frac{d}{dm}\frac{d}{dp} \quad .$$

The reader may wish to keep in mind the classical case

$$L = a(x)\frac{d^2}{dx^2} + b(x)\frac{d}{dx} \tag{1}$$

with $a(x) > 0$, a and b continuous. This corresponds to

$$dp = \frac{1}{r(x)}\, dx \ , \quad dm = \frac{r(x)}{a(x)}\, dx$$

where

$$r(x) = \exp\{\int_{x_0}^{x} \frac{b}{a}\, dx\} \ , \quad x_0 \in J \quad .$$

---

[1]Supported by NSF research contract MCS-74-05786 A02

ISBN 0-12-268380-3

We will describe the tail σ-field of X. This is equivalent to finding all bounded solutions $h(x,t)$ of

$$\left(\frac{\partial}{\partial t} + L\right)h = 0 \qquad (2)$$

on $J \times [-\infty,\infty)$. In particular the assertion that the tail σ-field is trivial is equivalent to the assertion that the only bounded solutions of (2) are $h \equiv$ constant.

Recently Rösler [2] obtained necessary and sufficient conditions for the tail σ-field to be trivial. His conditions, however, are not given explicitly in terms of L. In Theorem 1 we give an alternative proof of Rösler's result, and in Theorem 2 we find explicit conditions in terms of L. What interests us then is to specify the tail σ-field (or the bounded solutions of (1)) in the non-trivial case. Consider for example the case $J = [0,\infty)$, with L given by (1) and assume $X_t \to \infty$ with probability one. If $b(x) > c > 0$ for $x \geq x_1$ and (contrary to our assumptions) $a(x)$ vanishes for $x \geq x_1$, then once the process X_t gets to the right of x_1 it would just deterministically follow the path of

$$dX_t = b(X_t)dt \qquad (3)$$

and the entire future of X_t would be determined by the knowledge of T_{x_1}, the first time X_t hits x_1. In fact our assumptions rule out $a(x) = 0$, but since $a(x)$ need not be bounded below, and $b(x)$ need not be bounded above, the sort of thing just described can happen: ultimately the diffusion becomes overwhelmed by the drift, and X_t converges to a path of (3). Our results imply that this is exactly what always happens in the non-trivial case. Furthermore, it is all that happens: except for information about which path of (3) X_t will converge to, the future is random. For example, it can not happen that there exists positive probability that X_t will converge to the limiting path $x(t)$ from above (i.e. $X_t > x(t)$ for all big t), and also positive probability that X_t converges from below. The precise statement, for general L, is given in Theorem 3.

1. Rösler's Theorem. Let $\tilde{X}_t = (X_t, t_0 + t)$ be the space-time process associated with X. Since $(\frac{\partial}{\partial t} + L)$ is the generator for $\tilde{X}$, solutions of (2) will be referred to as harmonic functions of $\tilde{X}$. The connection between tail events of X and bounded harmonic functions of $\tilde{X}$ is very familiar: a tail event Λ of X is an invariant event of $\tilde{X}$, and

$$h(x,s) = P^{(x,s)}[\Lambda] \tag{4}$$

is a bounded harmonic function of $\tilde{X}$, where $P^{(x,s)}$ of course refers to the probability measure corresponding to $\tilde{X}$ started at (x,s), $x \in J$, $-\infty < s < \infty$.

As observed by Rösler [2], the general situation can easily be reduced to a special case. Let α and β be the left and right end-point, respectively of J. If X is recurrent the tail σ-field is easily seen to be trivial. In the transient case not both the harmonic functions of $\tilde{X}$ given by

$$h_\alpha(x,s) = P^{(x,s)}[\lim_{t\to\infty} X_t = \alpha], \quad h_\beta(x,s) = P^{(x,s)}[\lim_{t\to\infty} X_t = \beta]$$

can vanish. If neither of these functions vanishes it is easily seen that neither can be constant, and hence the tail σ-field is not trivial. This leaves the situation in which exactly one of these functions does not vanish, and by the symmetry of the situation we may restrict ourselves to $h_\alpha \equiv 0$, $h_\beta \equiv 1$. If β is accessible, that is X can reach β in finite time, it follows from $h_\alpha \equiv 0$ and X not being recurrent that β must be a trap: X remains at β once it gets there. If h is a bounded harmonic function of $\tilde{X}$, $H_t = h(\tilde{X}_t)$ is a bounded martingale, converging to a limit H_∞ with probability one. If β is a trap, H_∞ must equal a constant given by $\lim_{s\to\infty} h(\beta,s)$, and then $h(x,t) = E^{(x,t)}[H_\infty]$ is identically equal to that constant; the tail σ-field is trivial. So only the case β inaccessible has to be considered, with $X_t \to \beta$. In this case it is clear that the tail σ-field depends only on the behavior of p and m near β . One may therefore assume without losing any generality that α is accessible and instantaneously reflecting. Finally, by re-scaling one reduces to the situation $J = [0,1)$, the scale is the natural scale, $p(x) = x$, 0 is instantaneously reflecting, 1 is inacessible.

Say that X has a trivial tail σ-field if for every tail event Λ , the corresponding h defined by (4) is constant. The constant must be 0 or 1 , of course. Let

$$T_y = \inf\{t: X_t \geq y\} , \quad 0 \leq y < 1$$

be the first passage time of y .

Our first result extends a theorem of Rösler [2]. Rösler showed, by a different proof, that conditions (ii) and (iii) of Theorem 1 are equivalent.

Theorem 1. The following conditions are equivalent.

(i) There exist constant $c(y)$, $0 \leq y < 1$ and a random variable Θ such that $(T_y - c(y))$ converges to Θ as $y \uparrow 1$, $P^{(x,s)}$ - a.s. (almost surely) for all x,s .

(ii) $\lim_{y\uparrow 1} E^{(x,s)}[(T_y - E^{(x,s)}T_y)^2] < \infty$.

(iii) The tail σ-field of X is non-trivial.

(iv) $\lim_{s\to\infty} |s - c(X_s) - \Theta| = 0$, $P^{(x,s)}$ - a.s. for all x,s.

Before giving the proof of Theorem 1, we introduce some notation and recall some facts. Again let Λ be in the tail σ-field of X , and let $h(x,s) = P^{(x,s)}[\Lambda]$. Then $h(\tilde{X}_t)$ is a $P^{(x,s)}$ - martingale, for each x,s, and by the martingale convergence theorem $h(\tilde{X}_t)$ converges to I_Λ, the indicator function of Λ, $P^{(x,s)}$ - a.s. For fixed $\varepsilon \in (0,\frac{1}{2})$ introduce the following subsets of $J \times (-\infty,\infty)$:

$$\text{Red} = \{(x,t): h(x,t) > 1 - \varepsilon\}$$
$$\text{Blue} = \{(x,t): h(x,t) < \varepsilon\}$$
$$\text{Dark Red} = \{(x,t): P^{(x,t)}[\tilde{X}_u \in \text{Red for all } u \geq 0] > 1 - \varepsilon\}$$
$$\text{Dark Blue} = \{(x,t): P^{(x,t)}[\tilde{X}_u \in \text{Blue for all } u \geq 0] > 1 - \varepsilon\}$$
$$\text{Dark} = \text{Dark Red} \cup \text{Dark Blue}.$$

From the above it is clear that

$$P^{(x,s)}([\tilde{X}_t \in \text{Red for all sufficiently large } t] \cup [\tilde{X}_t \in \text{Blue for all sufficiently large } t]) = 1 .$$

In fact even

$$P^{(x,s)}[\tilde{X}_t \in \text{Dark for all sufficiently large } t] = 1 .$$

This can be seen as follows: let $\Gamma_t = [\tilde{X}_t \in \text{Red}]$. Then

$$\lim_{t \to \infty} P[\bigcap_{u=t}^{\infty} \Gamma_u \mid X_s, 0 \leq s \leq t] = I_{\bigcup_{v=1}^{\infty} \bigcap_{u=v}^{\infty} \Gamma_u}$$
$$= [\tilde{X}_t \in \text{Dark Red for all sufficiently large } t].$$

The first equality holding $P^{x,s}$ - a.s., as a consequence of a slight extension of the usual martingale convergence theorem (see for example Proposition 2.2 in Chapter 0 of [1]). Thus

$$P^{(x,s)}[\tilde{X}_t \in \text{Red for all sufficiently large } t]$$
$$= P^{(x,s)}[\tilde{X}_t \in \text{Dark Red for all sufficiently large } t],$$

and the same applies for blue.

Let Λ be a tail event, $h(x,s) = P^{(x,s)}[\Lambda]$. Suppose $h(x,s) = 1$, for some (x,s). We want to show that $h \equiv 1$. Note that $(x,s) \in$ Red, and since

$$P^{(x,s)}[\tilde{X}_t \notin \text{Red for some } t > 0] = 0$$

$(x,s) \in$ Dark Red. Let $s < s' < s''$. Then

$$P^{(x,s)}[X_{s'-s} \leq x,\ X_{s''-s} > x,\ \tilde{X}_u \in \text{Red},\ 0 \leq u \leq s''] > 0$$

$$P^{(x,s)}[X_{s'-s} > x,\ X_{s''-s} \leq x,\ \tilde{X}_u \in \text{Red},\ 0 \leq u \leq s''] > 0 .$$

Hence the point (x,s') is surrounded by a red curve. So $P^{(x,s')}$ - a.s the process will enter Red, and letting $\tau = \inf\{t: \tilde{X}_t \in \text{Red}\}$, $h(x,s') = E^{(x,s')}[h(\tilde{X}_\tau)] = 1$. It follows that for $y \in [0,1)$ there exists $s_0(y)$ such that $h(y,s_1) = 1$ for $s_1 > s_0(y)$, and $h(y,s_1) < 1$ for $s_1 < s_0(y)$. Suppose $h(x,s_1) < 1$. If there exist $y_n \uparrow 1$ with $s_0(y_n)$ bounded,

$$h(x,s_1) = E^{(x,s_1)}[h(\tilde{X}_{T_{y_n}})] \to 1$$

contradicting $h(x,s_1) < 1$. So $s(y) \to \infty$ as $y \uparrow 1$. Again suppose $h(x,s) = 1$ and choose y so that $s(y) > s$. Then $P^{(x,s)}[T_y < s(y)] > 0$ (since X is a time changed Brownian motion), and this leads to a contradiction.

So $h(x,s_1) < 1$ can not occur, and $h(y,t) \equiv 1$ follows. Let us call a tail event Λ <u>trivial</u> if the corresponding h is constant; this is equivalent to h ever assuming the values zero or one.

<u>Proof that (i) implies (iii)</u>. Assume $T_y - c(y) \to \Theta$, $P^{(x,s)}$ - a.s. for all (x,s). To see that Θ is a tail random variable, note that since $T_y \to \infty$ as $y \uparrow 1$, on setting $T_y^{(k)} = \inf\{t: t > k,\ X_t = y\}$ one has $(T_y^{(k)} - c(y)) \to \Theta$, $k = 1,2,\ldots$. Then $[\Theta < \frac{1}{2}]$ is a tail event which is clearly non-trivial.

<u>Proof that (iii) implies (i)</u>. Let Λ be a non-trivial tail event, $h(x,s) = P^{x,s}[\Lambda]$, and follow the terminology introduced before the proof of the Theorem. Let (y_n) be a

sequence of points in $[0,1)$ increasing to 1. Note that

$$T_{y_n} = (T_{y_1} - T_{y_0}) + (T_{y_2} - T_{y_1}) + \ldots + (T_{y_n} - T_{y_{n-1}})$$

is the nth partial sum of a series of independent random variables. Such a series is said to be essentially convergent if there exist constants $c(y_n)$ such that $T_{y_n} - c(y_n)$ converges a.s. If the series is not essentially convergent, then for every $k > 0$

$$\lim_{n\to\infty} P^{(x,s)}[u < T_{y_n} < u+k] = 0 \,, \quad \text{uniformly in } u \,.$$

From the discussion preceding the proof of the Theorem follows

$$(5) \quad P^{(x,s)}[\bigcup_{u=s}^{\infty} \bigcap_{t=u}^{\infty} [\tilde{X}_t \in \text{Dark Red}]] + P^{(x,s)}[\bigcup_{u=s}^{\infty} \bigcap_{t=u}^{\infty} [\tilde{X}_t \in \text{Dark Blue}]] = 1$$

and both terms on the left are positive. So there exist y, t_1 and t_2 with $(y,t_1) \in$ Dark Blue, $(y,t_2) \in$ Dark Red. Assume $t_1 > t_2$; the alternative case is treated in the same manner. Now consider $y < z < 1$, with z near one. Let m be the median of the distribution of T_z under $P^{(y,t_1)}$. Note that since the transition probabilities are stationary the distribution of T_z under $P^{(y,t_2)}$ is obtained from that under $P^{(y,t_1)}$ by translating by $(t_2 - t_1)$. Now under $P^{(y,t_1)}$ most paths are forever blue; and by (5) about half of all paths satisfy $T_z < m - (t_1 - t_2)$, so that there will be all blue paths connecting (y,t_1) to points of the form (z,s), with $s < m - (t_1 - t_2)$. A similar argument gives all red paths connecting (y,t_2) with points of the form (y,t), $t > m$. This leads to the intersection of red and blue paths, a contradiction. So after all $c(y_n)$ exist so that $(T_{y_n} - c(y_u)) \to \Theta$ $P^{(x,s)}$- a.s. Here we had an increasing sequence (y_n), but an obvious interpolation argument allows us to conclude that

$c(y)$ and Θ can be defined independently of (y_n). It is easy to see that $c(y)$ may be chosen to be increasing and continuous, and hence actually $(T_y - c(y)) \to \Theta$ as $y \uparrow 1$, $P^{(x,s)}$ - a.s. Finally, since $P^{(x,s)}[T_y - c(y) \to \Theta] = 1$, it again follows that this function is identically 1, so that the choice of $c(y)$ can be made independently of (x,s).

Proof that (ii) implies (i). Again let $y_n \uparrow 1$. For fixed (x,s) the 3 series theorem applies to show $(T_{y_n} - c(y_n))$ convergent, $P^{(x,s)}$ - a.s. where $c(y_n)$ may be taken as $E^{(x,s)}[T_{y_n}]$. Then proceed as in the previous part of the proof.

Proof that (i) implies (ii). Let $x \in [0,1)$, $x = c(x) + \theta$. Because of the stationarity of the transition probabilities there exist constants $m > 0$ and $\delta > 0$ not depending on x,s such that

$$P^{(x,s)}[\sup_{y>x} |T_y - (c(y) - \theta)| > m] < \delta < 1 .$$

By iterating (using the strong Markov property)

$$P^{(x,s)}[\sup_{y>x} |T_y - (c(y) - \theta)| > km] < \delta^k, \quad k = 1,2,\ldots$$

and (iii) follows.

Proof that (i) and (iv) are equivalent. Draw a picture. Clearly (iv) implies (i). For the converse consider $\Lambda = [\theta - \delta < \Theta < \theta + \delta]$. From the martingale convergence theorem $P^{(\tilde{X}_t)}[\Lambda]$ is near one for all large t, on the set Λ, and in particular this will hold for $t = T_z$ with z near one. Suppose $u > \delta$ and suppose $X_{T_z+u} = z$. Then, by the stationary transition probabilities,

$$P^{(z,T_z)}[\Lambda] = P^{(z,T_z+u)}[\theta - \delta + u < \Theta < \theta + \delta + u]$$

contradicting $P^{(z,T_z+u)}[\Lambda]$ near one. Hence (iv) follows.

Theorem 2. Each of the conditions (i) - (iv) of Theorem 1 is equivalent to

$$\int_0^1 (1-y)^2 m[0,y] m(dy) < \infty \qquad \text{(v)}$$

where $m(dy)$ is the speed measure.

Proof. It will be shown that (ii) of Theorem 1 is equivalent to (v) of the present Theorem.

Let $X^{(r)}$ be the process defined on $[0,r]$ which agrees with X up to T_r, and which is killed at time T_r. Let G_r be the corresponding potential operator,

$$G_r f(x) = E^x \int_0^\infty f(X^{(r)}(t)dt = \int_0^r G_r(x,y) f(y) m(dy) \quad .$$

Then

$$E^x[(\int_0^\infty f(X_t^{(r)} dt)^2] = 2 \int_0^\infty \int_0^u E^x[f(X_t^{(r)} f(X_u^{(r)})] dt\, du$$
$$= 2 \int_0^\infty \int_0^u E^x[f(X_t^{(r)}) G_r(X_t^{(r)})] dt = [\; G_r[f\, G_r f](x) \quad . \qquad (6)$$

Now G_r is easily obtained from the potential kernel $\overline{G}$ for the process $\overline{X}$ on $[-r,r]$, where $\overline{X}$ has natural scale and speed measure $\overline{m}$ determined by

$$\overline{m}[0,x] = \overline{m}[-x,0] = m[0,x] \;, \quad 0 < x < r$$

with $\overline{X}$ killed on reaching r or $-r$. Then

$$\overline{G_r}(x,y) = \begin{cases} \frac{1}{2r}(x+r)(r-y) & -r \le x \le y \le r \\ \frac{1}{2r}(y+r)(r-x) & -r \le y \le x \le r \end{cases} \quad .$$

Clearly $G_r(x,y) = \overline{G}_r(x,y) + \overline{G}_r(x,-y)$, giving

$$G_r(x,y) = r - (y \vee x) \quad 0 \le x < r \,, \quad 0 \le y < r \,.$$

Note

$$E^0[T_r^2] - (ET_r)^2 = E^0[(\int_0^\infty f(X_s^{(r)})ds)^2] - (E^0[\int_0^\infty f(X_s^{(r)})ds])^2$$

with $f \equiv 1$ and substituting into (6) and (7) gives

$$E^0[T_r^2] - (E^0T_r)^2 = 2\int_0^r (r-y)\int_0^r r - (y \vee z)m(dz)m(dy) - (\int_0^r (r-y)m(dy))^2$$

$$= \int_0^r (r-y)\{\int_0^r [2\, r-(y \vee z) - (r-z)]m(dz)\}m(dy)$$

$$= \int_0^r (r-y)(\int_0^y (r-2y+z)m(dz)m(dy) + \int_0^r (r-y)\int_y^r (r-z)m(dz)m(dy)$$

$$= \int_0^r (r-y)\int_0^y (r-2y+z)m(dz)m(dy) + \int_0^r (r-y)\int_0^y (r-z)m(dz)m(dy)$$

$$= 2\int_0^r (r-y)^2\int_0^y m(dz)m(dy) = 2\int_0^r (r-y)^2 m[0,y]m(dy)$$

$$\uparrow 2\int_0^1 (1-y)^2 m[0,y]m(dy)$$

as $r \uparrow 1$; (if $m(dz)m(dy)$ assigns positive mass to $y = z$, the integrals in which y appears as a limit should be interpreted so that half the mass $m\{y\}$ is counted).

Theorem 3. Suppose $(T_y - c(y)) \to \Theta$ as $y \uparrow 1$, $P^{(x,s)}$- a.s. for all (x,s) , (i.e. (i) of Theorem 1 holds). Then every tail random variable of X is Θ-measurable.

Proof. Let $m(x)$ be the median of Θ with respect to $P^{(x,c(x))}$. It is easily seen that the median is unique and $P^{(x,c(x))}[\Theta = m(x)] = 0$. Define

$$g_\theta(y) = c(y) + \theta - m(y) \quad .$$

Then

$$P^{(y,g_\theta(y))}[\Theta > \theta] = P^{(y,g_\theta(y))}[\Theta < y] = \frac{1}{2} \quad . \tag{8}$$

Now let Λ be a tail event, $h(x,s) = P^{(x,s)}[\Lambda]$, and use the terminology introduced preceding the proof of Theorem 1. Also define

$\Delta = \{\theta$: for every $\delta > 0$ there exists $y > 1 - \delta$ such that

$(y, g_\theta(y)) \in$ Dark$\}$.

Lemma. $\{\theta: \theta \notin \Delta$ is a Lebesque null set$\}$.

Proof. Let $\tilde{X}$ be started with an initial probability distribution μ concentrated on $\{x_0\} \times (-\infty, \infty)$ such that for each real Borel set A , the probability of $[\tilde{X}_0 \in \{x_0\} \times A]$ satisfies

$$P^\mu[\tilde{X}_0 \in \{x_0\} \times A] = \int_A p(s)ds$$

where p is a smooth probability density function, bounded away from zero on compact intervals. Since the distribution function of $T_y - (c(y) - m(y))$ under P^μ is obtained by convoluting a probability distribution with the initial distribution of the time component of $\tilde{X}$, it will also have a density which inherits the properties of p . As we know, $P^\mu[(y, T_y) \in \text{Dark}] \to 1$ as $y \uparrow 1$. Since $T_y - (c(y) - m(y)) \to \Theta$ there exist $k_{a,b} > 0$ independent of y such that

$$P^\mu[T_y \notin \text{Dark}] \geq P^\mu[T_y \notin \text{Dark},\ g_a(y) \leq T_y \leq g_b(y)] \geq$$
$$k_{a,b} |\ [g_a(y), g_b(y)] \setminus \text{Dark}|$$

where $|\cdot|$ denotes Lebesgue measure. As $y \uparrow 1$ the term on the left approaches zero. This implies that $|[a,b] \cap \Delta^c| = 0$, and the lemma follows.

Continuing with the proof of the Theorem, choose $\theta \in \Delta$. Suppose there exist y arbitrarily close to 1 with $(y, g(y)) \in$ Dark Red. For such a y then

$$P^{(y, g_\theta(y))}[X_t \in \text{Red for all } t] > 1 - \epsilon$$

and (8) holds. So there will exist two permanently red paths starting at $(y, g_\theta(y))$ with one path satisfying $t > c(X_t) + \theta$ for all sufficiently large t , the other satisfying the reversed inequality. Hence for any s, a space-time path $\tilde{X}_t(\omega)$ starting at (y,s) which is blue for all $t \geq 0$ can not cross these red paths, and so $\Theta(\omega) \neq \theta$. Setting

$$\Delta_R = \{\theta: \text{there exist } y \text{ arbitrarily near 1 with } (y, g_\theta(y)) \in \text{Dark Red}\}$$

we may conclude

$$P^{(x,s)}[\Lambda^c \cap \Theta \in \Delta_R] = 0$$

since on Λ^c the paths are a.s. blue from some point on. Likewise, if

$$\Delta_B = \{\theta: \text{there exist } y \text{ arbitrarily near 1 with } (y, g_\theta(y)) \in \text{Dark Blue}\}$$

then

$$P^{(x,s)}[\Lambda \cap \theta \in \Delta_B] = 0 .$$

By the lemma $\Delta^c = R \setminus \Delta$ is a Lebesgue null set. It only remains to argue $P^{(x,s)}[\Theta \in \Delta^c] \equiv 0$. Suppose otherwise, by the remarks after the statement of Theorem 1, $P^\mu[\Theta \in \Delta^c] > 0$ for all initial μ. In particular choose μ as in the proof of the lemma. The distribution function $P^\mu[\Theta \leq \theta]$ will again inherit the absolute continuity of the time component of $\tilde{X}_0$ and this gives the desired contradiction.

BIBLIOGRAPHY

[1] Revuz, D. Markov chains. North-Holland American Elsevier 1975, Amsterdam.

[2] Rösler, U. The tail σ-algebra for transient, stationary one dimensional diffusion and birth and death processes, Preprint.

THE PRINCIPAL EIGENVALUE FOR LINEAR SECOND ORDER ELLIPTIC EQUATIONS WITH NATURAL BOUNDARY CONDITIONS

Charles J. Holland[1,2]

Courant Institute of Mathematical Sciences
New York University
New York City, N.Y.

I. INTRODUCTION

We study the properties of the principal eigenvalue λ^* for the eigenvalue equation

$$-\nabla \circ (u_x a) + u_x b + cu = \lambda u \text{ in } \Omega, \tag{1}$$

with the natural boundary condition $u_x an = 0$ on $\partial\Omega$. Above a is a positive definite matrix for each $x \in \bar{\Omega}$, b is a vector, and n is the exterior normal. Throughout we assume that the functions a_{ij}, b_i, c are of class $C^2(\bar{\Omega})$ and the bounded domain is of class $C^{2+\alpha}$. Then the principal eigenvalue λ^* exists and is real and all other eigenvalues λ

[1]On leave from Purdue University.

[2]This research was supported by a grant to the Courant Institute from the Sloan Foundation and also by AFOSR 77-3286.

ISBN 0-12-268380-3

satisfy Re $\lambda \geq \lambda^*$, see [16]. Here a characterization of the principal eigenvalue is derived motivated by ideas from stochastic control theory. This characterization was derived earlier by Donsker-Varadhan [6] using different methods. The stochastic control theory interpretation suggests a possible numerical scheme for determining the principal eigenvalue and eigenfunction for (1), and this interpretation also gives a new characterization of the smallest eigenvalue for Schrodinger's equation. Finally we study the behavior of the principal eigenvalue λ^ε as $\varepsilon \downarrow 0$ when $a = \varepsilon I$ both for the Neumann boundary condition $\frac{\partial u}{\partial n} = 0$ and the mixed condition $\frac{\partial u}{\partial n} + \alpha^2 u = 0$. The above results are only outlined here. Some of these results have appeared in [10], [12].

II. CHARACTERIZATION

Define

$$F(\phi,V) = \phi_x a \phi_x^T + c\phi^2 + b^T(4a)^{-1}b\phi^2 - \frac{1}{2}\operatorname{div} b\, \phi^2$$
$$- \frac{1}{2}[(2a)^{-1}b - V_x^T]^T(2a)[(2a)^{-1}b - V_x^T]\phi^2. \quad (2)$$

Theorem 1. λ^* satisfies

$$\lambda^* = \min_{\phi\in\Phi} \max_{V\in C^1(\Omega)} \left[\int_\Omega F(\phi,V)\,dx + \int_{\partial\Omega} \frac{1}{2}(b\cdot n)\phi^2 ds\right] \quad (3)$$

where Φ is the set of functions ϕ of class $C^2(\Omega) \cap C(\bar{\Omega})$ with $\phi^2 > 0$ in $\bar{\Omega}$ and

$\int \phi^2 = 1$. Note that whenever $(2a)^{-1}b$ is a gradient, then V in (3) is chosen so that $V_x = (2a)^{-1}b$. In particular when $b = 0$ (3) reduces to Rayleigh-Ritz.

As we remarked earlier Donsker-Varadhan [6] utilizing their powerful methods established the analogue of (3) for the Dirichlet problem. Their method can be extended to cover this case. We established the formula (3) using stochastic control theory methods in [12] and outlined how to establish the result also for the Dirichlet problem.

Let us outline our derivation of (3). If λ^*, $u^* > 0$ represent the principal eigenvalue and normalized eigenfunction, then (1) can be rewritten as

$$-\text{tr}\ a u^*_{xx} + u^*_x(b+d) + cu^* = \lambda^* u^* \tag{4}$$

for an appropriate vector function d. Now $u^* > 0$ in $\overline{\Omega}$. Define $u^* = \exp(-\psi)$, then ψ satisfies the equation

$$\text{tr}\ a\psi_{xx} - \psi_x a \psi_x^T - \psi_x(b+d) + c = \lambda^* \tag{5}$$

with the boundary condition $\psi_x an = 0$. We recognize that (5) is the Hamilton-Jacobi equation for a stationary stochastic control problem and that it can be rewritten as

$$\text{tr}\ a\psi_{xx} + \min_{v \in R^n} [\psi_x v + (b+d+v)(4a)^{-1}(b+d+v) + c] = \lambda^* \tag{6}$$

A similar stochastic control theory interpretation

has been derived independently by Fleming [8] to obtain some Ventsel-Freidlin estimates.

The stochastic control problem is the following. For each Lipschitz function w, let p_w be the invariant measure associated with the Ito stochastic differential equation

$$d\xi = w(\xi)dt + \sigma(\xi)dz(t), \ \sigma\sigma^T = 2a, \tag{7}$$

with reflection in the direction $-an$ at the boundary. With each control w associate the cost

$$C(w) = \int_\Omega [(b+d+w)(4a)^{-1}(b+d+w)+c]p_w \, dx. \tag{8}$$

Then $\lambda^* = \underset{w}{\text{Min}}\ C(w)$. Since the min in (6) is obtained when $w = -b - d + 2(u^*)^{-1}(u_x^* a)^T$, then the Min $C(w)$ can be restricted to $\mathscr{W} = \{w: w = -b-d+2u^{-1}(u_x a)^T,$ $\int u^2 = 1, \ u \in C^2(\Omega) \cap C(\bar{\Omega})\}$.

Now the representation (3) is obtained by discovering the relationship between w and p_w for $w \in \mathscr{W}$. For any w, p_w satisfies the Fokker-Planck equation and boundary condition

$$\nabla \circ [(p_x a)^T - p(d+w)] = 0 \quad \text{in} \quad \Omega$$
$$((p_x a) - p(d+w)^T) \cdot n = 0 \quad \text{on} \quad \partial\Omega. \tag{9}$$

A little work shows that every $w \in \mathscr{W}$ can be written as

$$w = -d + 2u^{-1}(u_x a)^T + (-b+2(W_x a)^T) \tag{10}$$

where $p_w = u^2$ and u, W satisfy

$$\nabla \circ (u^2(-b + 2(W_x a)^T) = 0 \quad \text{in} \quad \Omega$$

$$(-b^T + 2W_x a) \cdot n = 0 \quad \text{on} \quad \partial\Omega. \tag{11}$$

Now W minimizes

$$I(V) = \int_\Omega ((2a)^{-1}b - V_x^T)^T (2a)((2a)^{-1}b - V_x^T) u^2 dx. \tag{12}$$

Substituting (10) into (8) and using the fact that W minimizes (12), one obtains the representation (3). This concludes the outline of the proof.

Corollary 1. $\underset{\Phi}{\text{Min}}\ \underset{C^1(\bar{\Omega})}{\text{Max}} = \underset{C^1(\bar{\Omega})}{\text{Max}}\ \underset{\Phi}{\text{Min}}$ in (3).

We shall outline a proof of this fact below. Note that the interchange of Min and Max gives the representation

$$\lambda^* = \underset{V \in C^1(\bar{\Omega})}{\text{Min}} \mu(V)$$

where $\mu(V)$ is the smallest eigenvalue to

$$-\text{tr}\, a\phi_{xx} + M(V)\phi = \mu\phi \ \text{in} \ \Omega, \tag{14}$$

$$M(V) = c + b^T(4a)^{-1}b - \frac{1}{2}\text{div}\, b$$
$$-\frac{1}{2}[(2a)^{-1}b - V_x^T]^T (2a)[(2a)^{-1}b - V_x^T],$$

with the boundary condition $\phi_x an + \frac{1}{2}(b \cdot n)\phi^2 = 0$ on $\partial\Omega$. It is not necessary that $\frac{1}{2}(b \cdot n) \geq 0$ to guarantee existence of the smallest eigenvalue. Now the fact that Min Max = Inf Sup = Sup Inf follows from the fact that the functional in (3) is strictly concave in V if we require the additional assumption that $V(0) = 0$. Convexity in the argument ϕ^2 follows from the following lemma which is the obvious modification of Lemma 2.2 in Donsker-Varadhan [6]. Note that if $q = \phi^2$, then $\int (q_x a q_x^T) q^{-1} dx = \int 4\phi_x a \phi_x^T$.

Lemma 1. Let $\mathscr{U} = \{p: p \in C^2(\Omega) \cap C^1(\overline{\Omega}),\ p > 0$ in $\overline{\Omega},\ p_x a n = 0$ on $\partial\Omega\}$. If

$q \in \mathscr{U}$, then $\int_\Omega (q_x a q_x^T) q^{-1} dx = -\inf_{p \in \mathscr{U}} \int_\Omega \frac{\nabla \circ (p_x a)}{p} q dx.$

Convexity in ϕ^2, concavity in V is sufficient to guarantee that Inf Sup = Sup Inf, see Theorem 4.1 in Sion [17]. The fact that Sup Inf = Max Min follows from the fact that $\mu(V)$ exists and Remark 2.3, p. 175 in Ekeland-Temam [7].

III. SCHRÖDINGER'S EQUATION

In [10] we considered the Schrödinger equation

$$ih \frac{\partial \psi}{\partial t} = H\psi \tag{15}$$

in bounded domains under the cases of periodic, zero Dirichlet, and zero Neumann boundary conditions. In (15) $H = -(h^2/2m)\Delta_x + K$, h is Planck's constant divided by 2π, and $K \geq 0$ is the potential function. Let us discuss only the Neumann case here. We look for solutions of (15) of the form $\psi(t,x) = \phi(x) \cdot \exp(-i\lambda h^{-1} t)$ in Ω satisfying $\frac{\partial \phi}{\partial n} = 0$ on $\partial\Omega$. Then ϕ satisfies

$$H\phi = \lambda\phi. \tag{16}$$

Let λ^* denote the smallest eigenvalue of (16)

and ϕ^* the corresponding normalized eigenfunction. We now formulate a stochastic control problem for λ^*, ϕ^*. Define $\phi^* = \exp(-Q/h)$. Then Q satisfies the equation

$$\frac{h}{2m}\Delta_x Q + \min_w [Q_x w + \tfrac{1}{2}m|w|^2 + K] = \lambda^* \tag{17}$$

For each control function v let p_v be the invariant density p_v generated by the diffusion process

$$d\xi = v(\xi)dt + (h/m)^{1/2} dz(t) \tag{18}$$

with reflection in the interior normal direction. Associate with each control v, the cost

$$C(v) = \int_\Omega [\tfrac{1}{2}m|v|^2 + K] p_v \, dx. \tag{19}$$

Theorem 2. $\lambda^* = \operatorname{Min}_v C(v)$ where the minimum is taken over Lipschitz functions $v: \overline{\Omega} \to R^n$. The minimum is obtained when $v = (h/m)(\phi_x^*/\phi^*)$ and $p_v = (\phi^*)^2$.

Note that v plays the role of "velocity" of the process (18). Thus λ^* can be thought of as the minimum energy of (19) and $(\phi^*)^2$ as the invariant probability distribution of the process. This interpretation corresponds to the standard physical interpretations. See Nelson [15] for another approach to (15) using stochastic differential equations. Of course, the interpretation (18), (19) for the self-adjoint problem $H\phi = \lambda\phi$ is Rayleigh-Ritz in disguise.

IV. NUMERICAL APPROXIMATIONS

The control theory interpretation in Section 2 also suggests an alternative numerical method for approximating the principal eigenvalue and eigenfunction for the Neumann problem (1). The procedure is to discretize the verification equation (6) obtaining the verification equations for stationary control of a Markov chain. The Markov chain problem is the discrete analogue of the stochastic control problem associated with (6). The Markov chain control problem has a special form which can be solved efficiently using an iterative method due to White [19]. See [10], Section 3 for details. We have done some numerical work on simple examples using this method, however, the convergence of this method has not been justified theoretically.

This type of numerical approximation can also be used for the periodic eigenvalue problem. However, the formal approximation suggested in [10] for the Dirichlet problem fails.

V. SINGULARLY PERTURBED PROBLEMS

Here the asymptotic behavior as $\varepsilon \downarrow 0$ of the principal eigenvalue to the singularly perturbed eigenvalue problem

$$-\varepsilon\Delta\phi - \phi_x(b(x)+\varepsilon\tilde{b}(x)) + ((c(x)+\varepsilon\tilde{c}(x))\phi = \lambda\phi \quad (20)$$

in a bounded convex domain Ω with boundary data $\frac{\partial\phi}{\partial n} + \alpha(x)\phi = 0$ on $\partial\Omega$ is studied. The behavior of λ^ε has been extensively studied for the Dirichlet problem with $c = 0$; see the papers [3], [9], [18]. There the behavior can vary from $\lambda^\varepsilon \to 0$ to $\lambda^\varepsilon \to \infty$ as $\varepsilon \downarrow 0$ depending upon the nature of the vector field b. For the zero Neumann problem such a wide range of behavior is not possible since one knows that $\min_{x\in\overline{\Omega}}(c(x)+\varepsilon\tilde{c}(x)) \le \lambda^\varepsilon \le \max_{x\in\overline{\Omega}}(c(x)+\varepsilon\tilde{c}(x))$. However, the behavior of the principal eigenvalue for Neumann and mixed problems is of interest, especially in the stability analysis of equilibrium states of nonlinear diffusion processes occurring in chemical reactors. See H. Keller [14] who treats the eigenvalue problem in one spatial dimension by transforming the problem to the standard self-adjoint Sturmian form.

The problem with the zero Neumann condition $\frac{\partial u}{\partial n} = 0$ is treated first. The mixed problem with $\alpha \not\equiv 0$ will then be solved by transforming the mixed problem into a Neumann problem. The properties of the principal eigenvalue λ^ε depend upon the nature of the vector field $b + \varepsilon\tilde{b}$, as is evident from (7)-(8). We need the following definition. Throughout let $\xi_x^0(t)$ be the solution to the differential equation

$$d\xi_x^o(t) = b(\xi_x^o(t))dt, \quad \xi_x^o(0) = x \in \bar{\Omega} \tag{21}$$

with reflection. Reflection means that if $y \in \partial\Omega$ and $b(y) \cdot n(y) \geq 0$, n the exterior normal as before, then $\xi_y^o(t)$ travels along the boundary with velocity equal to the projection of b on the tangent space to Ω at $\xi_y^o(t)$. Otherwise $\xi_y^o(t)$ is the solution to (21). This prevents the path from escaping from $\bar{\Omega}$. See Bensoussan-Lions [2] or Holland [11] for a more detailed discussion of the path.

First, the self-adjoint case $b = \tilde{b}$ is disposed of.

<u>Lemma 2</u>. If $b = \tilde{b} = 0$, then $\lambda^\varepsilon \to \min_{x \in \bar{\Omega}} c(x)$ as $\varepsilon \downarrow 0$.

Proof: This follows immediately from the Rayleigh-Ritz representation.

<u>Theorem 3</u>. Suppose that the equation (21) with reflection has a globally (for all x in $\bar{\Omega}$) asymptotically stable rest point $\bar{x}$. Then $\lambda^\varepsilon \to c(\bar{x})$ as $\varepsilon \downarrow 0$.

<u>Remark</u>. A rest point $\bar{x}$ is a point in $\bar{\Omega}$ such that $\xi_{\bar{x}}^o(t) = \bar{x}$ for all t. If $\bar{x} \in \Omega$, then $b(\bar{x}) = 0$. If $\bar{x} \in \bar{\Omega}$, then either $b(\bar{x}) = 0$ or else $b(\bar{x})$ is parallel to $n(\bar{x})$.

Outline of proof: Let λ^ε denote the principal eigenvalue and $\phi^\varepsilon > 0$ the corresponding eigenfunction. Define

$$u^{\varepsilon}(x,t) = \phi^{\varepsilon}(x)e^{-\lambda^{\varepsilon}t}. \tag{22}$$

Then u^{ε} satisfies

$$\varepsilon\Delta u + u_x(b+\varepsilon\tilde{b}) - (c+\varepsilon\tilde{c})u - u_t = 0 \tag{23}$$

with the boundary condition $\frac{\partial u}{\partial n} = 0$ on $\partial\Omega x[0,\infty)$.

Then using the Ito differential rule we have that

$$u^{\varepsilon}(x,t) = E\phi(\xi_x^{\varepsilon}(t))\exp\int_0^t -c(\xi_x^{\varepsilon}(s))ds \tag{24}$$

where $\xi_x^{\varepsilon}(t)$ is the solution to the Ito equation

$$d\xi_x^{\varepsilon}(t) = (b+\varepsilon\tilde{b})(\xi_x^{\varepsilon}(t))dt + (2\varepsilon)^{1/2}dz(t), \xi_x^{\varepsilon}(0) = x, \tag{25}$$

$z(t)$ Brownian motion as before, with reflection in the interior normal direction at the boundary of $\partial\Omega$. See [2] for a construction of the process.

The proof is then completed using the representations (22) and (24) for sufficiently large t. This type of approach follows ideas of Kac [13]. The proof requires the estimate that in convex domains

$$\Pr_{0\leq t\leq t'}[\|\xi_x^{\varepsilon}(\cdot)-\xi_x^{o}(\cdot)\| > \mu] \downarrow 0$$

as $\varepsilon \downarrow 0$ for any fixed $\mu > 0$. This estimate is derived in [11] (see equation (15) there) and depends upon estimates (3.13), (3.14) in [2] for reflecting processes in convex domains.

<u>Example</u>. If Ω is the unit disc in R^2 and $b = (b_1,b_2) = (1,-x_2^2)$, then $\lambda^{\varepsilon} \to c(1,0)$ as $\varepsilon \downarrow 0$.

<u>Theorem 4</u>. Suppose that the differential equation (21)

with reflection has a limit cycle Γ whose domain of attraction consists of all of $\overline{\Omega}$ except for isolated critical points in Ω. Further, at a critical point x^* assume that the eigenvalues of $b_x(x^*)$ have only positive real parts. Then

$$\lambda^\varepsilon \to \frac{1}{T^*}\int_0^{T^*} c(\xi_{\overline{x}}^o(t))dt \quad \text{as} \quad \varepsilon \downarrow 0$$

where T^* is the period of Γ and $\overline{x}$ is an arbitrary point on Γ.

Proof: The proof is similar to the proof of Theorem 3 except one has the added complication of possible critical points x^*.

We now consider the mixed problem for the cases considered in Theorems 3 and 4.

<u>Theorem 5</u>. Let H be a smooth solution to $\Delta H - H = 0$ in Ω with $\frac{\partial H}{\partial n} = \alpha$ on $\partial\Omega$. Suppose b satisfies the assumptions of Theorem 3. Then $\lambda^\varepsilon \to c(\overline{x}) + \nabla H(\overline{x})\cdot b(\overline{x})$ as $\varepsilon \downarrow 0$. Suppose b satisfies the assumptions of Theorem 4. Then

$$\lambda^\varepsilon \to \frac{1}{T^*}\int_0^{T^*} c(\xi_{\overline{x}}^o(t)) + \nabla H(\xi_{\overline{x}}^o(t))\cdot b(\xi_{\overline{x}}^o(t))dt \quad \text{as} \quad \varepsilon \downarrow 0.$$

Proof: The result follows by considering the eigenvalue problem for $v = e^{H(x)}u$.

<u>Example</u>. The limits for λ^ε in Theorem 5 can be evaluated in the following case. Suppose $\Omega \subset R^n$ and b satisfies the assumptions of Theorem 3 with

$\bar{x} \in \partial\Omega$. Then $b(\bar{x}) = |b(\bar{x})| n(\bar{x})$ and hence $\lambda^{\varepsilon} \to c(\bar{x}) + \alpha(\bar{x}) |b(\bar{x})|$. Thus if $\Omega = (0,1)$, $b > 0$ on $[0,1]$, then $\lambda^{\varepsilon} \to c(1) + \alpha(1)b(1)$ as $\varepsilon \downarrow 0$. This last remark is the content of the limit result (1.7) in [14].

ACKNOWLEDGMENT

The author would like to thank S. R. S. Varadhan for some stimulating discussions.

REFERENCES

1. Anderson, R., Orey, S., Small random perturbations of dynamical systems with reflecting boundary, Proc. Symposium Pure Math. Vol. XXXI, Univ. Ill., 1976, pp. 1-4, American Math Soc.
2. Bensoussan, A., Lions, J., Diffusion processes in bounded domains and singular perturbation problems for variational inequalities with Neumann boundary conditions in Probabilistic Methods in Differential Equations, Springer Verlag Lecture Notes in Mathematics #451(1975), 8-25.
3. Devinatz, A., Ellis, R., Friedman, A., The asymptotic behavior of the first real eigenvalue of second order elliptic operators with a small parameter in the highest derivatives II, Indiana Univ. Math. J. 23(1974) 991-1011.
4. Devinatz, A., Friedman, A., Asymptotic behavior of the principal eigenfunction for a singularly perturbed Dirichlet problem, preprint.
5. Donsker, M., Varadhan, S. R. S., Asymptotic evaluation of certain Wiener integrals for large time, in Proceedings of the International Conference on Integration in function spaces, Clarendon Press, Oxford (1974) 82-88.
6. Donsker, M., Varadhan, S. R. S., On the principal eigenvalue of second order differential operators, Comm. Pure. Appl. Math. 29(1976) 595-621.

7. Ekeland, I., Teman, R., Convex Analysis and Variational Problems. North-Holland (1976).
8. Fleming, W., Exit probabilities and optimal stochastic control, preprint.
9. Friedman, A., The asymptotic behavior of the first real eigenvalue of a second order elliptic operator with a small parameter in the highest derivatives, Indiana Univ. Math. J. 22(1973) 1005-1015.
10. Holland, C., A new energy characterization of the smallest eigenvalue of the Schrödinger equation Comm. Pure Appl. Math. 30(1977) 755-765.
11. Holland, C., The regular expansion in the Neumann problem for elliptic equations Comm. in Partial Differential Equations 1(1976) 191-213.
12. Holland, C., A minimum principle for the smallest eigenvalue for second order linear elliptic equations with natural boundary conditions, Comm. Pure Appl. Math., to appear.
13. Kac, M., On some connections between probability theory and differential and integral equations, Proc. of the 2nd Bekeley Symposium (1950) pp. 189-215.
14. Keller, H., Stability theory for multiple equilibrium states of a nonlinear diffusion process, SIAM J. Math. Anal. 4(1973) 134-140.
15. Nelson, E., Dynamical Theories of Brownian Motion, Princeton University Press, 1967.
16. Protter, M., Weinberger, H., On the spectrum of general second order operators, Bull. Amer. Math. Soc. 72(1966) 251-255.
17. Sion, M., On general minimax theorem, Pac. J. Math. 8(1958) 171-176.
18. Ventsel, A.,Freidlin, M., On small random perturbations of dynamical systems, Russ. Math. Surveys 25(1970) 1-56, Uspekhi Math. Nauk, 28(1970) 3-55.
19. White, D., Dynamic programming, Markov chains, and the method of successive approximations, J. Math. Anal. Appl. 6(1963) 373-376.

INVARIANCE PRINCIPLES FOR SOME INFINITE PARTICLE SYSTEMS

by

R. Holley[1] and D. Stroock[1]

0. Introduction

We present in this paper a survey (without detailed proofs) of our recent investigations into the invariance principle for infinite particle systems.

[1] Research supported in part by N.S.F. Grant MCS77-14881

ISBN 0-12-268380-3

The general set-up with which we will be dealing is the following. Let $\eta(t)$ be a Markov process describing the evolution of a system consisting of a countably infinite number of particles (either in R^d or in Z^d). Given a measurable set Γ, let $N_t(\Gamma)$ denote the number of particles in Γ at time t. The topic of our research is the problem of determining the limiting behavior of $N_\bullet$ after appropriate scaling and centering operations have been performed on $N_\bullet$. Exactly what we have in mind will be made clearer by an example.

Consider a countable number of mutually independent R^d-valued Brownian motions whose initial distribution is determined by a Poisson point process (i.e., initially the number of particles in two disjoint sets Γ_1 and Γ_2 has joint distribution $e^{-(|\Gamma_1|+|\Gamma_2|)}\dfrac{|\Gamma_1|^m}{m!}\dfrac{|\Gamma_2|^n}{n!}$, $m,n \geq 0$). Let $N_t(\Gamma)$ be the number of these Brownian particles in Γ at time t, and for $\alpha > 0$ set $N_t^{(\alpha)}(\Gamma) = \dfrac{N_{\alpha^2 t}(\alpha\Gamma) - \alpha^d|\Gamma|}{\alpha^{d/2}}$, where $|\Gamma|$ is the Lebesgue measure of Γ. Of course $N_t^{(\alpha)}$ is no longer a non-negative measure and so it is convenient to think of the state space of $N_\bullet^{(\alpha)}$ as being $\mathcal{S}'(R^d)$ (the space of tempered distributions). That is $N_t^{(\alpha)}$ is the tempered distribution given by:

$$N_t^{(\alpha)}(\varphi) = \frac{N_t(\varphi^{(\alpha)}) - \int \varphi^{(\alpha)}(x)dx}{\alpha^{d/2}} ,$$

where $\varphi^{(\alpha)}(x) = \varphi(x/\alpha)$, for test functions $\varphi \in \mathcal{S}(R^d)$. It is easy to see that if $f \in C_0^\infty(R^1)$, then

$$f(N_t^{(\alpha)}(\varphi)) - \frac{1}{\alpha^2}\int_0^{\alpha^2 t} f'(N_{s/\alpha^2}^{(\alpha)}(\varphi))N_{s/\alpha^2}^{(\alpha)}(1/2\,\Delta\varphi)ds$$
$$- \frac{1}{2\alpha^2}\int_0^{\alpha^2 t} f''(N_{s/\alpha^2}^{(\alpha)}(\varphi))\frac{N_s(|\nabla\varphi|^2)^{(\alpha)}}{\alpha^d} ds$$

is a martingale. Thus

$$f(N_t^{(\alpha)}(\varphi)) - \int_0^t f'(N_s^{(\alpha)}(\varphi))N_s^{(\alpha)}(1/2\,\Delta\varphi)ds$$
$$- 1/2\int_0^t f''(N_s^{(\alpha)}(\varphi))\frac{N_{\alpha^2 s}((|\nabla\varphi|^2)^{(\alpha)})}{\alpha^d} ds$$

is a martingale. After a little computation one can show that for $\psi \in C_0(R^d)$:

$$\sup_{0\le s\le t}\left|\frac{N_{\alpha^2 s}(\psi^{(\alpha)})}{\alpha^d} - \int \psi(x)ds\right| \to 0$$

as $\alpha \uparrow \infty$. Thus it is reasonable to suppose that if $N_\bullet^{(\alpha)}$ has a limit $N_\bullet^{(\infty)}$, then

$$f(N_t^{(\infty)}(\varphi)) - \int_0^t f'(N_s^{(\infty)}(\varphi))N_s^{(\infty)}(1/2\,\Delta\varphi)ds$$
$$- 1/2\int_0^t \||\nabla\varphi|\|^2 f''(N_s^{(\infty)}(\varphi))ds$$

is a martingale, where $\|\psi\|^2 = \int |\psi(x)|^2 dx$. In other words,

$$\text{(0.1)} \qquad \text{"}dN_t^{(\infty)} = d\mathfrak{w}_t^{\nabla} + 1/2\,\Delta N_t^{(\infty)}dt\text{"},$$

where $\mathfrak{w}_t^{\nabla}$ is the Siegel process with covariance $s\wedge t\,\langle\nabla\varphi,\nabla\psi\rangle$. It turns out that good sense can be made out of all this and that a rigorous proof can be given that $N_\bullet^{(\alpha)}$ does indeed tend in distribution to the Ornstein-Uhlenbeck process described formally by (0.1). The first version of this fact was given by Martin-Löf in [8]. For details of the approach just outlined, see [6].

A slightly more interesting example of the sort just given is studied in detail in [6]. In this example the Brownian motions are still independent and initially Poisson distributed. However, they are now branching Brownian motions which, when they branch, vanish or double with equal probability. Assigning N_t the same meaning as in the preceding, we now define:

$$N_t^{(\alpha)}(\Gamma) = \begin{cases} \dfrac{N_{\alpha^2 t}(\alpha\Gamma)}{\alpha^{(d+2)/2}} & \text{if } d = 1 \text{ or } 2 \\[2ex] \dfrac{N_{\alpha^2 t}(\alpha\Gamma) - \alpha^d|\Gamma|}{\alpha^{(d+2)/2}} & \text{if } d \geq 3 . \end{cases}$$

If $d = 1$, one can show that $N_{\bullet}^{(\alpha)} \to 0$ as $\alpha \uparrow \infty$. If $d = 2$, then $N_{\bullet}^{(\alpha)}$ tends to a continuous state branching process on non-negative tempered measures. Finally, if $d \geq 3$, then $N_{\bullet}^{(\alpha)}$ tends to the Ornstein-Uhlenbeck process described by

$$\text{(0.2)} \qquad \text{"}dN_t^{(\infty)} = d\mathfrak{w}_t^0 + 1/2\,\Delta N_t^{(\infty)}\,dt\text{"}$$

where $\mathfrak{w}_t^0$ is now the Siegel process with covariance $s \wedge t(\varphi,\psi)$. The origin of our interest in this example came from a problem posed to us by F. Spitzer.

Although these two examples provide some insight into the nature of this type of limit theorem, neither one involves systems in which the particles interact. Obviously the introduction of interaction opens up the possibility for the appearance of new phenomena (as well as new complications). The rest of this article is devoted to the study of two types of interacting systems in which the interaction is simple enough to be tractible but complicated enough to be interesting. The first example is based on the "voter models" of Holley and Liggett [2]. Here the interaction is quite tame (i.e., not too long range) and the limit phenomena resemble those coming from the independent cases described above. For the second example we look at the so called "nearest neighbor" interactions proposed by F. Spitzer [9]. In spite of their name, this interaction is of indeterminant range because one does not know how far away one's "nearest neighbors" reside. In a certain sense this example stands at the opposite extreme from the "voter model"; and for considering it we are rewarded with genuinely new phenomena. In particular, we find that the limiting process will not be Markovian unless the nearest neighbor interaction is trivial.

1. Rescaling the Voter Model

The voter model is an example of a spin flip process. Generally a spin flip process is a type of Markov process with state space $E = \{-1,1\}^{Z^d}$. If $\eta \in E$ and $k \in Z^d$ we interpret $\eta(k)$ as the spin (or in the case of the voter model the opinion) of the particle (or person) at site k. For each site k, there is a positive continuous function, c_k, on E called the flip rates. c_k governs the rate that the spin at site k flips, and no two sites flip simultaneously. More precisely, if η_t is the spin flip process, then

$$(1.1)\quad P(\eta_u(k) = -\eta_t(k) \text{ for some } t \leq u \leq +h \mid \eta_s, 0 \leq s \leq t) = c_k(\eta_t)h + o(h)$$

and if $k \neq j$

$$(1.2)\quad P(\eta_u(k) = -\eta_t(k) \text{ for some } t \leq u \leq t+h \text{ and } \eta_v(j) = -\eta_t(j) \text{ for some } t \leq v \leq t+h \mid \eta_s, 0 \leq s \leq t) = o(h).$$

If one adds some uniformity assumptions to (1.1) and (1.2) and some regularity assumptions on the functions c_k then it can be shown that there is a unique strong Markov process η_t which satisfies (1.1) and (1.2) (see [3] or [7]).

In the case of the voter model the c_k's have the form

(1.3) $$c_k(\eta) = 1 - a\eta(k) - b\eta(k) \sum_{j \in Z^d} p(k,j)\eta(j) \quad ,$$

where $p(k,j)$ is the transition function for a Markov chain, $0 \le a,b$, and $a+b \le 1$. The interpretation is that a measures the interaction of person k with his conscience and $bp(k,j)$ measures the importance that person k attaches to the opinion of person j . Note the if person j agrees with person k then person k is less likely to change his opinion in the near future than if they disagree.

For simplicity we will restrict ourselves to the case in which $p(k,j)$ is the transition function for simple random walk on Z^d . Thus each person just samples the opinions of his 2d neighbors.

The proofs of the results in this section rely on the following property of any spin system with flip rates c_k . For any cylinder function f on E

(1.4) $$f(\eta_t) - \int_0^t \sum_{k \in Z^d} c_k(\eta_s)(f({}_k\eta_s) - f(\eta_s))ds$$

is a martingale, where $${}_k\eta(j) = \begin{cases} \eta(j) & \text{if } j \neq k \\ -\eta(j) & \text{if } j = k \end{cases}$$

(see [3]).

When rescaling the voter model we are interested in how many of the people in a large region $\Lambda \subset Z^d$ hold the opinion $+1$. Thus if $\Lambda \subset Z^d$ we will be interested in quantities such as

$$\sum_{k \in \Lambda} \frac{1+\eta(k)}{2} = \frac{1}{2} \sum_{k \in \Lambda} \eta(k) + \frac{|\Lambda|}{2} \quad .$$

In most of the cases considered by us, the mean of the above expression will be $\frac{|\Lambda|}{2}$. Thus after subtracting the mean we are left with $\frac{1}{2} \sum_{k \in \Lambda} \eta(k)$. As a function of Λ this is no longer a measure, and for technical reasons we must again pass to tempered distributions. Thus if $\varphi \in \mathcal{S}(R^d)$ we will define

$$\eta(\varphi) = \sum_{k \in Z^d} \varphi(k)\eta(k) \quad .$$

Of course before rescaling we may still write

$$\eta(\varphi) = \sum_{k \in Z^d} (1+\eta(k))\varphi(k) - \sum_{k \in Z^d} \varphi(k) \quad ,$$

and think of η as the difference of two measures. However, in the limit this is no longer possible, and we must interprete the limiting process as having values in the tempered distributions (or at least a subspace thereof).

In order to rescale the voter model we define, for $\varphi \in \mathcal{S}(R^d)$ and $\alpha \geq 1$,

$$\eta^{(\alpha)}(\varphi) = \sum_{k \in Z^d} \varphi(\tfrac{k}{\alpha})\eta(k)/N(\alpha,a,b,d) \quad , \tag{1.5}$$

where $N(\alpha,a,b,d)$ is chosen according to the various cases described below. A superscript α on an η will mean that we are thinking of $\eta^{(\alpha)}$ as a tempered distribution.

For the limit theorems described below we will always assume that the initial distribution of the voter model (i.e., the distribution of η_0) is product measure with $\text{Prob}(\eta_0(k) = 1) = \frac{1}{2}$ for all $k \in Z$. Any product measure having equal factors would do just as well; however, this one permits us to take the centering to be zero and therefore simplifies the expressions.

<u>Case 1</u>. $a > 0$.

In this case we take $N(\alpha,a,b,d) = \alpha^d$, and the processes

$\eta_t^{(\alpha)}$ then converge weakly to the deterministic process N_t on the tempered distributions given by

$$N_t = (1 - e^{-2at}) \text{ (Lebesgue measure on } R^d) .$$

Case 2. $a = 0,\ b < 1$

In this case we take $N(\alpha,a,b,d) = \alpha^{d/2}$ and the processes $\eta_t^{(\alpha)}(\cdot)$ converge weakly to the process N_t characterized as follows: i) N_0 has characteristic function $e^{-1/2(\varphi,\varphi)}$, and ii) for all $\varphi \in \mathcal{S}(R^d)$ and all $F \in C_0^\infty(R)$

$$F(N_t(\varphi)) + 2(1-b)\int_0^t F'(N_s(\varphi))N_s(\varphi)ds$$
$$- 2\|\varphi\|^2 \int_0^t F''(N_s(\varphi))(1 - b\rho(s))ds$$

is a martingale. Here

$$\rho(s) = P^u(\tau_0 \le 2s) \quad , \tag{1.6}$$

where P^u is the probability measure of a continuous time simple random walk on Z^d starting at a unit vector u and having jump rate 1, and τ_0 is the hitting time of 0.

Case 3. $a = 0$ and $b = 1$

This case is divided into two subcases. The first one works in all dimensions and the second only works in dimensions three or greater.

Case 3a.

Take $N(\alpha,a,b,d) = \alpha^{d/2}$. Then $\eta_t^{(\alpha)}$ converges weakly to the process N_t which satisfies i) N_0 has characteristic function $e^{-1/2(\varphi,\varphi)}$, and ii) for all $\varphi \in \mathcal{S}(R^d)$ and $F \in C_0^\infty(R)$

$$F(N_t(\varphi)) - 2\|\varphi\|^2\int_0^t F''(N_s(\varphi))(1 - \rho(s))ds$$

is a martingale. Here $\rho(s)$ is as in (1.6).

<u>Case 3b</u>. $a = 0$, $b = 1$ and $d \geq 3$.

Take $N(\alpha,a,b,d) = \alpha^{(d+2)/2}$. In this case we must also speed up the time to get a nontrivial limit. Since this is the most interesting case we state the result as a theorem.

(1.7) <u>Theorem</u>. <u>If</u> $d \geq 3$, $a = 0$, $b = 1$, <u>and</u> $N(\alpha,a,b,d) = \alpha^{(d+2)/2}$ <u>then as</u> α <u>goes to infinity the processes</u> $\eta^{(\alpha)}_{\alpha^2 t}(\cdot)$ <u>converge weakly to the process</u> N_t <u>determined by</u> $P(N_t = 0) = 1$ <u>and for all</u> $\varphi \in \mathcal{S}(R^d)$ <u>and all</u> $F \in C_0^\infty(R)$

$$F(N_t(\varphi)) - \frac{1}{d}\int_0^t F'(N_s(\varphi))N_s(\Delta\varphi)ds$$
$$- 2(1-\rho(\infty))\|\varphi\|^2 \int_0^t F''(N_s(\varphi))ds$$

<u>is a martingale</u>, <u>where</u> $\rho(\infty) = \lim_{s\to\infty} \rho(s)$ <u>and</u> ρ <u>is as in</u> (1.6).

The proofs of these results rely on the duality between the voter model and certain finite particle systems (see [2] or [4]). This duality has also been used by Bramson and Griffeath [1] in studying the renormalized equilibrium measure for the voter model in Case 3b. It was their results which led us to guess that Theorem (1.7) was true.

We now sketch a proof of Theorem (1.7). The proofs of the results in the other cases are similar.

First, since $E[(\sum_{k\in Z^d} \varphi(\frac{k}{\alpha})\eta_0(k))^2] = \sum_{k\in Z^d} \varphi^2(\frac{k}{\alpha})$,

it is clear that $\eta^{(\alpha)}_{\alpha^2 0}(\cdot)$ converges in distribution to zero as α goes to infinity.

Now from (1.3) and (1.4) we see that for all $\varphi \in \mathcal{S}(R^d)$ and all $F \in C_0^\infty(R)$

$$F(\eta_t(\varphi)) - \int_0^t \sum_{k\in Z^d} (1-\eta_s(k)\frac{1}{2d} \sum_{|j-k|=1} \eta_s(j))[F({}_k\eta_s(\varphi)) - F(\eta_s(\varphi))]ds$$

$$= F(\eta_t(\varphi)) - 2\int_0^t F'(\eta_s(\varphi)) \sum_{k\in Z^d} \eta_s(k) \sum_{|j-k|=1} \frac{1}{2d}(\varphi(j) - \varphi(k)ds$$

$$+ 2\int_0^t F''(\eta_s(\varphi)) \sum_{k\in Z^d} (1-\eta_s(k) \sum_{|j-k|=1} \frac{1}{2d}\eta_s(j))\varphi^2(k)$$

$$+ tO(\sum_{k\in Z^d} |\varphi(k)|^3)$$

is a martingale.

Replacing $\varphi(\cdot)$ by $\varphi(\frac{\cdot}{\alpha})/\alpha^{(d+2)/2}$ and t by $\alpha^2 t$ we see that

(1.8) $$F(\eta^{(\alpha)}_{\alpha^2 t}(\varphi))$$

$$- 2\int_0^{\alpha^2 t} F'(\eta^{(\alpha)}_s(\varphi)) \sum_{k\in Z^d} \eta_s(k) \sum_{|j-k|=1} \frac{1}{2d}(\varphi(\frac{j}{\alpha}) - \varphi(\frac{k}{\alpha}))/\alpha^{(d+2)/2} ds$$

$$+ 2\int_0^{\alpha^2 t} F''(\eta^{(\alpha)}_s(\varphi)) \sum_{k\in Z^d} (1-\eta_s(k)) \sum_{|j-k|=1} \frac{1}{2d}\eta_s(j))\varphi^2(\frac{k}{\alpha})/\alpha^{d+2} ds$$

$$+ \alpha^2 t\, O(\sum_{k\in Z^d} |\varphi(\frac{k}{\alpha})|^3/\alpha^{(3d+6)/2})$$

is a martingale.

Now $\sum_{|j-k|=1} \frac{1}{2d}(\varphi(\frac{j}{\alpha}) - \varphi(\frac{k}{2})) \approx \frac{1}{2d}\alpha^{-2}\Delta\varphi(\frac{k}{\alpha})$, and by using the duality between the voter model and finite particle systems one can show that for all positive s ,

$$\sum_{k \in Z^d} (1 - \eta_{\alpha^2 s}(k)) \sum_{|j-k|=1} \frac{1}{2d} \eta_{\alpha^2 s}(j)) \varphi^2(\tfrac{k}{\alpha}) / \alpha^d$$

converges in L^2 to $(1 - \rho(\infty))\|\varphi\|^2$ as α goes to infinity.

Using these two observations in (1.8), making the change of variables $s' = \alpha^2 s$ and letting α go to infinity we see that any limit of the $\eta^{(\alpha)}_{\alpha^2 t}$ processes should have the property that for all $\varphi \in \mathcal{S}(R^d)$ and all $F \in C_0^\infty(R)$

$$(1.9) \quad F(N_t(\varphi)) - \frac{1}{d}\int_0^k F'(N_s(\varphi))N_s(\Delta\varphi)ds - 2(1-\rho(\infty))\|\varphi\|^2 \int_0^t F''(N_s(\varphi))ds$$

is a martingale.

To make all this rigorous one has to worry about compactness and show that (1.9) and $P(N_0 = 0) = 1$ uniquely determine the process N_t. The uniqueness is proved in [6] and the proof of compactness is messy but can be modeled on the approach taken in [6]. There are a few added complications here since we are on Z^d instead of R^d.

Section 5 of [6] also contains a discussion of the ergodic properties of the process determined by (1.9). In particular, it can be shown that $E[e^{iN_t(\varphi)}]$ converges to $e^{-d(1-\rho(\infty))(\varphi,(-\Delta)^{-1}\varphi)}$ as t goes to ∞, where N_t is the process determined by (1.9) and satisfying $P(N_0 = 0) = 1$.

2. Nearest Neighbor Birth and Death Processes

The ideas in this section rely heavily on F. Spitzer's [9] and our [5].

Let E be the set of $\mu \in (\{0,1\})^Z$ such that $\sum_0^\infty \mu(k) = \sum_{-\infty}^0 \mu(k) = \infty$, where $\mu(k) \in \{0,1\}$ is the kth coordinate of μ. We think of $\mu \in E$ as describing the distribution of a population on Z: $\mu(k) = 1$ means that someone is present at site k and $\mu(k) = 0$ means that no one is at site k. A nearest neighbor birth and death process on Z is a Markov process μ_t with state space E which evolves as follows:

$$P((\exists k \in [-n,n])\mu_{t+h}(k) \neq \mu_t(k) \mid \sigma(\mu_s : s \leq t)) = \sum_{-n}^{n} c_k(\mu_t)h + o(h)$$

where

$$c_k(\mu) = \begin{cases} b_k(\ell_\mu(k), r_\mu(k)) & \text{if } \mu(k) = 0 \\ d_k(\ell_\mu(k), r_\mu(k)) & \text{if } \mu(k) = 1 \end{cases}$$

and

$$\ell_\mu(k) = k - \max\{\ell < k: \mu(\ell) = 1\}$$
$$r_\mu(k) = \min\{\ell > k: \mu(\ell) = 1\} - k .$$

In keeping with our intuitive picture of μ, we think of b_k's as being the birth rates and d_k's as the death rate. Assuming that the b_k's are uniformly bounded, one can show that there is a unique Markov family of probability measures $\{P_\mu : \mu \in E\}$ on $D([0,\infty), E)$ corresponding to b and d (cf. section 3 of [5]). Moreover, if $b_k \equiv b$ and $d_k \equiv d$ where b and d are positive functions, then F. Spitzer [9] has shown that this Markov process admits a reversible initial distribution if and only if there is a probability distribution $\{f(n) : n \geq 1\}$ on Z^+ such that $\sum_1^\infty nf(n) < \infty$ and

$$\frac{b(\ell,r)}{d(\ell,r)} = \frac{f(\ell)f(r)}{f(\ell+r)} , \quad \ell,r \in Z^+ .$$

Finally, if such an f exists, then it is unique and the one and only reversible distribution for $\{P_\mu : \mu \in E\}$ is the renewal measure m_f on E determined by f.

In the present discussion, we are going to assume that $d \equiv 1$ and that

(2.1) $$b(\ell,r) = \frac{f(\ell)f(r)}{f(\ell+r)} , \quad \ell,r \in Z^+ ,$$

where f is a positive probability distribution on Z^+ having first moment

(2.2) $$m = \sum_1^\infty nf(n)$$

and variance

(2.3) $$\sigma^2 = \sum_1^\infty n^2 f(n) - m^2 .$$

Let $\{P_\mu : \mu \in E\}$ denote the associated Markov family. Then the renewal measure m_f determined by f is the unique reversible measure for $\{P_\mu : \mu \in E\}$. Denote by P the process having m_f as its initial distribution (i.e., $P = \int_E P_\mu m_f(d\mu)$).

Reversibility means that

(2.4) $$E^P[\varphi(\mu_s)\psi(\mu_t)] = E^P[\varphi(\mu_t)\psi(\mu_s)]$$
$$E^P[\varphi(\mu_0)\psi(\mu_{t-s})]$$

for all $0 \le s \le t$ and $\varphi,\psi \in B(E)$.

At the crux of all our reasoning is the next important observation. Let

(2.5) $$X_t = \min\{k > 0: \mu_s(k) = 1 \text{ for } 0 \le s \le t\}$$

and for $n \ge 1$ define:

(2.6) $$X_t^{(n)} = \begin{cases} X_t & \text{if } n = 1 \\ \min\{k > X_t^{(n-1)}: \mu_s(k) = 1 \text{ for } 0 \le s \le t\} & \text{if } n \ge 2. \end{cases}$$

Next define $\mathcal{F}_t^{(n)} = \sigma(\mu_s(k): 0 \le s \le t$ and $X_t^{(n)} \le k \le X_t^{(n+1)})$. Then for each $t \ge 0$ the σ-algebras $\{\mathcal{F}_t^{(n)}: n \ge 1\}$ are mutually independent under P. Moreover, for any $N \ge 1$, $\Phi: (\{0,1\})^N \to R^1$, and $\{s_n\}_1^N \subseteq [0,t]$, the distribution of

$$(2.7) \quad \Phi(\mu_{s_1}(X_t^{(n)}+1), \mu_{s_2}((X_t^{(n)}+2) \wedge X_t^{(n+1)}), \ldots,$$
$$\mu_{s_N}((X_t^{(n)} + N) \wedge X_t^{(n+1)}))$$

under P is the same for all $n \ge 1$. In fact, let $\{P_\mu^0 : \mu \in E\}$ be the family associated with the coefficients

$$b_k = \begin{cases} 0 & \text{if } k \le 0 \\ b & \text{if } k > 0 \end{cases}$$

$$d_k = \begin{cases} 0 & \text{if } k \le 0 \\ 1 & \text{if } k > 0, \end{cases}$$

and define m_f^0 on E to be the "one-sided" renewal measure determined by f (that is, $m_f^0(\mu : \mu(k) = 1$ for $k \le 0) = 1$ and and for all $n \ge 1$ and $1 \le k_1 < \ldots < k_n$

$$m_f^0(\{\mu : (\forall 0 \le k \le k_n)\mu(k) = 1 \text{ iff } k \in \{k_1, \ldots, k_n\}\})$$
$$= f(k_1)f(k_2 - k_1) \ldots f(k_n - k_{n-1}).$$

Set $P^0 = \int P_\mu^0 \, m_f^0(d\mu)$. Then for all $n \ge 1$ the distribution of the quantity in (2.7) under P coincides with the distribution of

$$(2.8) \quad \Phi(\mu_{s_1}(1), \mu_{s_2}(2 \wedge X_t), \ldots, \mu_{s_N}(N \wedge X_t))$$

under P^0. These assertions are intuitively reasonable in that so long as a given site is occupied, what goes on on the right of that site cannot tell what is going on to the left of it. Indeed, it is this intuition which underlies the results in section 3 of [5], and a slight modification of the methods used there yields the facts needed here.

We now make several observations based on the remarks in the preceding paragraph. First note that for any $n \in Z$ the distribution of $\mu_{\cdot}$ under P is the same as the distribution of $\tau_n \mu_{\cdot}$, where $\tau_n \mu$ is that element of E given by $(\tau_n \mu)(k) = \mu(k+n)$. Thus if we think of $\mu \in E$ as being the measure on R^1 defined by

$$(2.9) \qquad \mu((x,y)) = \sum_{k \in (x,y) \cap Z} \mu(k) \;, \quad x < y \;,$$

then $E^P[\mu_t([x,y))] = |[x,y)| E^P[\mu_t(0)]$, where $|[x,y)| = \text{card}\{k \in Z\colon k \in [x,y)\}$. But $\mu_{\cdot}$ is stationary under P and so $E^P[\mu_t(0)] = m_f(\{\mu : \mu(0) = 1\}) = 1/m$. Thus

$$(2.10) \qquad E^P[\mu_t([x,y))] = |[x,y)|/m \;.$$

Next define

$$(2.11) \qquad \Xi_T^{(n)}(t) = \mu_t([X_T^{(n)}, X_T^{(n+1)})) - \frac{X_T^{(n+1)} - X_T^{(n)}}{m}$$

for $n \geq 1$ and $0 \leq t \leq T$. Then, for fixed $T > 0$, the $\Xi_T^{(n)}(\cdot)$, $n \geq 1$, are $\mathcal{F}_T^{(n)}$ - measurable and are distributed under P in the same way as is

$$(2.12) \qquad \Xi_T(\cdot) = \mu_{\cdot}([0,X_T)) - \frac{X_T}{m}$$

under P^0. Furthermore, if

$$(2.13) \qquad N_T(y) = \min\{n \geq 1\colon X_T^{(n)} \geq y\} \;, y > 0 \;,$$

then

$$\mu_{\cdot}([0,y)) - \frac{|[0,y)|}{m}$$
$$= (\mu_{\cdot}([0,X_T) - \frac{X_T}{m}) + \sum_{1}^{N_T(y)-1} \Xi_T^{(n)}(\cdot) - (\mu_{\cdot}([y,Y_T) - \frac{|[y,Y_T)|}{m})$$

where $Y_T = \min\{k > y: \mu_\cdot(k) = 1 \text{ for } 0 \le t \le T\}$. By translation invariance, $\mu_\cdot([0,X_T)) - \frac{X_T}{m}$ and $\mu_\cdot([y,Y_T)) - \frac{|[y,Y_T)|}{m}$ have the same distribution under P, and so

$$0 = E^P[\mu_t([0,X_T) - \frac{X_T}{m}] = E^P[\sum_1^{N_T(y)-1} \Xi_T^{(n)}(t)] \quad .$$

Noting that $N_T(y) - 1$ is a stopping time relative to $\{\mathcal{F}_T^{(n)}: n \ge 1\}$ and the $\Xi_T^{(n)}(\cdot)$ are $\mathcal{F}_T^{(n)}$-measurable, we arrive at

$$E^P[N_T(y) - 1]E^{P^0}[\Xi_T(t)] = 0 \ .$$

In particular, since

$$E^P[N_T(2)] = P(X_T \ge 2) + 2P(X_T = 1) = 1 + P(X_T = 1) = 1 + \frac{e^{-T}}{m} \quad ,$$

we have now shown that:

$$E^{P^0}[\Xi_T] = 0 \ , \ T > 0 \ . \tag{2.14}$$

Next note that

$$|\Xi_T(\cdot)| \le 2X_T$$

and so

$$E^{P^0}[\sup_{0 \le t \le T} |\Xi_T(t)|^2 \le 4E^{P^0}[X_T^2] < \infty \ . \tag{2.15}$$

We can therefore apply the invariance principle to the $\Xi_T^{(n)}$'s both as real-valued random variables (i.e., for a fixed $t \in [0,T]$) and as $L^2([0,T])$-valued random variables and thereby conclude that if

$$S_{n,T}(t,y) = \frac{1}{n} \sum_1^{[n^2 y]} \Xi_T^{(\ell)}(t) \quad ,$$

then $S_{n,T}(\cdot,y)$ tends in distribution as $n \to \infty$ to the $L^2([0,T])$ - valued Brownian motion $B(\cdot,y)$ governed by $\mathfrak{w}_T$ on $C([0,\infty), L^2([0,T]))$ such that

$$(2.16)\quad E^{\mathfrak{w}_T}[\langle \varphi , B(\cdot,y)\rangle^2] = y\int_0^T\int \rho_T(t-s)\varphi(s)\varphi(t)dsdt$$

for $\varphi \in L^2([0,T])$, where

$$(2.17)\quad \rho_T(t) = E^{P^0}[\Xi_T(0)\Xi_T(t)] \quad , \quad t \in [0,T] .$$

See Remark (2.33).

We now want to apply the preceding considerations to study the following scaling. For $\alpha > 0$ define

$$(2.18)\quad \mu_t^{(\alpha)}([x,y)) = (\mu_t([\alpha^2 x,\alpha^2 y)) - \frac{|[\alpha^2 x,\alpha^2 y)|}{m} / \alpha$$

for $x < y$. Then

$$(2.19)\quad \mu_t^{(\alpha)}([0,y)) = (\mu_t([0,X_T)) - \frac{X_T}{m})/\alpha + \frac{1}{\alpha}\sum_1^{N_T(\alpha^2 y)-1} \Xi_T^{(n)}(t)$$

$$- (\mu_t([\alpha^2 y,X_T^{(N_T(\alpha^2 y))})) - \frac{|[\alpha^2 y,X_T^{(N_T(\alpha^2 y))})|}{m})/\alpha .$$

From renewal theory one can show that

$$(2.20)\quad \lim_{\alpha \uparrow \infty} E^P[\sup_{0 \le y \le Y} |\frac{N_T(\alpha^2 y)}{\alpha^2} - \frac{ye^{-T}}{m}|] = 0$$

for all $T > 0$ and $Y > 0$. Combining this with the preceding, one can easily show that the $L^2_{loc}([0,\infty))$ - valued process $y \to \mu_\cdot^{(\alpha)}([0,y))$ tends in distribution to the Brownian motion $\mathfrak{w}$ on $C([0,\infty), L^2_{loc}([0,\infty)))$ such that for each $T > 0$ and $\varphi \in L^2([0,T))$:

$$(2.21)\quad E^{\mathfrak{w}}[\langle \varphi , B(\cdot,y)\rangle^2] = y \int_0^T\int \rho(t-s)\varphi(s)\varphi(t)dsdt \quad ,$$

where

$$(2.22)\quad \rho(t) = \frac{e^{-T}}{m}\rho_T(t) , 0 \le t \le T .$$

We now want to discuss this limit Brownian motion a little bit. In the first place, it is clear that another expression for $\rho(t)$ is:

$$\rho(t) = \lim_{\alpha \uparrow \infty} E^P[\mu_0^{(\alpha)}([0,1))\mu_t^{(\alpha)}([0,1))] \quad . \tag{2.23}$$

Now observe that by the description of the original process:

$$\begin{aligned}\frac{d}{dt} & E^P[\mu_0^{(\alpha)}([0,1))\mu_t^{(\alpha)}([0,1))] \\ &= \frac{1}{\alpha} E^P[\mu_0^{(\alpha)}([0,1))(\sum_{0 \le k < \alpha^2} (b_k(\mu_t)(1-\mu_t(k))-\mu_t(k))] ;\end{aligned}$$

and by time reversibility:

$$\begin{aligned}\frac{d}{dt} & E^P[\mu_0^{(\alpha)}([0,1))\mu_t^{(\alpha)}([0,1))] \\ &= \frac{1}{\alpha} E^P[\sum_{0 \le k < \alpha^2} (b_k(\mu_0)(1-\mu_0(k)) - \mu_0(k))\mu_t^{(\alpha)}([0,1))] .\end{aligned} \tag{2.24}$$

Proceeding as we did before, we can now show that

$$\frac{d}{dt} \rho(t) = \frac{e^{-T}}{m} E^{P^0} [\Xi_T'(0)\Xi_T(t)] \quad , \quad 0 \le t \le T \quad , \tag{2.25}$$

where

$$\Xi_T'(t) = \sum_{0 \le k < X_T} (b_k(\mu_t)(1 - \mu_t(k)) - \mu_t(k)) \quad . \tag{2.26}$$

If we repeat this line of reasoning, starting from (2.24) we can also arrive at:

$$\frac{d^2}{dt^2} \rho(t) = \frac{e^{-T}}{m} E^{P^0} [\Xi_T'(0)\Xi_T'(t)] \; , \; 0 \le t \le T \; . \tag{2.27}$$

From (2.25) it is immediate that the measure $\mathfrak{w}$ is in fact concentrated on $C([0,\infty)^2,R^1)$ and that $B(t,y)$ under $\mathfrak{w}$ is a continuous two parameter Gaussian process which for fixed $t \ge 0$ is a Brownian motion in y having diffusion coefficient

$\rho(0)$ and for fixed $y > 0$ is a stationary Gaussian process with covariance $y\rho(\cdot)$.

We now wish to investigate when the limiting process is Markovian in time. Invoking Doob's characterization of stationary Gaussian Markov processes, we know that a necessary and sufficient condition is that $\rho(t)$ be exponential. This will be the case if $b(\mu) \equiv b$ is constant; or equivalently if $f(n) = qp^{n-1}$ with $0 < p < 1$, $q = 1-p$, and $b = q/p$. Indeed, (2.22) plus (2.25) then yield:

$$\frac{d}{dt}\rho(t) = \frac{e^{-T}}{m} E^{P^0}[bX_T - (b+1)\mu_t([0,X_T)]$$

$$= -\frac{e^{-T}}{m}\frac{1}{p} E^{P^0}[(\mu_t([0,X_T)) - qX_T)]$$

$$= -\frac{1}{p}\rho(t)$$

since $m = 1/q$. What is slightly more intriguing is that $\rho(t)$ is <u>not</u> exponential unless $f(n)$ is geometric. To see this, suppose that ρ is exponential. Then

$$\frac{\rho(0)}{\frac{d\rho}{dt}(0)} = \frac{\frac{d\rho}{dt}(0)}{\frac{d^2\rho}{dt^2}(0)} . \tag{2.28}$$

After a little work using (2.22), (2.25), and (2.26), one can see that (2.27) implies

$$\sum_{2}^{\infty} \frac{(f*^2(n))^2}{f(n)} = \frac{\sigma^2 + m^2}{\sigma^2} . \tag{2.29}$$

On the other hand, for any $\lambda \in R^1$:

$$|2m+\lambda| = |\sum_{1}^{\infty} (n+\lambda)f*^2(n)| = |\sum_{1}^{\infty} (n+\lambda)\frac{f*^2(n)}{f(n)} f(n)|$$

$$\leq (\sum_{1}^{\infty} (n+\lambda)^2 f(n))^{1/2} (\sum_{2}^{\infty} \frac{(f*^2(n))^2}{f(n)})^{1/2} ,$$

and equality hold only if for some $c \neq 0$:

$$f^{*2}(n) = c(n+\lambda)f(n) \ , \ n \geq 1 \ . \tag{2.30}$$

That is, for any λ:

$$\sum_{2}^{\infty} \frac{(f^{*2}(n))^2}{f(n)} \geq \frac{(2m+\lambda)^2}{\sigma^2+(m+\lambda)^2}$$

and equality obtains only if (2.30) holds. Taking $\lambda = \frac{\sigma^2 - m^2}{m}$, we conclude that

$$\sum_{2}^{\infty} \frac{(f^{*2}(n))^2}{f(n)} \geq \frac{\sigma^2+m^2}{\sigma^2} \tag{2.31}$$

and equality implies that

$$f^{*2}(n) = c\left(n + \frac{\sigma^2 - m^2}{m}\right)f(n) \ , \ n \geq 1 \ . \tag{2.32}$$

From (2.32) with $n = 1$ we now see that (2.29) implies that $\frac{\sigma^2 - m^2}{m} = -1$ and that $f^{*2}(n) = c(n-1)f(n)$, $n \geq 1$. It is easy to deduce from this that $f(n)$ must be geometric.

(2.33) Remark: The sense in which this convergence takes place is actually much better. In fact the distributions of $S_n(\cdot,y)$ converge as measures on $C([0,\infty),D([0,T]))$. The proof of this fact is not entirely elementary.

References

[1] Bramson, M., and Griffeath, D., Renormalizing the 3-dimensional Voter Model, to appear.

[2] Holley, R., and Liggett, T., Ergodic Theorems for Weakly Interacting Infinite Systems and the Voter Model, Ann. of Prob. 3, 643-663 (1975).

[3] Holley, R. and Stroock, D., A Martingale Approach to Infinite Systems of Interacting Processes, Ann. of Prob. 4, 195-228 (1976).

[4] __________, Dual Processes and their Application to Infinite Interacting Systems, to appear in Adv. in Math.

[5] __________, Nearest Neighbor Birth and Death Processes on the Real Line, to appear in Acta Math.

[6] __________, Generalized Ornstein-Uhlenbeck Processes and Infinite Particle Branching Brownian Motions, to appear.

[7] Liggett, T., Existence theorems for Infinite Particle Systems, Trans. Amer. Math. Soc. 165, 471-481 (1972).

[8] Martin-Löf, A., Limit Theorems for Motion of a Poisson System of Independent Markovian Particles with High Density, Z. Wahr. verw. Geb. 34, 205-223 (1976).

[9] Spitzer, F., Stochastic Time Evolution of One-Dimensional Infinite Particle Systems, Bull. Amer. Math. Soc. 83, 880-890 (1977).

STOCHASTIC INTEGRAL OF DIFFERENTIAL FORMS AND ITS APPLICATIONS

Nobuyuki Ikeda
Shojiro Manabe

Department of Mathematics
Osaka University
Toyonaka, Japan

I. INTRODUCTION

Let us consider a d-dimensional σ-compact oriented C^∞ manifold M. Let $X = (X_t, F_t, P_x)$ be a diffusion process on M satisfying some reasonable conditions. We consider a C^2 differential 1-form α on M. Recently, in (5), we have defined the integral of α along X_t, $t \geq 0$ and investigated its properties. In some special cases, a number of authors discussed closely related subjects (cf. (3), (6), (10), (16)). In the case when M is a Riemannian manifold some geometrical properties of paths of X can be expressed in terms of integrals of harmonic forms along X_t, $t \geq 0$ (cf. (12), (13)). Also using integrals of coclosed differential 1-forms we can analyze the relation between the theory of ordinary differential equations and that of stochastic differential equations (cf.(6)). In this article we will survey some of them and indicate some extensions of the methods and results in (5). The details of the proofs will be omitted occasionally. See Ikeda-Manabe (5) for details.

ISBN 0-12-268380-3

II. DEFINITION

Let M be a d-dimensional σ-compact oriented C^∞ manifold. Let us consider a diffusion process $X = (X_t, F_t, P_x)$ on M, that is, X is a Hunt process whose almost all sample functions : $t \longmapsto X_t$ are continuous (cf. Kunita-Watanabe (9)). Throughout this article, we assume that X satisfies the following conditions :

Condition 1. X is conservative.

Condition 2. For every C^∞ function f on M, $f(X_t)$, $t \geq 0$, is a continuous semi-martingale, that is, $f(X_t)$ is expressed as

$$f(X_t) = f(X_0) + M_t[f] + A_t[f], \tag{2.1}$$

where $M_t[f]$ is a continuous locally square-integrable (F_t)-martingale and $A_t[f]$ is a continuous (F_t)-adapted process whose almost all sample functions $t \longmapsto A_t[f]$ are of bounded variation on each finite interval.

Now we will introduce a notation. For any continuous mapping $c : [s, t] \longrightarrow M$, we denote by $c[s, t]$ the curve in M joining $c(s)$ and $c(t)$ defined by $c[s, t] = \{c(u) ; s \leq u \leq t\}$. Then $c[s, t]$ can be regarded, in the usual way, as a singular 1-chain c on M (cf. (15)). We are now in a position to define an integral of a differential 1-form along the curve $X[0, t]$.

Definition 2.1. Let α be a C^2 differential 1-form on M. An additive functional[1] A_t of X is called the integral of α

[1] See Kunita-Watanabe (9).

along the path X_t, if for every coordinate neighborhood U and local coordinate $(x^1, x^2, \cdots, x^d)$ on U,

$$A_t = \sum_{i=1}^{d} \int_0^t \alpha_i(X_s) \circ dX_s^i, \qquad 0 \leq t < \sigma_U, \tag{2.2}$$

where $(X_t^1, X_t^2, \cdots, X_t^d)$ is the coordinate of X_t,

$$\alpha = \sum_{i=1}^{d} \alpha_i(x) dx^i, \qquad \text{on } U,$$

and σ_U is the first exit time from U. Here the symbol $\circ$ denotes Stratonovich's symmetric multiplication in the sense of Itô (7).

We will denote the integral of α along the path X_t as

$$\int_{X[0,t]} \alpha$$

or simply $I(t\ ;\ \alpha)$.

Remark 2.1. By using Itô's formula, it is easy to see that the right hand side of (2.2) is independent of a particular choice of the local coordinate $(x^1, x^2, \cdots, x^d)$.

Now we choose locally finite open coverings $\{W_n\}$, $\{U_n\}$ and $\{V_n\}$ of M satisfying the following conditions :

(i) For any n, W_n is a coordinate neighborhood.

(ii) For any n, $\bar{U}_n \subset V_n \subset \bar{V}_n \subset W_n$.

We define the sequences $\{\sigma_{n,k}\}$ and $\{\tau_{n,k}\}$ of stopping times by the following relations : for $n = 1, 2, \cdots$ and $k = 0, 1, \cdots$,

$$\sigma_{n,-1} = 0,$$
$$\tau_{n,k} = \inf \{t \; ; \; t > \sigma_{n,k-1}, \; X_t \in U_n\}^{2},$$
$$\sigma_{n,k} = \inf \{t \; ; \; t > \tau_{n,k}, \; X_t \notin V_n\}.$$

Then, as an easy consequence of the definition, $I(t \; ; \alpha)$ can be written as

$$I(t \; ; \alpha) = \sum_{n=1}^{\infty} \sum_{k=0}^{\infty} \sum_{i=1}^{d} \int_{\tau_{n,k}\wedge t}^{\sigma_{n,k}\wedge t} (\psi_n \alpha)_i (X_s) \circ dX_s^i, \tag{2.3}$$

where $\{\psi_n\}$ is a partition of unity subordinate to $\{U_n\}$. By using (2.3) and Itô's formula, it is easy to show the following :

Proposition 2.1. (i) For every C^2 differential 1-form α, $I(t \; ; \alpha)$ is a continuous semi-martingale.

(ii) Let α and β be C^2 differential 1-forms such that there exists a function v on M satisfying $\alpha = v\beta$. Then

$$I(t \; ; \alpha) = \int_0^t v(X_s) \circ dI(s \; ; \beta).$$

(iii) If α is an exact C^2 differential 1-form, i.e. there exists a function u on M such that $\alpha = du$, then

$$I(t \; ; \alpha) = u(X_t) - u(X_0), \qquad \text{for every } t \geq 0. \tag{2.4}$$

A case which is of some importance is that in which the differential 1-form is closed. In this case we have

Theorem 2.1. Let α be a closed C^2 differential 1-form, i.e. $d\alpha = 0$ and c be a piecewise smooth curve joining X_0 and X_t. If $c[0, t]$ is homologous to $X[0, t]$, then

---

[2] $\inf \phi = \infty$.

$$\int_{X[0,t]} \alpha = \int_{c[0,t]} \alpha , \tag{2.5}$$

where the integral in the right hand side of (2.5) is the ordinary integral of α along $c[0,t]$.

Proof. To prove this, let us consider a subdivision $\delta(m) = \{0 = t_{m,0} < t_{m,1} < \cdots\}$ of $[0,\infty)$ such that $|t_{m,k} - t_{m,k-1}| \le 1/m$, $k = 1, 2, \cdots$. Let $\tilde{\delta}(m) = \{0 = s_{m,0} < s_{m,1} < \cdots\}$ be the refinement of $\delta(m)$ obtained by adding $\{\sigma_{m,k}\}$, $\{\tau_{m,k}\}$ and t. Then it follows from (2.4) that for sufficiently large m there exists a piecewise smooth curve $\tilde{c}$ joining X_0 and X_t satisfying the following conditions :

(i) $\tilde{c}(s_{m,k}) = X_{s_{m,k}}$, $k = 0, 1, \cdots, q$, where $s_{m,q} = t$.

(ii) For every k, $0 \le k \le q - 1$, $\tilde{c}[s_{m,k}, s_{m,k+1}]$ is homologous to $X[s_{m,k}, s_{m,k+1}]$.

(iii)

$$\int_{X[0,t]} \alpha = \int_{\tilde{c}[0,t]} \alpha .$$

Since c and $\tilde{c}$ are homologous,

$$\int_{c[0,t]} \alpha = \int_{\tilde{c}[0,t]} \alpha,$$

which completes the proof.

Now we will give a remark which plays an important role in applications of results presented here.

Remark 2.2. Let M_1 be an open submanifold of M and α be a C^2 differential 1-form on M_1. If $M \setminus M_1$ is a polar set relative to X, then X can be restricted on M_1 and we can define the integral of α along the path X_t with $X_0 \in M_1$.

As a simple application of this, we have

Example 2.1. Let M be the complex plane C and X be a diffusion process on C such that the origin $\{0\}$ is polar relative to X. We consider a C^∞ differential 1-form α on $C\backslash\{0\}$ given by $\alpha = (a^2 + b^2)^{-1}(adb - bda)$, for $x = a + ib \in C\backslash\{0\}$. Set $\theta_t = I(t ; \alpha)$. Since $d\alpha = 0$ on $C\backslash\{0\}$, it follows from (2.5) that θ_t is equal to $\arg(X_t)$ (cf. Yor (16)).

III. THE CASE OF A RIEMANNIAN MANIFOLD

In the remainder of the article we always assume that (M, g) is a d-dimensional connected complete Riemannian manifold. Let O(M) be the bundle of orthonormal frames on M. Let us consider a Brownian motion $X = (X_t, F_t, P_x)$ on M, i.e. X is a minimal diffusion process on M corresponding to $\Delta/2$, where Δ is the Laplace-Beltrami operator. Let $r = (r_t, F_t, P_r)$ be the lifted diffusion on O(M) of X, i.e. the horizontal Brownian motion on O(M) (cf. Malliavin (11)). Let ω^i, $i = 1, 2, \cdots, d$, be the solder 1-forms in the sense of Bishop-Crittenden (1), pp. 90 ~ 93. For every C^2 differential 1-form α, let $\pi^*\alpha$ be the pull-back of α by π, where $\pi : O(M) \ni r = (x, e_1, e_2, \cdots, e_d) \longmapsto \pi(r) = x \in M$. Then $\pi^*\alpha$ can be written as

$$\pi^*\alpha = \sum_{i=1}^{d} \bar{\alpha}_i(r)\omega^i . \tag{3.1}$$

Set

$$B_t^i = \int_{r[0,t]} \omega^i, \qquad i = 1, 2, \cdots, d. \tag{3.2}$$

Then $B_t = (B_t^1, B_t^2, \cdots, B_t^d)$, $t \geq 0$, is a d-dimensional (F_t)-Brownian motion starting from $0 \in R^d$. Now we have

Theorem 3.1. Let α be a C^2 differential 1-form on M. Then

$$\int_{X[0,t]} \alpha = \sum_{i=1}^{d} \int_0^t \bar{\alpha}_i(r_s)dB_s^i - \frac{1}{2}\int_0^t \delta\alpha(r_s)ds, \tag{3.3}$$

where $\bar{\alpha}_i$, $i = 1, 2, \cdots, d$, are the components of $\pi^*\alpha$ given by (3.1) and δ is the adjoint operator of the exterior differential operator d.

The details of the proof can be found in (5).

Remark 3.1. (i) In the case when α is an exact C^2 differential 1-form, (3,3) is nothing else than Itô's formula. See (5) for details.

(ii) The first term in the right hand side of (3.3) :

$$\sum_{i=1}^{d} \int_0^t \bar{\alpha}_i(r_s)dB_s^i$$

is uniquely determined from $I(t\ ;\ \alpha)$ and is called the martingale part of $I(t\ ;\ \alpha)$.

We will use $M(t\ ;\ \alpha)$ to denote the martingale part of $I(t\ ;\ \alpha)$. An immediate corollary may be mentioned.

Corollary (i) Let α be a C^2 differential 1-form on M. If $\delta\alpha = 0$ on M, $I(t\ ;\ \alpha)$ is a local (F_t)-martingale.

(ii) For any C^2 differential 1-forms α and β,

$$d\langle M(\cdot\ ;\ \alpha), M(\cdot\ ;\ \beta)\rangle_t = \langle\alpha, \beta\rangle(X_t)dt \quad ,$$

where $\langle\alpha, \beta\rangle(x)$ is the inner product of α and β in $T_x^*(M)$.

Let $\{W_n\}$, $\{U_n\}$ and $\{V_n\}$ be the locally finite open coverings of M given in §2. We further assume that for any n and $x, y \in W_n$, there exists a unique minimal geodesic γ such that $\gamma(0) = x$, $\gamma(1) = y$ and $\{\gamma(s) ; 0 \leq s \leq 1\} \subset W_n$ (cf. Helgason (4)). Let $\delta(m)$ be the subdivision of $[0, \infty)$ defined in the proof of Theorem 2.1 and $\tilde{\delta}(m)$ be the refinement of $\delta(m)$ obtained by adding $\{\sigma_{n,k}\}$ and $\{\tau_{n,k}\}$. Let us consider the broken geodesic $X_m(t)$ such that the restriction of X_m on $[s_{m,k}, s_{m,k+1}]$ is the minimal geodesic joining $X_{s_{m,k}}$ and $X_{s_{m,k+1}}$. We are now in a position to state an approximation theorem.

<u>Theorem</u> 3.2. Let α be a C^2 differential 1-form. Then, for every compact set K in M and $T > 0$,

$$\lim_{m\uparrow\infty} E_x[\sup_{0\leq t\leq T\wedge\sigma_k} |\int_{X_m[0,t]} \alpha - \int_{X[0,t]} \alpha|^2] = 0, \tag{3.4}$$

where σ_k is the first exit time from K.

The details of the proof can be found in (5).

Remark 3.2. Throughout this remark, let $\{X_m(t)\}$ be a sequence of piecewise smooth approximations to X_t such that, as $m \longrightarrow \infty$, $X_m(t) \longrightarrow X_t$, a.s., uniformly on every compact set. If α is closed, then

$$\int_{X_m[0,t]} \alpha \longrightarrow \int_{X[0,t]} \alpha, \qquad \text{a.s.,} \qquad \text{as } m \longrightarrow \infty,$$

by Theorem 2.1. In (14), McShane gave a simple example which will show what may happen in the case when α is not closed. Let $M = R^2$ and X be a 2-dimensional Brownian motion. We consider

α given by

$$\alpha = \frac{1}{2}(a\,db - b\,da), \qquad x = (a, b) \in R^2.$$

Then $d\alpha = da \wedge db$ and $\delta\alpha = 0$. For every $k \in R^1$, there exists a sequence $\{X_m(t)\}$ of approximations to X_t such that

(i)

$$E_0[|\int_{X_m[0,t]} \alpha - (\int_{X[0,t]} \alpha + kt)|^2] \longrightarrow 0, \qquad \text{as } m \longrightarrow \infty,$$

(ii) for every C^∞ differential 1-form β on M with compact support,

$$E_0[|\int_{X_m[0,t]} \beta - (\int_{X[0,t]} \beta + k\int_0^t g(X_s\ ;\ \beta)ds)|^2] \longrightarrow 0,$$

$$\text{as } m \longrightarrow \infty,$$

where $d\beta = g(x\ ;\ \beta)da \wedge db$. See (6) for details.

IV. APPLICATIONS

First, as an application of the results in §2 and §3, we will describe the following :

Example 4.1. Let M be the open unit disk $D : x = a + ib$, $a^2 + b^2 < 1$, equipped with the Riemannian metric $g_{i,j} = 4(1 - a^2 - b^2)^{-2}\delta_{i,j}$, $i,j = 1, 2$. We consider a Brownian motion X on M. Setting $\alpha = 2(1 - a^2 - b^2)^{-1}(a^2 + b^2)^{-1}(a\,db - b\,da)$, we have $d\alpha = 4(1 - a^2 - b^2)^{-2}da \wedge db$. Hence, noting (3.4) and Stokes' theorem if $X_0 = 0 \in D$ we may regard $I(t\ ;\ \alpha)$ as a stochastically defined area enclosed by a Brownian curve on M up to moment t and the minimal geodesic joining X_t and 0 (cf. (6), (10)). Next, let $r : M \rightarrow R$ be the geodesic distance between x and 0. Set $\beta = dr$. Let

$$\gamma = 2(1 - a^2 - b^2)^{-1}(a^2 + b^2)^{-1/2}(adb - bda).$$

Then we have

$$\langle\beta, \gamma\rangle = 0, \qquad \langle\beta, \beta\rangle = \langle\gamma, \gamma\rangle = 1, \qquad \delta\gamma = 0$$

and

$$\delta\beta = -\cosh r(x)/\sinh r(x) .$$

Hence, by setting $B_t^1 = M(t ; \beta)$ and $B_t^2 = I(t ; \gamma)$, it follows from the corollary in §3 that (B_t^1, B_t^2), $t \geq 0$, is a 2-dimensional (F_t)-Brownian motion starting from the origin in R^2. By Proposition 2.1 and Theorem 3.1, we have

$$dI(t ; \alpha) = (\cosh \frac{r_t}{2} / \sinh \frac{r_t}{2}) dB_t^2$$

and

$$\begin{cases} dr_t = dB_t^1 + \frac{1}{2}(\cosh r_t/\sinh r_t)dt \\ d\theta_t = (\sinh r_t)^{-1} dB_t^2 \end{cases}$$

where $r_t = r(X_t)$ and $\theta_t = \arg (X_t)$ (cf. (8), (16)).

In a forthcoming paper, (Manabe (12)), it is shown that a number of geometrical properties of diffusion processes on M can be expressed as properties of the integrals of harmonic forms along X_t, $t \geq 0$. Here we will confine ourselves to a simple example.

Example 4.2, (Manabe (12)). From now on, let (M, g) be a compact Riemannian surface with the genus k and X be a Brownian motion on M. Let (A_i, B_i), $i = 1, 2, \cdots, k$, be the canonical basis of the 1-dimensional homology group $H_1(M)$ of M and α^i, β^i, $i = 1, 2, \cdots, k$, be the harmonic 1-forms corresponding, in usual way, to A_i, B_i, $i = 1, 2, \cdots, k$, respectively. Let $c : [0, t] \longrightarrow M$ be a smooth curve joining X_0 and X_t such that c meets neither A_i nor B_i, $i = 1, 2, \cdots, k$. Set

$$I[X[0, t] ; A_i] = \mathrm{Re} \int_{X[0,t]-c} \alpha^i, \qquad I[X[0, t] ; B_i] = \mathrm{Re} \int_{X[0,t]-c} \beta^i, \qquad i = 1, 2, \cdots, k .$$

By Theorem 2.1, we may regard $I[X[0, t] ; A_i]$ $(I[X[0, t] ; B_i])$ as a stochastic version of the intersection number of $X[0, t]$ and A_i (B_i, respectively) (cf. (2), (12)).

REFERENCES

1. Bishop, R. L. and Crittenden, R. J., Geometry of manifolds, Academic Press, 1964.
2. de Rham, G., Variétés différentiables, Hermann, 1960.
3. Gaveau, M. B., Acta Math., 139 (1977).
4. Helgason, S., Differential geometry and symmetric spaces, Academic Press, 1962.
5. Ikeda, N. and Manabe, S., Integral of differential forms along the path of diffusion processes, to appear.
6. Ikeda, N., Nakao, S. and Yamato, Y., Publ. RIMS. Kyoto Univ., 13 (1977).
7. Itô, K., Appl. Math. Opt. 1 (1975).
8. Itô, K. and McKean, H. P., Diffusion processes and their sample paths. Springer, 1965.
9. Kunita, H. and Watanabe, S., Nagoya Math. Jour., 30 (1967).
10. Lévy, P., Processus stochastiques et mouvement brownien, Gauthier-Villars, 1948.
11. Malliavin, P., Jour. Funct. Anal., 17 (1974).
12. Manabe, S., On the intersection number between diffusion and chains, to appear.
13. McKean, H. P., Stochastic integrals, Academic Press, 1969.
14. McShane, E. J., Proc. 6-th Berkeley Symp. on Math. Statist. and Prob., 3 (1970).
15. Spanier, E. H., Algebraic topology, McGraw-Hill, 1966.
16. Yor. M., Bull. Soc. Math. France, 105 (1977).

STOCHASTIC ANALYSIS IN INFINITE DIMENSIONS

Kiyosi Itô

Research Institute for Mathematical Sciences
Kyoto University
Kyoto, Japan

I. INTRODUCTION

Stochastic analysis in infinite dimensions is appearing in several fields, i.g. random vibrations, random particle systems, etc. An abstract theory of stochastic differential equations has recently been established by an important work of H. H. Kuo [1]. However, it is also interesting from the view-point of applications to discuss stochastic differential equations in the Schwartz space $\mathcal{S}' = \mathcal{S}'(\mathbb{R}^r)$. In Section II of this paper we will review some known facts on $\mathcal{S}'$. In Section III we will use Kuo's theory to introduce $\mathcal{S}'$-valued Wiener processes and related stochastic integrals. In Section IV we will mention an interesting concrete example of stochastic differential equations. We thank H. Tanaka and D. W. Stroock for their valuable suggestions on this example.

II. THE SCHWARTZ SPACE $\mathcal{S}'$

Let $g(x)$ denote the standard Gauss density on $\mathbb{R}$ and $H_n(x)$, $n=0,1,2,\ldots$ the Hermite polynomials of $x \in \mathbb{R}$, i.e.

ISBN 0-12-268380-3

$$g(x) = \frac{1}{\sqrt{2\pi}} e^{-\frac{x^2}{2}}$$

and

$$H_n(x) = 2^{-n/2}(-1)^n e^{x^2} \frac{d^n(e^{-x^2})}{dx^n}, \quad n=0,1,2,\ldots$$

Then the functions

$$e_n(x) = \sqrt{g(x)}\, H_n\left(\frac{x}{\sqrt{2}}\right) c_n, \quad n=0,1,2,\ldots, \quad c_n = (2^n n!\sqrt{\pi})^{-\frac{1}{2}}$$

form a complete orthonormal system in $L^2 = L^2(\mathbb{R})$.

Let $\mathcal{F} = \mathcal{F}(\mathbb{R})$ be the space of all functions $f(x)$ of the form:

$$f(x) = \sqrt{g(x)}\, P(x), \qquad P(x) : \text{a polynomial.}$$

Then every $f \in \mathcal{F}$ is expressed uniquely as follows:

$$f = \sum_n f_n e_n \tag{2.1}$$

where $f_0, f_1, \ldots$ are constants vanishing except for a finite number of n's. It is easy to see that $\mathcal{F}$ is a pre-Hilbert space with the Hilbertian norm:

$$\|f\|_p^2 = \sum_n |f_n|^2 (n+\tfrac{1}{2})^{2p} \tag{2.2}$$

for every $p \in \mathbb{R}$. It is obvious that $\| \ \|_0$ is the usual L^2-norm.

The completion of $\mathcal{F}$ with respect to the norm $\| \ \|_p$ is denoted by $\mathcal{S}_p = \mathcal{S}_p(\mathbb{R})$. Then

$$\mathcal{S}_0 = L^2 \quad \text{and} \quad \mathcal{F} \subset \mathcal{S}_p \subset \mathcal{S}_q \quad (p \geqq q). \tag{2.3}$$

Set

$$\mathcal{S}_{\infty} = \bigcap_p \mathcal{S}_p \quad \text{and} \quad \mathcal{S}_{-\infty} = \bigcup_p \mathcal{S}_p. \tag{2.4}$$

It is easy to see that $\mathcal{S}_{\infty}$ coincides with the Schwartz space $\mathcal{S} = \mathcal{S}(\mathbb{R})$ of all rapidly decreasing functions. $\mathcal{S}$ is $\| \ \|_p$-dense in $\mathcal{S}_p$, so $\mathcal{S}_p$ is the $\| \ \|_p$-completion of $\mathcal{S}$. The Schwartz topology on $\mathcal{S}$ is the same as the topology determined by the system of norms $\| \ \|_p$, $p \in \mathbb{R}$ or equivalently by the countable system of norms $\| \ \|_p$, $p \in \mathbb{N}$. Defining the coupling $\langle f,h \rangle$ by

$$\begin{aligned} \langle f,h \rangle &= \sum_n f_n h_n \qquad (f \in \mathcal{S}_{-p},\ g \in \mathcal{S}_p) \\ &= (f,h)_0 \qquad (f,\ h \in \mathcal{S}) \end{aligned}$$

we can regard $\mathcal{S}_{-p}$ as the dual space of $\mathcal{S}_p$. In view of these facts we can easily see that $\mathcal{S}_{-\infty}$ coincides with the dual space of $\mathcal{S}$, i.e. the Schwartz space $\mathcal{S}' = \mathcal{S}'(\mathbb{R})$ of tempered distributions. The strong topology in $\mathcal{S}'$ coincides with the topology whose restriction to $\mathcal{S}_p$ is the $\| \ \|_p$-topology for every p. $\mathcal{S}_{-p}$ is often denoted by $\mathcal{S}'_p$.

The well-known properties of Hermite polynomials ensures that both $e_n'(x)$ and $xe_n(x)$ are expressed as linear combinations of $e_{n-1}(x)$ and $e_{n+1}(x)$ with coefficients of order $n^{1/2}$. Hence the maps from $\mathcal{S}$ into $\mathcal{S}$:

$$\frac{d}{dx} : \phi(x) \longmapsto \frac{d}{dx}\phi(x) \qquad \text{and} \qquad x : \phi(x) \longmapsto x\phi(x)$$

can be extended to continuous linear maps from $\mathcal{S}_p$ into $\mathcal{S}_{p-1/2}$ (i.e. from $\mathcal{S}'_p$ into $\mathcal{S}'_{p+1/2}$). Also it follows from the definition of $\| \ \|_p$ that the map from $\mathcal{S}$ into $\mathcal{S}$:

$$D : \phi(x) \longmapsto D\phi(x) \equiv \left(\frac{d^2}{dx^2} - \frac{x^2}{4}\right)\phi(x)$$

can be extended to a norm-preserving linear map from $\mathcal{S}_p$ into

$\mathcal{S}_{p-1}$ (i.e. from $\mathcal{S}'_p$ into $\mathcal{S}'_{p+1}$), because

$$De_n = (n + \frac{1}{2})e_n, \quad n = 0,1,2,\ldots$$

Let $f \in \mathcal{S}$. It is obvious that $\| f\phi \|_0 \leqq c\| \phi \|_0$ (c: constant). Noting that

$$\begin{aligned} \|f\phi\|_{p+1} &= \|D(f\phi)\|_p = \|f''\phi + 2f'\phi' + f\phi'' - \frac{1}{4}fx^2\phi\|_p \\ &\leqq c'\|\phi\|_{p+1} \qquad (c'\text{: constant}) \end{aligned}$$

we can use the induction argument on p to prove that $\|f\phi\|_p \leqq c_p\|\phi\|_p$ (c_p: constant) for $p = 0,1,2,\ldots$, which implies that the map

$$f : \phi \longmapsto f\phi$$

can be extended to a linear continuous map from $\mathcal{S}_p$ into $\mathcal{S}_p$ for $p = 0,1,2,\ldots.$ This is also true for $p = -1,-2,\ldots,$ as is seen from

$$\|f\phi\|_{-p} = \sup\{(f\phi,\psi)_0 : \psi \in \mathcal{S}_0, \ \|\psi\|_0 \leqq 1\} \quad (\phi \in \mathcal{S}).$$

All the facts mentioned above hold in the r-dimensional case with the obvious modifications:

$$\begin{aligned} &x = (x_1,x_2,\ldots,x_r), && g(x) = g(x_1)g(x_2)\ldots g(x_r), \\ &n = (n_1,n_2,\ldots,n_r), && e_n(x) = e_{n_1}(x_1)e_{n_2}(x_2)\ldots e_{n_r}(x_r), \\ &\alpha = (\alpha,\alpha,\ldots,\alpha) && (\alpha = \pm 1, \pm 1/2), \\ &p = (p_1,p_2,\ldots,p_r), && (n + 1/2)^p = \Pi_{i=1}^r (n_i + 1/2)^{p_i}. \end{aligned}$$

Denoting $(\delta_{i1},\delta_{i2},\ldots,\delta_{ir})$ by δ_i, we have that for

$$\partial_i = \frac{\partial}{\partial x_i} : \phi(x) \longmapsto \frac{\partial}{\partial x_i}\phi(x)$$

$$x_i : \phi(x) \longmapsto x_i\phi(x)$$

can be extended to continuous linear maps from $\mathcal{S}_p$ into $\mathcal{S}_{p-\frac{1}{2}\delta_i}$ and that

$$D_i : \phi(x) \longmapsto D_{x_i}\phi(x) = \left(\frac{\partial^2}{\partial x_i^2} - \frac{x_i^2}{4}\right)\phi(x)$$

can be extended to a norm-preserving linear map from $\mathcal{S}_p$ into $\mathcal{S}_{p-\delta_i}$. Also the map

$$f : \phi \longmapsto f\phi \qquad (f \in \mathcal{S})$$

can be extended to a continuous linear map from $\mathcal{S}_p$ into itself for every integer p.

III. WIENER $\mathcal{S}'$-PROCESS

Let $\mathcal{S}' = \mathcal{S}'(\mathbb{R}^r)$. The topological σ-algebra on $\mathcal{S}'$ is denoted by $\mathcal{B}(\mathcal{S}')$. $\mathcal{B}(\mathcal{S}')$ coincides with the σ-algebra on $\mathcal{S}'$ generated by the sets

$$E_{\phi,a} = \{\xi \in \mathcal{S}' : \langle\xi,\phi\rangle < a\}, \quad \phi \in \mathcal{S}, \quad a \in \mathbb{R}.$$

It is obvious that $\mathcal{B}(\mathcal{S}'_p) = \mathcal{B}(\mathcal{S}') \cap \mathcal{S}'_p$ for every p.

$\mathcal{S}'$-valued random variables (<u>$\mathcal{S}'$-variables</u>) and $\mathcal{S}'$-valued stochastic processes (<u>$\mathcal{S}'$-processes</u>) are defined in the same way as in the real-valued case. Similarly for $\mathcal{S}'_p$-variables and $\mathcal{S}'_p$-processes. A sample-continuous $\mathcal{S}'$-process $B = \{B_t\}$ is called a <u>Wiener $\mathcal{S}'$-process</u> if the following conditions are satisfied:

(i) $B_0 \equiv 0$,

(ii) B_t has independent increments,

(iii) $E(e^{i\langle B_t,\phi\rangle}) = e^{-\frac{t}{2}Q(\phi)}, \quad t \in [0,\infty), \quad \phi \in \mathcal{S} = \mathcal{S}(\mathbb{R}^r).$

where $Q(\phi)$ is a continuous positive definite quadratic functional.

It is easy to see that

$$0 \leqq Q(\phi) \leqq c\|\phi\|_p^2$$

for some $c \geqq 0$ and some $p \in \mathbb{R}^r$. Using the Minlos theorem [2], we can see that $B_t \in \mathcal{S}'_{p+1}$ a.s., because

$$(\varepsilon_n, \varepsilon_m)_{p+1} = \delta_{nm} \quad \text{and} \quad \sum_n |\varepsilon_n|_p^2 < \infty$$

for $\varepsilon_k = (k+1/2)^{-p-1} \cdot e_k$. Hence there exists a unique (in law) Wiener $\mathcal{S}'_{p+1}$-process for every continuous positive definite quadratic functional $Q(\phi)$ satisfying (3.1).

The Wiener $\mathcal{S}'$-process corresponding to $Q(\phi) = \|\phi\|_0^2$ is called a standard Wiener $\mathcal{S}'$-process, denoted by $b = \{b_t\}$. b is regarded as a Wiener $\mathcal{S}'_1$-process. Hence

$$\Delta b_t = \Delta b_t(\phi) \equiv b_t(\Delta\phi), \qquad \Delta = \sum_{i=1}^r \partial_i^2,$$

is a Wiener $\mathcal{S}'_2$-process by virtue of the last statement of Section II. The quadratic functional $Q(\phi, \Delta b)$ corresponding to $\Delta b = \{\Delta b_t\}$ is as follows:

$$Q(\phi, \Delta b) = \|\Delta\phi\|_0^2. \tag{3.1}$$

Similarly $\partial_i b_t = \langle \partial_i b_t, \phi \rangle$ $(= \langle b_t, -\partial_i \phi \rangle)$ is a Wiener $\mathcal{S}_{1+\frac{1}{2}\delta_i}$-process with

$$Q(\phi, \partial_i b) = \|\partial_i \phi\|_0^2. \tag{3.2}$$

Let $\mathbf{b}_t = (b_t^1, b_t^2, \ldots, b_t^r)$, where the component processes b_t^i, $i=1,2,\ldots,r$, are independent standard Wiener $\mathcal{S}'$-processes. Then

$$\operatorname{div} \mathbf{b}_t = \sum_{i=1}^r \partial_i b_t^i$$

is a Wiener $\mathcal{S}'_{3/2}$-process with

$$Q(\phi, \operatorname{div} \mathbf{b}_t) = \sum_{i=1}^{r} \|\partial_i \phi\|_0^2 = -(\Delta\phi, \phi) \tag{3.3}$$

Now we will explain <u>stochastic integrals</u> of the form:

$$X_t = \int_0^t a_s dB_s \tag{3.4}$$

where B_t is a Wiener $\mathcal{S}'_p$-process and a_t is an operator-valued process such that a_t is a bounded linear operator from $\mathcal{S}'_p$ to $\mathcal{S}'_q$. Note that a_t may be a differential operator, in which case q must be larger than p. We assume that $a_t = a_t(\omega)$ is measurable in (t,ω) and adapted to the process b. Measurability and adaptability are defined with respect to the uniform norm topology on the space of bounded linear operators: $\mathcal{S}'_p \to \mathcal{S}'_q$. The stochastic integral (3.4) can be defined in the same way as in the real-valued case; see Kuo [1] for general discussions. Needless to say, the process $\{X_t\}$ defined by (3.4) is a sample-continuous $\mathcal{S}'_q$-process.

In the special case where a_t is deterministic (=independent of ω) the process X_t defined by (3.4) is a sample-continuous Gaussian $\mathcal{S}'_q$-process with mean 0 which has independent increments. Also we can easily show that

$$E(\langle X_t, \phi\rangle^2) = \int_0^t Q(\alpha_s^* \phi, B)\, ds \tag{3.5}$$

where $\alpha_t^* : \mathcal{S}'_q \to \mathcal{S}'_p$ is the dual operator of α_t.

IV. AN INFINITE SYSTEM OF INDEPENDENT BROWNIAN PARTICLES

Let $\{W_t^k\}$, $k=1,2,\ldots$ be independent r-dimensional Wiener processes with $W_0^k = 0$. In the physical terminology $\{\{W_t^k\}, k=1,2,\ldots\}$ is a <u>countable system of independent Brownian particles</u>. Let $N_t^n(E)$ denote the number of the first n particles lying in a Borel subset E of $\mathbb{R}^r$ at time t, i.e.

$$N_t^n(E) = \sum_{k=1}^{n} 1_E(W_t^k) \quad (1_E = \text{the indicator of } E) \tag{4.1}$$

$\{N_t^n\}$ is regarded as a measure-valued stochastic process which is equivalent to the following $\mathcal{S}'$-process:

$$\langle N_t^n, \phi \rangle = \int_{\mathbb{R}^r} \phi(x) N_t^n(dx) = \sum_{k=1}^{n} \phi(W_t^k). \tag{4.2}$$

Define a process X_t^n by

$$X_t^n = n^{-1/2}[N_t^n - E(N_t^n)] \tag{4.3}$$

i.e.

$$\langle X_t^n, \phi \rangle = n^{-1/2} \sum_{k=1}^{n} [\phi(W_t^k) - E(\phi(W_t^k))] \tag{4.4}$$

$\phi(W_t^k)$, $k=1,2,\ldots$ are independent random variables bounded uniformly and the variance

$$V(\phi(W_t^k)) = \int_{\mathbb{R}^r} \phi(x)^2 g_t(x)dx - \left(\int_{\mathbb{R}^r} \phi(x) g_t(x)dx\right)^2$$

is independent of k and will be denoted by $V_t(\phi)$. Hence the central limit theorem ensures that $\langle X_t^n, \phi \rangle$ converges in law to a Gaussian variable $X_t(\phi)$ with mean 0 and variance $V_t(\phi)$ for t, ϕ fixed. Using the central limit theorem in several dimensions, we can similarly prove that $\{X_t(\phi)\}_{t,\phi}$ is a Gaussian system. By taking an appropriate version we can regard $\{X_t\}$ as a Gaussian $\mathcal{S}'$-process ($\mathcal{S}' = \mathcal{S}'(\mathbb{R}^r)$) with mean 0, so we denote $X_t(\phi)$ by $\langle X_t, \phi \rangle$ from now on. Since

$$E[\langle X_t, \phi \rangle^2] = V_t(\phi),$$

we have

$$E[e^{i\langle X_t,\phi\rangle}] = e^{-\frac{1}{2}V_t(\phi)}$$

Since $V_t(\phi) \leq c\|\phi\|_0^2$ (c: constant), $\{X_t\}$ is an $\mathcal{S}_1'$-process. Since

$$V_t(\phi) = (A_t\phi,\phi)_0 \quad \text{where} \quad A_t\phi(x) = g_t(x)\phi(x) - (g_t,\phi)_0\, g_t(x)$$

and since A_t is not a trace operator in $\mathcal{S}_0$, the Sazonov theorem [2] ensures that $\{X_t\}$ is not an $\mathcal{S}_0'$-process.

Theorem The limit $\mathcal{S}_1'$-process $\{X_t\}$ obtained above satisfies the following stochastic differential equation:

$$dX_t = \sum_{i=1}^{r}\left((\partial_i\sqrt{g_t})+\sqrt{g_t}\,\partial_i\right)db_t^i + \frac{1}{2}\Delta X_t dt, \qquad X_0 = 0 \tag{4.5}$$

where b_t^i, $i=1,2,\ldots,r$ are independent standard Wiener $\mathcal{S}'$-processes.

We will sketch the proof of this fact. It follows from (4.2) that

$$\begin{aligned}\langle N_t^n,\phi\rangle &= \int_0^t d\langle N_s^n,\phi\rangle = \sum_{k=1}^{n}\int_0^t d[\phi(W_t^k)] \\ &= \sum_{k=1}^{n}\int_0^t \nabla\phi(W_s^k)\cdot dW_s^k + \frac{1}{2}\sum_{k=1}^{n}\int_0^t \Delta\phi(W_s^k)\,ds, \\ &= \sum_{k=1}^{n}\langle Y_t^k,\phi\rangle + \frac{1}{2}\int_0^t \langle N_s^n,\Delta\phi\rangle ds\end{aligned}$$

where

$$\langle Y_t^k,\phi\rangle = \int_0^t \nabla\phi(W_s^k)\cdot dW_s^k. \tag{4.6}$$

This, combined with (4.4), implies that

$$\langle X_t^n, \phi \rangle = n^{-1/2} \sum_{k=1}^{n} \langle Y_t^k, \phi \rangle + \frac{1}{2} \int_0^t \langle X_s^n, \Delta\phi \rangle ds \tag{4.7}$$

Observing that $\langle Y_t^k, \phi \rangle$, $k=1,2,\ldots$ are independent and that

$$E(\langle Y_t^k, \phi \rangle) = 0, \quad E(\langle Y_t^k, \phi \rangle^2) = \int_0^t (|\nabla\phi|^2, g_s)_0 ds$$

$$(| \ | : \text{ Euclidean norm in } \mathbb{R}^r)$$

and using the central theorem as above, we can see that the first term M_t^n of the right hand side of (4.7) converges in law to a Gaussian $\mathcal{S}'$-process M_t. Since

$$E(\langle M_s, \phi \rangle \langle M_t - M_s, \psi \rangle) = \lim_{n\to\infty} E(\langle M_s^n, \phi \rangle \langle M_t^n - M_s^n, \psi \rangle) = 0$$

for $s < t$, M_t has independent increments. Since $\{\langle Y_t^k, \phi \rangle\}$, $k=1,2,\ldots$ are independent, we have

$$E(\langle M_t, \phi \rangle^2) = \int_0^t (|\nabla\phi|^2, g_s) ds$$

$$= \sum_{i=1}^{r} \int_0^t \| \sqrt{g_s}(\partial_i \phi) \|_0^2 ds \tag{4.8}$$

Let $\{b_t^i\}$, $i=1,2,\ldots,n$ be independent standard Wiener $\mathcal{S}'$-processes. Then

$$\tilde{M}_t = \sum_{i=1}^{r} \int_0^t \big((\partial_i \sqrt{g_s}) + \sqrt{g_s}\,\partial_i \big) db_s^i$$

is also a Gaussian $\mathcal{S}'$-process with independent increments. By (3.5) we have

$$E(\langle \tilde{M}_t, \phi \rangle^2) = \sum_{i=1}^{r} \int_0^t \| \big((\partial_i \sqrt{g_s}) + \sqrt{g_s}\,\partial_i \big)^* \phi \|_0^2 ds$$

$$= \sum_{i=1}^{r} \int_0^t \| -\sqrt{g_s}(\partial_i \phi) \|_0^2 ds,$$

which, combined with (4.8), implies that

$$E(\langle\tilde{M}_t,\phi\rangle^2) = E(\langle M_t,\phi\rangle^2) \tag{4.9}$$

Since both $\{\tilde{M}_t\}$ and $\{M_t\}$ are Gaussian $\mathcal{S}'$-processes with mean 0 and have independent increments, $\{\tilde{M}_t\}$ is regarded as a version of $\{M_t\}$. Taking the limits (in law) of both sides of (4.7), we have

$$X_t = \sum_{i=1}^{r} \int_0^t \big((\partial_i \sqrt{g}_s) + \sqrt{g}_s \partial_i\big) db_s^i + \frac{1}{2}\int_0^t \Delta X_s ds,$$

proving (4.5).

REFERENCES

1. Kuo, H. H., Gaussian Measures in Banach Spaces, Lecture Note in Mathematics 463, Springer-Verlag (1975).
2. Gelfand, I. M. and Vilenkin, N. Ja., Generalized functions IV, Academic Press, 1934.

C^k-HYPOELLIPTICITY WITH DEGENERACY

Paul Malliavin

Institut Henri Poincaré
Paris, France

Let $A_1,\ldots,A_n$ be C^∞ vector fields on a compact manifold V of dimension m. We want to study the heat kernel $p_t(v_0,dv)$ associated to the semi-elliptic operator

$$0.1 \quad S = \frac{1}{2}\sum_{k=1}^{n} \mathcal{L}^2_{A_k}.$$

We consider the case with degeneracy, that is, the case where the Lie algebra generated by the A_k do not span at each point the tangent space, therefore we are out of the range of Hörmander's Theorem [4]. In [8] we have given a condition which implies absolute continuity relative to the Lebesgue measure dv

$$p_t(v_0,dv) = p_t(v_0,v)\, dv.$$

We want here to give more stringent conditions which will imply that $p_t(v_0,v)$ is a C^k function on V×V. The method used can be described as follows: We denote by $\Omega_1(R^n)$ the probability space of the brownian $b^k_\omega(\tau)$ on $\mathbb{R}^n$, for $\tau \in [0,1]$. We denote by μ the Wiener measure on $\Omega_1(\mathbb{R}^n)$. Then there is a natural map

$$0.2 \quad g:\Omega_1(R^n) \to V$$

defined by $g(\omega) = v_\omega(1)$ where $v_\omega(t)$ is the solution of the

ISBN 0-12-268380-3

Cauchy's Problem:

$$0.3 \qquad \begin{cases} dv_\omega = A_k d_s b_\omega^k & (d_s = \text{Stratanovitch differential}) \\ v_\omega(0) = v_0 \end{cases}$$

Let $g_* \mu$ be the direct image by g of μ, then

$$p_1(v_0, v)dv = g_* \mu .$$

The problem of the smoothness of p will be lifted through the map g to a problem of the estimation of the divergence of μ under the action of some "flows" defined on $\Omega_1(\mathbb{R}^n)$. Resources of the ordinary differential calculus are not intrinsically available on $\Omega_1(\mathbb{R}^n)$. Therefore the precise meaning of "flows" will be subordination of some semigroup defined on $\Omega_1(R^n)$ to the Ornstein-Uhlenbeck process which leaves μ invariant. This construction can be summarized by saying that by the map g we lift an hypoelliptic problem on V to an elliptic problem in infinite dimensions on $\Omega_1(R^n)$.

CONTENTS

* * *

1. COMPUTATION OF A DIVERGENCE BY LIFTING TO THE WIENER SPACE

1.0 Notation and Known Results

We denote $\Omega_1(\mathbb{R}^n)$ by X; the Wiener's measure on X is denoted by μ. Let on $\mathbb{R}^n$ the O.U. (Ornstein-Uhlenbeck) process be associated to the elliptic operator

$$\frac{1}{2} \Delta_{\mathbb{R}^n} - \xi \cdot \nabla_{\mathbb{R}^n} .$$

We denote by W_s ($0 \leq s$, $s \in \mathbb{N}$) an enumerable set of independent copies of the probability space of the O.U. process on $\mathbb{R}^n$. Define

$$W = \prod_s W_s \qquad \text{(equipped with the product measure).}$$

Now define a map $u:[0,1] \times R^+ \times W \to R^n$ by

1.0.1 $$u(\tau,t,w) = \tau G_{w_0}(t) + \sum_{q=1}^{+\infty} \frac{\sin(\pi q \tau)}{q\pi 2^{-\frac{1}{2}}} G_{w_q}(t).$$

1.0.2 <u>Theorem</u> ([8] p. 246 and p. 203)

For most w, the map $(w,t) \to u(\cdot,t,w)$ is a continuous map in t with values in α-Hölderian functions ($\alpha < 1/2$). This map defines a stochastic process,

$$(w,t) \to x_w(t)$$

which has the Wiener space as states spaces; it will be called the O.U. process on X. The O.U. process has μ for invariant measure.

We shall define regularity for a real function f defined on X in terms of the O.U. process [8], p. 206:

1.0.3 $$f \in \mathcal{C} \text{ iff } \mu(\{x;\ P_x(|f(x_w(t)) - f(x)|) > \varepsilon\}) \to 0$$

$$(\varepsilon \text{ being fixed, } t \to 0)$$

(It is quite possible that all measurable functions are in $\mathcal{C}$.)

1.0.4 $f \in \mathcal{C}^0$ iff a.s. in w $t \to f(x_w(t))$ is continuous.

1.0.5 $f \in \mathcal{C}^1$ iff $f \in \mathcal{C}^0 \cap L^1(\mu)$ and there exist functions

$$u \in \mathcal{C} \cap L^1(\mu) \qquad v \in \mathcal{C} \cap L^2(\mu),\ v > 0$$

such that

1.0.6 $$M_f(w,t) = f(x_w(t)) - \int_0^t u(x_w(\lambda))d\lambda$$

is an L^2-martingale having for increasing process

1.0.7 $\langle M_f, M_f\rangle_t = \int_0^t [v(x_w(\lambda))]^2 d\lambda.$

We shall denote

1.0.8 $u = Lf, \quad v = \| \nabla f \|.$

Then $\mathcal{C}$, $\mathcal{C}^0$, $\mathcal{C}^1$ are vector spaces and given $f_1, f_2 \in \mathcal{C}^1$ the Kunita-Watanabe [7] bilinear form $\langle M_{f_1}, M_{f_2}\rangle_t$ defines a scalar product $\nabla f_1 \cdot \nabla f_2$ by

1.0.9 $(\nabla f_1 \cdot \nabla f_2)(x_w(t)) = \frac{d}{dt} \langle M_{f_1}, M_{f_2}\rangle_t.$

1.0.10 Theorem ([8] p. 249). *Let g be as in 0.2; let ϕ be a smooth function defined on V. Denote $f = \phi \circ g$. Then*

$$f \in \mathcal{C}^1.$$

Given a local chart $v^1, \ldots, v^n$ on V, we shall always extend the coordinate function to smooth functions defined on V. We consider in this chart the semi-definite matrix

1.0.11 $\sigma^{i,j}(x) = (\nabla g^i \cdot \nabla g^j)(x).$

Finally we define

1.0.12 $\mathcal{C}^1_4 = \{f \in \mathcal{C}^1 \cap L^2(\mu);\ Lf \in L^2(\mu) \text{ and } \|\nabla f\| \in L^4(\mu)\}.$

Let F be a C^2 map from R^q to R bounded with its first and second derivatives; let $f_1 \cdots f_q \in \mathcal{C}^1_4$, then $F(f_1 \cdots f_q) \in \mathcal{C}^1_4$ and LF, $\| \nabla F \|$ can be computed using Itô's differential calculus. Therefore

1.0.13 *The class $\mathcal{C}^1_4$ is stable by symbolic calculus.*

We define the class $\mathcal{C}^1_\pi$ as

1.0.14 $\mathcal{C}^1_\pi = \{f \in \mathcal{C}^1;\ f, \|\nabla f\| \text{ and } Lf \text{ belongs to } L^p(\mu) \text{ for all } p < \infty\}.$

We introduce the following assumption:

(✯) $\sigma^{i,j}(x)$ positive definite μ a.e. on $\{x;\ g(x) \in \text{Domain of the chart } v^i\}$.

Under (✯) it can be proved [7] that $g_*\mu$ is absolutely continuous relative to the Lebesgue measure.

We want to get more regularity for $p_1(v_0,dv)$ and we will proceed on some stronger assumptions than assumption (✯).

Remark. In ordinary differential calculus, the hypothesis (✯) is equivalent to the surjectivity of the differential of g.

1.1 Lifting Vector Fields on V to Martingales on X

Let $v^1 \cdots v^m$ be a local chart near $v \in V$. Then, by hypothesis (✯), the matrix 1.0.12 is positive definite if $g(x)$ is sufficiently close to v^1.

A vector field Z is given on V. For our convenience we shall suppose that Z is zero outside of the domain of the local chart $v^1 \cdots v^n$; we will write

1.1.1 $$Z = Z^k(v) \frac{\partial}{\partial v^k}.$$

We shall associate to Z the martingale

1.1.2 $$B_Z(w,t) = \sum_k \xi_k(x_w(0)) M_{g^k}(w,t)$$

where the coefficients ξ_k are chosen so that

1.1.3 $$\left[\frac{d}{dt} \langle B_Z, M_{g_k} \rangle\right]_{t=0} = Z^k(g(x_w(0))).$$

Using the matrix $\sigma^{i,j}$, 1.1.3 can be written

1.1.4 $$\sigma^{k,j}(x)\xi_j(x) = Z^k(g(x)).$$

By (✯) we have a unique solution when $g(x) \in$ support of Z; otherwise we shall take $\xi^k(x) = 0$.

1.2 Divergences on V

Consider on V a Radon measure ρ and a vector field Z. We denote by U_t^Z the flow on V associated to Z. Then we want to compute

1.2.1 $$\frac{d}{dt}\left[\left(U_t^Z\right)_* \rho\right]_{t=0} = D_Z(\rho).$$

The derivative on the left hand side will be taken in the sense of distribution. Then 1.2.2 may be written for a smooth test function ϕ

1.2.2 $$\int (\alpha_Z \phi)(v)\rho(dv) = \langle \phi, D_Z(\rho) \rangle$$

where $\alpha_Z \phi$ denote the derivative of ϕ along Z.

1.3 Theorem

Given $q \in L^1(\mu)$ *and* Z *a vector field on* V, *define*

$\rho = g_*(q\mu)$. *Suppose*

1.3.1 $$g \in \mathcal{C}^1_\pi, \qquad \sigma^{i,j} \in \mathcal{C}^1_\pi.$$

Define γ *as the inverse matrix of* σ. *Suppose that for some* $\varepsilon > 0$ *we have*

1.3.2 $$\|\gamma\| \in L^{6+\varepsilon}(\mu)$$

1.3.3 $$q \in \mathcal{C}^1 \cap L^{2+\varepsilon}(\mu), \quad (Lq) \in L^{2+\varepsilon}(\mu), \; \|\nabla q\| \in L^{2+\varepsilon}(\mu).$$

Then

1.3.4 $$\begin{cases} D_Z \rho = g_*(q_1 \mu), \quad \text{where} \\ q_1 = \gamma_{s,k} z^k \nabla g^s \cdot \nabla q + aq \quad \text{with } z^k = Z^k \circ g \quad \text{and} \\ a = -2\gamma_{s,k} z^k L(g^s) - \gamma_{s,\ell}\gamma_{m,k} z^k \nabla \sigma^{\ell,m} \cdot \nabla g^s + \sum_k \frac{\partial Z^k}{\partial v^k} \circ g. \end{cases}$$

Remark. We can remark by Hölder that $(\gamma_{s,k} z^k \nabla g^s \cdot \nabla q) \in L^{\frac{3}{2}+\varepsilon}(\mu)$ and that $a \in L^{3+\varepsilon}(\mu)$. We get therefore

$$q_1 \in L^{\frac{3}{2}+\varepsilon}(\mu).$$

1.4 Infinitesimal Girsanov's Formulae

We shall use the approach of [8] p. 200 but the majorations of pp. 210-212 have to be worked again in our case.

The first observation is that the O.U. process is invariant under time reversal, if we take as initial measure the invariant measure μ ([8] p. 204).

Define

1.4.1 $$\tilde{B}(w,t) = \xi_k(x_w(t))\{g^k(x_0) - g^k(x_w(t)) - \int_0^t (Lg^k)(x_w(\xi))\,d\xi\}$$

$$= -\xi_k(x_w(t))[M_{g^k} + 2\int_0^t Lg^k].$$

Let $\phi \in C^\infty(v)$ denote $f = \phi \circ g$, and B being defined in 1.1.2 we have

1.4.2 Lemma

$$E_\mu\{q(x_0)E_{x_0}[(1+B)f(x_w(t)) - f(x_0))]\}$$
$$= E_\mu\{f(x_0)E_{x_0}[(1+\tilde{B})q(x_w(t)) - q(x_0)]\}.$$

Proof. Time reversal as in [8] p. 200.

1.4.3 Lemma

Denote $f_Z = (\alpha_Z\phi) \circ g$, then

$$\lim_{t\to 0} E_\mu\{q(x_0)\left|E_{x_0}\left[\frac{(1+B)f(x_w(t)) - f(x_0)}{t} - (Lf)(x_0) - f_Z(x_0)\right]\right|\} = 0.$$

Proof: We write for $f(x_w(t))$ Itô's formula. Introducing $f_k = \partial\phi/\partial v^k \circ g$, we get

$$f(x_w(t)) = f(x_0) + \int_0^t (Lf)(x_w(\lambda))\,d\lambda + \int_0^t f_k\, dM_{g^k}.$$

Therefore the $|E_{x_0}|$ is majorized by

$$|I_1| + \|\gamma\|\, I_2 + \|\gamma\|\, I_3$$

where

$$I_1 = \left|\int_0^1 [(Lf)(x_\omega(t\lambda)) - (Lf)(x_0)]d\lambda\right|$$

$$I_2 = \sum_k \left|\int_0^1 [\sigma^{k,s}(x_w(t\lambda))f_k(x_w(t\lambda)) - \sigma^{k,s}(x_0)f_k(x_0)]d\lambda\right|$$

$$I_3 = \sum_k \left|M_{g^k}(w,t) \int_0^1 (Lf)(x_w(t\lambda))d\lambda\right|.$$

We can use according to 1.3.3 the Schwarz inequality to get out from q, and by using 1.3.2 the Schwarz inequality makes it possible to eliminate $\|\gamma\|$. Therefore we have to majorize:

$$E_\mu(|I_1|^2), \quad E_\mu(|I_2|^4), \quad E_\mu(|I_3|^4).$$

The two first expectations tend to zero by 1.3.1 and [8] p. 208. The last can be majorized again by Schwarz's inequality:

$$E_\mu(|I_2|^4) \leq (E_\mu(|M_{g^k}|^8)E_\mu(|Lf|^8))^{\frac{1}{2}}$$

and 1.3.1 with [8] p.   show that the right member is $O(t^2)$.

1.4.4 <u>Lemma</u>

<u>Denote</u>

$$G = Lq - \xi_k \nabla g^k \cdot \nabla q - q[\nabla \xi_k \cdot \nabla g^k + 2\xi_k L g^k].$$

<u>Then</u>

$$\lim_{t\to 0} E_\mu\left\{\left|E_{x_0}\left[\frac{(1+\tilde{B})q(x_w(t)) - q(x_0)}{t} - G\right]\right|\right\} = 0.$$

<u>Proof</u>: We replace on the left hand side the functions of t by their Itô integral representation; we have then to study

1.4.4.1 $$\left[1 + (\xi_k(x_0) + M_{\xi_k} + \int_0^t L\xi_k)(-M_{g^k} - 2\int_0^t Lg^k)\right](q(x_0) + M_q + \int_0^t Lq).$$

The terms which will contribute to G are:

$$\int_0^1 (Lq)(x_w(t\lambda))d\lambda \ ;$$

(we dispose of this term by 1.3.3 and [8] p. 208)

1.4.4.2 $$-q(x_0)t^{-1} E_{x_0}(\langle M_{\xi_k}, M_{g^k}\rangle) = -q(x_0)\int_0^1 (\nabla\xi_k\cdot\nabla g^k)(x_w(\lambda t))d\lambda$$

$$-2q(x_0)\xi_k(x_0) \int_0^1 Lg^k(x_\omega(t\lambda))d\lambda$$

$$-\xi_k(x_0)t^{-1}E_{x_0}(\langle M_{g^k}, M_q\rangle) = -\xi_k(x_0)\int_0^1 (\nabla g^k\cdot\nabla q)(x_w(\lambda t))d\lambda.$$

We dispose of these last two terms by 1.3.1, 1.3.2, 1.3.3 and [ ] p. 208. For 1.4.4.2 we need

1.4.5 <u>Lemma</u>

$\xi_k \in \mathcal{C}^1$ <u>with</u> $L\xi_k \in L^{2+\varepsilon}(\mu)$, $\nabla\xi_k \in L^{3+\varepsilon}(\mu)$.

<u>Proof</u>: We have

$$\xi_k = \gamma_{k,s}\, z^s \quad \text{with } z^s \in \mathcal{C}^1_\pi \ \text{ by 1.3.1.}$$

Therefore we have only to see that $\gamma_{s,k}$ satisfies

$$\gamma_{s,k} \in \mathcal{C}^1, \quad L\gamma_{s,k} \in L^{2+\varepsilon}(\mu), \quad \nabla\gamma_{s,k} \in L^{3+\varepsilon}(\mu).$$

This comes from the stochastic differentiation of the relation $\gamma = \sigma^{-1}$, plus the fact $\sigma \in \mathcal{C}^1_\pi$ and $\|\gamma\| \in L^{6+\varepsilon}$.

We can now majorize 1.4.4.2 plus $q\nabla\xi_k\cdot\nabla g^k$ by

$$E_\mu\Big(\int_0^1 [\nabla\xi_k\cdot\nabla g^k)(x_w(t\lambda)) - \nabla\xi_k\cdot\nabla g^k(x_w(0))]^2 d\lambda\Big).$$

Remark now that

$$\| \nabla\xi_k \cdot \nabla g^k \|_{L^2(\mu)} \leq \| \nabla\xi_k \|_{L^3(\mu)} \| \nabla g^k \|_{L^6(\mu)} < +\infty$$

and apply [4] p. 208.

1.4.6 Lemma

$$\left(E_\mu(|M_{g^k}(w,t)|^p)\right)^{1/p} \leq c_p \, t^{\frac{1}{2}}$$

$$\left(E_\mu(|\int_0^t Lg^k|^p)\right)^{1/p} \leq c_p t.$$

Proof: The second inequality results from 1.3.1, which implies also the first by [8]. p.

1.4.7 Conclusion of the proof of Lemma 1.4.4.

Now we resume the proof of the domination of the terms of 1.4.4.1 which do not contribute to G and therefore have to tend to zero. Using 1.4.6 we can forget in this computation $(M_{g^k} + \int_0^t Lg_k)$ if we get majorization in $L^{1+\varepsilon}$ of the remaining product. More precisely we want to prove

$$E_\mu(|q(x_0) \int_0^t L\xi_k|^{1+\varepsilon}) < ct$$

$$E_\mu(|\xi_k(x_0) \int_0^t Lq|^{1+\varepsilon}) < ct$$

$$E_\mu(|M_{\xi_k} M_q|^{1+\varepsilon}) < ct$$

$$E_\mu(|M_{\xi_k} \int_0^t Lq|^{1+\varepsilon}) < ct$$

$$E_\mu(|M_q \int_0^t L\xi_k|^{1+\varepsilon}) < ct$$

$$E_\mu(|\int_0^t Lq|^{1+\varepsilon} |\int_0^t L\xi_k|^{1+\varepsilon}) < ct.$$

All these majoration results from 1.3.3 and 1.4.5 combine with Schwarz' inequalities and the estimation [1] page of the norm L^v of a martingale

$$E(|M_q|^{2+\eta}) \le E(\int_0^t \|\nabla q\|^{2+\eta}) \qquad \eta > 0.$$

1.5 Conclusion of the Proof of Theorem 1.3

We use 1.4.2, multiply by t^{-1} and pass to the limit using 1.4.3 and 1.4.4 to get

$$E_\mu(q(Lf - f_Z)) = E_\mu(fG).$$

Now remark that by [8] page 208

$$E_\mu(q\,L\,f) = E_\mu(f\,L\,q).$$

Therefore

$$E_\mu(qf_Z) = E_\mu(f(Lq - G))$$

which looking at 1.2.2 can be read as

$$D_Z(\rho) = g_*(q_1\mu)$$

with

$$q_1 = Lq - G = \xi_k \nabla g^k \cdot \nabla q + q[\nabla\xi_k \cdot \nabla g^k + 2\xi_k L(g^k)].$$

Now computing $\nabla\xi_k = \nabla(\gamma_{k,s} z^s)$ we get 1.3.4.

2. W_r^1 ESTIMATES FOR THE HEAT KERNEL

2.1 Iterated Divergences

We want to get regularity for the image ρ of the Wiener measure by Itô's map g. It is therefore sufficient to differentiate along coordinate vector fields $Z = \partial/\partial v^i$. We introduce the operators

2.1.0 $B_j q = \gamma_{s,j} \nabla g^s \cdot \nabla q + a_j q$

with

$$a_j = -2\gamma_{s,j} L(g^s) - \gamma_{s,\ell}\gamma_{m,j} \nabla\sigma^{\ell,m} \cdot \nabla g^s.$$

We start with

$$q_0 = 1.$$

Given a multi index $\vec{j} = (j_1, j_2, \ldots, j_r)$, $1 \leq \gamma_k \leq m$, denote $|\vec{j}| = r$ and define

2.1.1 $q_{\vec{j}} = B_{j_r} B_{j_{r-1}} B_{j_1} q_0.$

(We mean that $q' = B_{j_1} q_0$ has to be defined; that is, that q_0 fulfilled conditions of Theorem 1.3. At the second step q' has to fulfill the hypothesis of Theorem 1.3. Define $q'' = B_{j_2} q_0'$, and so on.)

Denote by W_m^1 the space of functions in L^1 having all their first m-derivatives (in the distributional sense) belonging to the space of Radon measure of finite total mass.

Proposition. *Suppose that*

2.1.2 $B_{\vec{j}} q_0$ *is defined for all* $\vec{j}$, $|\vec{j}| \leq r$. *Then*

2.1.3 $g_*(\mu) \in W_r^1.$

Proof. Iterative applications of Theorem 1.3.

2.2 Recursion Formulae

We want to make more explicit the hypotheses 2.1.1. Let us consider the increasing sequence C_k of a set of functions on X defined by

$$C_0 = \{Lg^i, \sigma^{i,j}, \nabla\sigma^{i,j} \cdot \nabla g^s\}$$

$$C_r = C_{r-1} \cup \{\nabla g^s \cdot \nabla u;\ u \in C_{r-1},\ s = 1, \ldots, m\}.$$

We make the temporary assumption

2.2.1 For every $r > 0$ all functions $u \in C_r$ belong to $\mathcal{C}^1_\pi$.

Then the ^ recursive construction defining B_k from B_{k-1} can always be done.

Proposition. Given r, there exists a finite number of polynomials with numerical coefficients $P_1,\ldots,P_s$ such that

2.2.2 $$q_{\vec{j}} = \Sigma\ \gamma P_{i_1}(u)\gamma P_{i_2}(u),\ldots,\gamma P_{i_j}(u),\quad |\vec{j}| < r$$

where $u \in C_k$, and where at most $2r$ products of the matrix γ appear in each monomial.

Proof. The recursion formula 2.1.0.

2.3 Theorem

Suppose

2.3.1 $$\|\gamma\| \in L^{4r+\varepsilon}(\mu),\ r \geq 2, \quad \text{or} \quad \|\gamma\| \in L^{6+\varepsilon}(\mu) \text{ for } r = 1.$$

Then

2.3.2 $$(g_*\mu) \in W^1_r.$$

Proof: We have to prove that the hypotheses of Theorem 1.3 are fulfilled for all $g_{\vec{j}}$ with $|\vec{j}| < r$. We shall proceed under the temporary assumption 2.2.1. First for $q_0 = 1$ we have only to check 1.3.2, which is implied by 2.3.1. Now if $|j| \leq (r-1)$ we have by 2.2.2

$$q_{\vec{j}} = \alpha w, \quad w \in L^p(\mu) \text{ for all } p \text{ and } |\alpha| < \|\gamma\|^{2r-2}.$$

Computing the stochastic differential of 2.2.2 we get

$$\|\nabla q_{\vec{j}}\| = \beta_1 w_1, \qquad L q_{\vec{j}} = \beta_2 w_2$$

with $w_1, w_2 \in L^p(\mu)$ for all p and $|\beta_1| < \|\gamma\|^{2r-1}$, $|\beta_2| < \|\gamma\|^{2r}$.

Therefore

$$q_{\vec{j}},\ \|\nabla q_{\vec{j}}\|,\quad Lq_{\vec{j}} \in L^{2+\varepsilon}(\mu)$$

and Theorem 1.3 can be applied, which gives 2.3 under the extra assumption 2.2.1.

2.4 Proof of 2.2.1.

We use the mechanism of prolongation as [8] p. 242. At each step to go from C_s to C_{s+1} we make two successive prolongations in the sense of differential geometry, that is, if the computation of the invariant at the level C_s has been done, solving a system of stochastic differential equations having for coefficients

$$a_\alpha^s(v),$$

We consider the prolongation which consists of working with

$$a_\alpha^2(v),\quad \frac{\partial}{\partial v^i} a_\alpha^2(v),\quad \frac{\partial^2}{\partial v^i \partial v^j} a_\alpha^r.$$

Then solving stochastic differential equations with these new coefficients we can compute as in [8] p. 237 the invariants of the stochastic variation of a function in C_s and in particular compute the function in C_{s+1}. Then we get by [8] p. 246 that the Itô invariant of a function in C_s belongs to L^p for all p. Therefore by the same proof used in [8] Theorem 5.6, p. 249, all the functions in C_s are in $\mathcal{C}_\pi^1$. We can resume the same proof starting now from functions of C_{s+1} and then get 2.2.1.

3. STATEMENT FOR C^∞ HYPOELLIPTICITY WITH DEGENERACY

We introduce GL(V) the bundle of linear frames on V. Then the vector fields A_k have a prolongation $\tilde{A}_k$ to GL(V) defined in a local chart by

$$\tilde{A}_k = (A_k^i,\ \partial_j A_k^i).$$

The process $v_\omega(t)$ can be lifted to GL(V) (cf. [8] p. 231) in a process $r_\omega(t)$ defined by

3.0.1 $$\begin{cases} dr_\omega(t) = \tilde{A}_k d_s b_\omega^k \\ r_\omega(0) = r_0 \in p^{-1}(v_0) \qquad (r_0 \text{ chosen arbitrarily}) \end{cases}$$

The vector field A_k on V can be read as a function defined on GL(V) with values in $\mathbb{R}^m$, let f_A. Then we have [8] p. 240

3.0.2 $$q_\omega(z) = \sum_{k=1}^{n} \int_0^1 |\langle \ell, f_{A_k}(r_\omega(\xi))\rangle|^2 d\xi$$

with

3.0.3 $$z \in T_{v_\omega(1)}(V), \qquad \ell = r_\omega^*(1)z.$$

3.1 Lemma. The condition 2.3.1

$$\|\gamma\| \in L^{4r+\varepsilon}(\mu)$$

is equivalent to

$$\|\beta\| \in L^{4r+\varepsilon}(\mu)$$

where β *is the inverse matrix of* α *defined by*

3.1.1 $$\alpha^{i,j}(\omega) = \int_0^1 \sum_k f_{A_k}^i(r_\omega(\xi)) f_{A_k}^j(r_\omega(\xi)) d\xi$$

$r_\omega(\xi)$ *being the solution of* 3.0.1.

Proof: We remark by 3.0.2 that

$$\|\gamma\| \leq \|r_\omega(1)\| \, \|\beta\|.$$

As $\|r_\omega(1)\|$ and $\|r_\omega^{-1}(1)\| \in L^p(\mu)$ for all p (by [8] p. 244) we get the equivalence.

3.2 Theorem. *Given an integer* s, *denote by* Σ_s *the set of vector fields constituted by* A_k, $[A_{k_1}, A_{k_2}] \cdots [A_{k_1}[\cdots A_{k_s}]$. *Let us choose on* V *a volume form and define*

3.2.1 $\phi_s(v) = \Sigma|\det(B_v \cdots B_v^m)|^2$

the summation being taken on all systems of vectors $B^1 \ldots B^m \in \Sigma_s$.

Define

3.2.2 $u(\varepsilon) = \sup_{v_0} P_{v_0}(\phi_s(v_\omega(t))) < \varepsilon, \quad t \in [0,1]$.

Then

3.2.3 $\lim \varepsilon^{-k} u(\varepsilon) = 0$ *for all* k

implies that for all $t > 0$ *fixed,*

3.2.4 $p_t(v_0,v)$ *is* C^∞ *in* v, *uniformly in* v_0.

Rescaling the time, we define

$$u_{t_0}(\varepsilon) = \sup_{v_0} P_{v_0}(\Phi_s(v_\omega(t)) < \varepsilon, \quad t \in [0,t_0]).$$

Then u_{t_0} can be evaluated going to $A_k' = t_0^{1/2} A_k$ and therefore 3.2.3 implies

(3.2.5) $\lim_{\varepsilon \to 0} \varepsilon^{-k} u_{t_0}(\varepsilon) = 0$ for all $k > 0$.

Therefore it is sufficient to prove 3.2.4 for $t = 1$. Before proceeding to the proof of the theorem, we shall collect in the following paragraph several lemmas. These appear in Part II, beginning on page 327 of this volume.

BOUNDARY BEHAVIOR OF BRANCHING TRANSPORT PROCESSES *

George C. Papanicolaou
Courant Institute, New York University
New York, New York 10012

TABLE OF CONTENTS

INTRODUCTION

Consider a particle starting from position x with velocity v and moving along the line x+vt until a random time τ_1 which is exponentially distributed. At that time the particle disappears and a random number of particles appear at $x+v\tau_1$ with randomly distributed velocities. Each new particle now moves along and splits as before independently of the other particles. Let B be

*This research was supported by the National Science Foundation, Grant No. NSF-MCS75-09837-A01.

ISBN 0-12-268380-3

a subset of phase space (i.e. position and velocity space) and let $\nu_t(B)$ denote the number of particles in B at time t. As the set B varies we get a process ν_t which for each t is an integer-valued random measure; this is (roughly) the branching transport process (BTP) that we shall analyze.

Branching transport processes have been introduced primarily as models for fission phenomena in neutron transport theory [1]-[3]. They fall within the scope of the theory of branching processes [4], [5] and of point processes [6], [7], [8]. However, we are primarily interested here in asymptotic approximations of branching transport processes by measure diffusion processes [9], [10] and their connection with the usual asymptotic approximations [11]-[14] (discussed in Sections 7 and 12).

In Section 1 we define branching transport processes using the Laplace functional of the transition probabilities. Equations for moments are given in Section 2. Branching transport processes with absorption at the boundary are defined in Section 3. The asymptotic analysis begins in Section 4 and continues until Section 13.

We do not give proofs here; they will appear in a forthcoming monograph by the author. However, a typical argument (for Theorem 3 here) is given in [14] in detail while most of the methods are presented in [13].

1. BRANCHING TRANSPORT PROCESSES

Let $S \subset R^n$ be the set of velocities and denote by $E = R^n \times S$ the phase space. Let $0 \leq q(x,v) \leq C < \infty$ be given and smooth. For each $k = 0,1,2,\ldots$ and each $(x,v) \in E$ let $\pi_k(x,v,A_1,\ldots,A_k)$ be a bounded measure on $S\times\cdots\times S$ (k times). Here $A_1,A_2,\ldots,A_k$ are Borel subsets of S and π_k depends smoothly on (x,v). We assume that

$$\sum_{k=0}^{\infty} \pi_k(x,v,S,\ldots,S) = 1, \quad \sum_{k=1}^{\infty} k\pi_k(x,v,S,\ldots,S) < \infty . \qquad (1.1)$$

Let $\lambda(x,v) \geq 0$ be bounded and measurable. Replacing f by $\exp(-\lambda)$ in (1.3) yields

$$u(t,x,v) = E\left\{\exp\left\{\sum_{k=1}^{N(t)} \lambda(x_k(t),v_k(t)\right\}\right\}$$

$$= E\left\{\exp\left\{-\int_E \lambda(\bar{x},\bar{v})\ \nu_t(d\bar{x}\ d\bar{v})\right\}\right\} \tag{1.6}$$

Here we have introduced the random measure $\nu_t(B)$, $B \subset E$ such that $\nu_t(B) = 0,1,2,\ldots$ (integer-valued i.e. a point measure) is the random number of particles in the subset B of phase space at time t. We write for short

$$(\lambda,\nu_t) = \int_E \lambda(\bar{x},\bar{v})\ \nu_t(d\bar{x}\ d\bar{v}) \tag{1.7}$$

Let $M^+(E)$ be the set of Radon measures on E with the topology of vague convergence (cf. [7]); it is a separable, locally compact and metrizable space. Let $M_p^+(E) \subset M^+(E)$ be the closed subset of point measures on E. We denote by $X = D([0,\infty);M^+(E))$ (or $X_p = D([0,\infty);M_p^+(E))$ the space of right-continuous functions on $[0,\infty) \to M^+(E)$ (or $M_p^+(E)$) endowed with the Skorohod metric. X and X_p are complete separable metric spaces.

The BPT is a probability measure on X_p which is, more precisely, a Markov process with state space $M_p^+(E)$. If $P(t,\mu,d\nu)$ denotes the transition function of this process then, on account of the branching property,

$$\int_{M_p^+(E)} e^{-(\lambda,\nu)}\ P(t,\mu,d\nu) = E_\mu\left\{e^{-(\lambda,\nu_t)}\right\}$$

$$= \exp\left\{\int_E \log u(t,x,v;\lambda)\ \mu(dx\ dv)\right\}$$

$$= \exp\left\{-\ (w(t;\lambda),\mu)\right\} \tag{1.8}$$

$$w(t,x,v;\lambda) = -\log u(t,x,v;\lambda)\ . \tag{1.9}$$

The branching transport process (BTP) is defined as in the Introduction where if τ_1 is the first splitting time

$$P\left\{\tau_1 \le t \text{ and } k \text{ particles are created with velocities } v_1(\tau_1) \in A_1, \ldots, v_k(\tau_1) \in A_k\right\}$$
$$= \int_0^t \pi_k(x+vs, v, A_1, \ldots, A_k) q(x+vs,v) \exp\left\{-\int_0^s q(x+v\gamma, v)\, d\gamma\right\} ds, \tag{1.2}$$

given that a single particle starts from (x,v). After the splitting each particle moves independently of the others the same way and splits, etc.

Let $N(t) = 0,1,2,\ldots$ be the number of particles at time t and let $(x_k(t), v_k(t))$, $k = 1,2,\ldots,N(t)$, denote their positions and velocities. Under the condition $q \le C < \infty$ and (1.1) $N(t) < \infty$ with probability one. Let $f(x,v)$ be a measurable function on E with $0 \le f(x,v) \le 1$ and consider

$$u(t,x,v) = E\left\{f(x_1(t), v_1(t)) \ldots f(x_{N(t)}(t), v_{N(t)}(t))\right\} \tag{1.3}$$

with the integrand on the right interpreted as one when $N(t) = 0$. By the usual renewal argument u satisfies the following renewal equation which is a nonlinear integral equation.

$$u(t,x,v) = f(x+vt,v) \exp\left\{-\int_0^t q(x+vs,v)\, ds\right\}$$
$$+ \int_0^t \Phi(u)(t-s, x+vs, v) q(x+vs,v) \exp\left\{-\int_0^s q(x+v\gamma,v)\, d\gamma\right\} ds \tag{1.4}$$

Here Φ stands for the nonlinear integral operator

$$\Phi(u)(t,x,v) = \sum_{k=0}^{\infty} \int_S \cdots \int_S u(t,x,v_1) \ldots u(t,x,v_k) \cdot \pi_k(x,v,dv_1,\ldots,dv_k) \tag{1.5}$$

Note that Φ acts on u as a function of v only. To emphasize parametric dependence on x we sometimes write $\Phi_x(\cdot)$.

Under (1.1) and $0 \le q \le C < \infty$, it is easily shown that (1.4) has a unique continuous solution u with $0 \le u \le 1$ if f is continuous and $0 \le f \le 1$. Thus the expectation (1.3) is well defined.

Here $\mu \in M_p^+(E)$ is the initial distribution of particles, i.e. $\nu_0 = \mu$, $u(t,x,v;\lambda)$ is the solution of (1.4) with $f = \exp(-\lambda)$ and E_μ denotes expectation conditional on $\nu_0 = \mu$. The transition function $P(t,\mu,d\nu)$ is completely characterized by its Laplace functional, which is given by (1.8) via the solution of (1.4), and it is Feller continuous. Thus, the BTP is well defined as a family of probability measures Q_μ on X_p for each $\mu \in M_p^+(E)$ and constitutes a strong Markov process.

The infinitesimal generator of the BTP ν_t has the following simple expression on functions of the form $\exp(-(\lambda,\mu))$ on $M_p^+(E)$, with $\lambda \in C_0^+(E)$ and λ differentiable in x.

$$Le^{-(\lambda,\mu)} = \frac{d}{dt} E_\mu\left\{e^{-(\lambda,\nu_t)}\right\}\Big|_{t=0} = e^{-(\lambda,\mu)}\left(N(\lambda),\mu\right) \tag{1.10}$$

where $N(\cdot)$ is the nonlinear operator

$$N(\lambda)(x,v) = -v\cdot \frac{\partial\lambda(x,v)}{\partial x} + q(x,v)\left[e^{\lambda(x,v)}\ \Phi(e^{-\lambda})(x,v)-1\right] \tag{1.11}$$

2. MOMENTS

It is convenient to write in differential form the defining equation for $w = -\log u$ where u satisfies (1.4). We have that

$$\begin{aligned} &\frac{\partial w}{\partial t} = v\cdot\frac{\partial w}{\partial x} - q\left[e^{+w}\ \Phi(e^{-w}) - 1\right], && t > 0, \\ &w(0,x,v;\lambda) = \lambda(x,v), && (x,v) \in E . \end{aligned} \tag{2.1}$$

It will always be understood that equations such as (2.1) are to hold in their integral form.

To obtain equations for moments of (λ,ν_t), we replace λ by $\alpha\lambda$ where $\alpha > 0$ is a parameter, differentiate with respect to α and then set $\alpha = 0$. For the first moment we have

$$E_\mu\left\{(\lambda,\nu_t)\right\} = (w^{(1)}(t;\lambda),\mu) \tag{2.2}$$

where $w^{(1)}(t,x,v;\lambda)$ satisfies

$$\frac{\partial w^{(1)}}{\partial t} = v \cdot \frac{\partial w^{(1)}}{\partial x} + Mw^{(1)}, \qquad t > 0$$

$$w^{(1)}(0,x,v;\lambda) = \lambda(x,v), \qquad (x,v) \in E. \tag{2.3}$$

Here M is the *mean operator* defined by

$$(Mw)(v) = q(x,v)\left[(\Phi^{(1)}(1)w)(x,v) - w(x,v)\right] \tag{2.4}$$

where

$$\Phi^{(1)}(1)w = \int_S \pi^{(1)}(x,v,dv_1)\, w(v_1), \tag{2.5}$$

$$\pi^{(1)}(x,v,A) = \sum_{k=1}^{\infty} \sum_{j=1}^{k} \pi_k(x,v,S,\ldots,\overset{\longleftarrow j \longrightarrow}{A},\ldots,S), \quad A \subset S. \tag{2.6}$$

Note that x is merely a parameter in (2.4)-(2.6).

Equation (2.3) is the adjoint or backward version of the usual linear transport equation.

Let us assume that

$$\sum_{k=1}^{\infty} k^2 \pi_k(x,v,S,\ldots,S) < \infty. \tag{2.7}$$

We replace λ in (1.8) by $\alpha_1\lambda_1 + \alpha_2\lambda_2$ where λ_1 and λ_2 are in $C_0^+(E)$ (nonnegative continuous functions on E with compact support). Differentiating with respect to α_1 and α_2 and setting $\alpha_1 = \alpha_2 = 0$ we see that

$$E_\mu\left\{(\lambda_1,\nu_t)(\lambda_2,\nu_t)\right\} - E_\mu\left\{(\lambda_1,\nu_t)\right\}E_\mu\left\{(\lambda_2,\nu_t)\right\}$$

$$= \left(w^{(2)}(t;\lambda_1,\lambda_2),\mu\right) \tag{2.8}$$

where

$$w^{(2)}(t,x,v;\lambda_1,\lambda_2) = -\frac{\partial^2}{\partial\alpha_1\partial\alpha_2} w(t,x,v;\alpha_1\lambda_1+\alpha_2\lambda_2)\Big|_{\alpha_1=\alpha_2=0} \tag{2.9}$$

Thus, differentiating (2.1) we find that $w^{(2)}$ satisfies the linear equation (the covariance equation)

$$\frac{\partial w^{(2)}}{\partial t} = v \cdot \frac{\partial w^{(2)}}{\partial x} - Mw^{(2)} + q\left(\Phi^{(2)}(1)w^{(1)}(t;\lambda_1), w^{(1)}(t;\lambda_2)\right)$$
$$+ q\Big| w^{(1)}(t;\lambda_1)w^{(1)}(t;\lambda_2) - w^{(1)}(t,\lambda_1)\Phi^{(1)}(1)w^{(1)}(t;\lambda_2)$$
$$+ \Phi^{(1)}(1)w^{(1)}(t;\lambda_1)\; w^{(1)}(t;\lambda_2)\Big|, \qquad t > 0, \tag{2.10}$$

$$w^{(2)}(0,x,v;\lambda_1,\lambda_2) = 0 .$$

Here $\Phi^{(2)}(1)$ is the bilinear operator defined by

$$\left(\Phi^{(2)}(1)u,w\right) = \int_S \int_S u(v_1)\; w(v_1)\; \pi^{(2)}(x,v,dv_1,dv_2) \tag{2.11}$$

where $\pi^{(2)}$ is given by

$$\pi^{(2)}(x,v,A_1,A_2) = \sum_{k=1}^{\infty} \sum_{\substack{j=1 \\ j\neq \ell}}^{k} \sum_{\ell=1}^{k} \pi_k(x,v,S,\ldots,\overset{\longleftarrow j \longrightarrow}{A_1},\ldots,\overset{\longleftarrow \ell \longrightarrow}{A_2},\ldots,S). \tag{2.12}$$

3. BOUNDARY PROCESS

Let D be a bounded open subset of R^n with smooth boundary ∂D. Let $\hat{n}(x)$ be the unit outward normal to ∂D at x and define

$$S^{\pm}(x) = \left\{ v \in S \;\middle|\; \hat{n}(x)\; v \gtrless 0 \right\}, \qquad x \in \partial D . \tag{3.1}$$

$S^+(x)$ $(S^-(x))$ is the set of all outgoing (incoming) velocities at the boundary point x. Let

$$E = (D \times S) \cup \left\{ (x,v) \;\middle|\; x \in \partial D,\; v \in S^-(x) \right\} \tag{3.2}$$

and

$$E_B = \left\{ (x,v) \;\middle|\; x \in \partial D,\; v \in S^+(x) \right\} . \tag{3.3}$$

Let $(x_k(t), v_k(t))$ be the trajectory of the k-th particle when we start with a single particle at (x,v). Let T_k be the first time $x_k(t)$ reaches ∂D with $T_k = +\infty$ if $x_k(t)$ becomes extinct or never reaches ∂D. Let f be a bounded measurable function on $E \cup E_B$ with $0 \leq f \leq 1$ and consider

$$u(t,x,v) = E\left\{ \prod_{k=1}^{N(t)} f(x_k(t \wedge T_k),\ v_k(t \wedge T_k)) \right\} \tag{3.4}$$

where $t \wedge T = \min(t,T)$ and the integrand is interpreted as zero when $N(t) = 0$.

Let $\tau = \tau(x,v)$ be the (deterministic) time to reach ∂D starting from (x,v) i.e.

$$\tau = \inf\{\ t \geq 0 \mid x+vt \in \partial D\}$$

By the usual renewal argument we find that u of (3.4) satisfies the renewal equation

$$\begin{aligned} u(t,x,v) = {} & f(x+vt,v)\chi_{\tau>t}(x,v)\exp\left\{-\int_0^t q(x+vs,v)\ ds\right\} \\ & + f(x+v\tau,v)\chi_{\tau\leq t}(x,v)\ \exp\left\{-\int_0^\tau q(x+vs,v)\ ds\right\} \\ & + \int_0^{t\wedge\tau} \Phi(u)(t\wedge\tau-s, x+vs,v)q(x+vs,v) \\ & \quad \exp\left\{-\int_0^s q(x+v\gamma,v)\ d\gamma\right\} ds \end{aligned} \tag{3.5}$$

Here $\chi_A(x,v)$ is the indicator function of the set A i.e., $\chi_A = 1$ if $(x,v) \in A$ and zero otherwise. It is easily seen that for each $t < \infty$, (3.5) has a unique bounded measurable solution u with $0 \leq u \leq 1$.

Let $\lambda(x,v) \geq 0$ be a bounded measurable function on E_B, identically zero in E, and let $f = \exp(-\lambda)$ in (3.4). Then we obtain

$$u(t,x,v) = E\left\{\exp\left\{-\sum_{k=1}^{N(t)} \lambda(x_k(t \wedge T_k),\ v_k(t \wedge T_k))\right\}\right\} \tag{3.6}$$

$$= E\left\{\exp\left\{-\int_{E_B} \lambda(\bar{x},\bar{v})\ \nu_t^B(d\bar{x}\ d\bar{v})\right\}\right\}$$

where ν_t^B is an integer-valued random measure on E_B. In fact if A is a Borel subset of E_B then $\nu_t^B(A)$ is the number of particles that reach E_B and fall into A by time t.

If $\mu \in M_p^+(E)$ then, as above, for each $\lambda \in C_0^+(E_B)$,

$$E_\mu\left\{e^{-(\lambda,\nu_t^B)}\right\} = \exp\left\{-\ (w^B(t;\lambda),\mu)\right\}, \tag{3.7}$$

where $w^B(t,x,v;\lambda)$ satisfies the integral equation version of

$$\frac{\partial w^B}{\partial t} = v\cdot\frac{\partial w^B}{\partial x} - q[e^{w^B}\ \Phi(e^{-w^B}) - 1]\ ,\quad t > 0,\ (x,v)\in E, \tag{3.8}$$

$$w^B(0,x,v) = 0\ ,\qquad (x,v) \in E,$$

$$w^B(t,x,v) = \lambda(x,v)\ ,\qquad t > 0,\quad (x,v) \in E_B\ .$$

The boundary process ν_t^B is well defined for each $t < \infty$ and it is nondecreasing with t i.e. $\nu_t^B(A)$ is nondecreasing for each $A \subset E_B$ as $t \uparrow \infty$. Thus there exists a limiting point process ν^B on E_B which may be $+\infty$. We have, for $\lambda \in C_0^+(E_B)$,

$$E_\mu\left\{e^{-(\lambda,\nu^B)}\right\} = \exp\left\{-\ (w^B(\lambda),\mu)\right\}\ ,\qquad \mu \in M_p^+(E)\ , \tag{3.9}$$

where, *formally*, $w^B(x,v;\lambda)$ is given by the solution of (the integral equation version of)

$$- v\cdot\frac{\partial w^B}{\partial x} + q\left[e^{w^B}\ \Phi\left(e^{-w^B}\right) - 1\right] = 0\ ,\ (x,v) \in E,$$
$$w^B(x,v) = \lambda(x,v)\ ,\qquad (x,v) \in E_B\ . \tag{3.10}$$

The nonlinear boundary value problem (3.10) does not, in general have a solution or a unique solution. Therefore, the Laplace functional of ν^B cannot be identified, in general, in the manner of (3.9). We need an existence and uniqueness theory for (3.10) and, of course, some hypotheses beyond $q \leq C$ and (1.1) which were enough up to now. We return to this in Section 9.

The formal boundary value problems for moments of ν^B are obtained the same way as in Section 2. For example

$$E\left\{(\nu^B,\lambda)\right\} = \left(w^{(1)}(\lambda),\mu\right) \tag{3.11}$$

where $w^{(1)}(x,v;\lambda)$ satisfies

$$\begin{aligned} v\cdot \frac{\partial w^{(1)}}{\partial x} + Mw^{(1)} &= 0 \;, \qquad (x,v) \in E \;, \\ w^{(1)}(x,v) &= \lambda(x,v) \;, \qquad (x,v) \in E_B . \end{aligned} \tag{3.12}$$

4. SCALINGS

An objective, as mentioned in the Introduction, is to recover at first the usual diffusion approximation results of transport theory [11]-[13] within the context of BTP and then to analyze the fluctuations due to branching. Recall that conventional transport theory deals only with the *average* density of particles in phase space and diffusion approximations deal also only with average particle densities. We will limit discussion to the simplest case of one-group diffusion approximations. The multi-group case requires a more complicated setup but otherwise amounts to the same thing.

For conventional diffusion theory the crucial parameter is the mean free time between collisions or splittings which is very small compared to macroscopic time scales in reactor theory, for example. In suitable dimensionless variables the ratio of these time scales is denoted by $\varepsilon > 0$ which is small. In the basic equation (2.1) for w (cf. (1.8)) this ε appears on the right side. The drift term is multiplied by ε^{-1} and the collision term (the bracket) by ε^{-2}.

Let $\tilde{\nu}_t^{\varepsilon}$ be the $M_p^+(E)$-valued process obtained by this scaling. For the diffusion approximation to make sense it is necessary that the average number of particles produced in each splitting be approximately one (near criticality). We shall make this precise in the next section. It is also necessary that there be a large number of particles around so that the *average* density of particles (in suitable units) represents reasonably well the random number of particles (suitably scaled). This means that the diffusion limit must be coupled to a *high-density* limit [15], [16] as follows.

For each $\mu \in M^+(E)$ let $\mu^{\varepsilon,\alpha} \in M_p^+(E)$ be such that $\varepsilon^\alpha \mu^{\varepsilon,\alpha} \to \mu$ weakly as $\varepsilon \to 0$ with $\alpha > 0$ fixed i.e., for each $\lambda \in C(E)$

$$(\varepsilon^\alpha \mu^{\varepsilon,\alpha}, \lambda) \to (\mu, \lambda) \tag{4.1}$$

Fix a particular way of constructing the $\mu^{\varepsilon,\alpha}$. Let $\nu_t^{\varepsilon,\alpha}$ be the BTP obtained from $\tilde{\nu}_t^\varepsilon$ by

$$\nu_t^{\varepsilon,\alpha} = \varepsilon^\alpha \tilde{\nu}_t^\varepsilon, \qquad \nu_0^{\varepsilon,\alpha} = \mu^{\varepsilon,\alpha} . \tag{4.2}$$

Note that $\nu_t^{\varepsilon,\alpha}$ is a process in $M^+(E)$ and not in $M_p^+(E)$. For the Laplace functional of $\nu_t^{\varepsilon,\alpha}$ we have

$$E_{\mu^{\varepsilon,\alpha}}\left\{e^{-(\lambda,\nu_t^{\varepsilon,\alpha})}\right\} = \exp\left\{-\left(w^{\varepsilon,\alpha}(\lambda), \varepsilon^\alpha \mu^{\varepsilon,\alpha}\right)\right\} \tag{4.3}$$

where $w^{\varepsilon,\alpha}(t,x,v;\lambda)$ satisfies

$$\frac{\partial w^{\varepsilon,\alpha}}{\partial t} = \frac{1}{\varepsilon} v \cdot \frac{\partial w^{\varepsilon,\alpha}}{\partial x} - \frac{1}{\varepsilon^2} q \left[\frac{e^{\varepsilon^\alpha w^{\varepsilon,\alpha}} \Phi(e^{-\varepsilon^\alpha w^{\varepsilon,\alpha}}) - 1}{\varepsilon^\alpha}\right] \tag{4.4}$$

$$w^{\varepsilon,\alpha}(0,x,v) = \lambda(x,v), \qquad (x,v) \in E .$$

For the boundary process the scaling is the same.

It will become clear in later sections that the following three cases are of interest.

$$\begin{array}{lll} \text{(i)} & \alpha < 2, & \text{Branching dominates diffusion,} \\ \text{(ii)} & \alpha = 2, & \text{Branching and diffusion balance,} \\ \text{(iii)} & \alpha > 2, & \text{Diffusion dominates branching.} \end{array} \tag{4.5}$$

For reactor calculations case (iii) is undoubtedly the most significant.

There are many other scalings of interest. For example one may consider the high density limit without any diffusion approximation. The scalings chosen here seem to us to be the most appropriate ones for reactor problems.

5. LOCAL ERGODICITY PROPERTIES

We shall introduce now the relevant hypotheses necessary for the asymptotic analysis. These hypotheses are not optimal in any sense but do have a local character that is, they specify the local splitting mechanism only.

We assume first that the splitting distribution has third moments

$$\sum_{k=1}^{\infty} k^3 \pi_k(x,v;S,\ldots,S) \leq C < \infty , \qquad (x,v) \in E. \qquad (5.1)$$

Next we assume that the local splitting frequency $q(x,v)$ is uniformly bounded from below

$$0 \leq q_\ell \leq q(x,v) \leq q_u < \infty . \qquad (5.2)$$

Let $\pi^{(1)}(x,v,A)$ be the mean splitting kernel defined by (2.6). We assume that there is a reference probability measure γ on S such that $\pi^{(1)}$ has a density $\tilde{\pi}^{(1)}$ relative to γ which is continuous and

$$0 < \pi_\ell \leq \tilde{\pi}^{(1)}(x,v,\bar{v}) \leq \pi_u < \infty , \qquad (x,v) \in E. \qquad (5.3)$$

Under (5.2) and (5.3) the mean operator M given by (2.4) has an *isolated* maximal real eigenvalue with strictly positive right and left eigenfunctions $\phi(v;x)$, $\tilde{\phi}(v;x)$, respectively ($x \in R^n$ is a parameter here) [4, p. 67].

We assume that the maximal eigenvalue of M is zero. (5.4)

This means that *locally* the BTP is assumed to be critical. Actually the maximal eigenvalue could deviate a little (depending on ε) from zero without any changes in what follows.

Since the x-dependence of everything is smooth and the maximal eigenvalue is isolated it follows that ϕ and $\tilde{\phi}$ are smooth functions of x. We assume that they satisfy the following centering conditions.

$$\int_S \tilde{\phi}(v;x)\ \phi(v;x)v\ \gamma(dv) = 0\ , \tag{5.5}$$

$$\int_S \tilde{\phi}(v;x)\ v\cdot \frac{\partial\phi(v;x)}{\partial x}\ \gamma(dv) = 0\ . \tag{5.6}$$

The above hypotheses, referred to collectively as the local ergodicity properties, will be in force in all that follows.

6. LAW OF LARGE NUMBERS AND DIFFUSION LIMIT (NO BOUNDARIES)

We fix attention to the case $\alpha > 2$ in (4.5) and consider the asymptotic limit of $\nu_t^{\varepsilon,\alpha}$ as $\varepsilon \to 0$. First we define the limiting diffusion operator L as follows. For $f \in C_0^\infty(R^n)$ we set

$$L\ f(x) = \int_S \int_S \tilde{\phi}(v;x)v\cdot \frac{\partial}{\partial x}\left[\frac{\phi(v;x)\psi(v,d\bar{v};x)}{\phi(\bar{v};x)}\ \bar{v}\cdot \frac{\partial}{\partial x}\left(\phi(\bar{v};x)f(x)\right)\right] \cdot\gamma(dv) \tag{6.1}$$

where $\psi(v,d\bar{v};x)$ is the kernel of the operator $-Q^{-1}$ where $Qw \equiv \phi^{-1}M(\phi w)$, with M the mean operator. The operator Q is the infinitesimal generator of a jump Markov process on S which is ergodic with invariant measure $\tilde{\phi}(v;x)\phi(v;x)\gamma(dv)$. This invariant measure is approached exponentially fast as time goes to infinity so $-Q^{-1}$ is well defined and its kernel is actually the recurrent potential kernel of the process generated by Q.

It is easily seen that the operator L of (6.1) is elliptic. We will assume in the sequel that

$$L \text{ of (6.1) is uniformly elliptic with smooth coefficients.} \tag{6.2}$$

Actually uniform ellipticity is needed only for the analysis of the boundary process (Section 9 and following). Assume the hypotheses of Section 5 hold.

THEOREM 1. *For each* $t > 0$ *and finite, for each* $\alpha > 2$ *and for each* $\lambda(x,v) \geq 0$, *continuous and of compact support in* x,

$$- \lim_{\varepsilon \downarrow 0} \log E_{\mu^{\varepsilon,\alpha}} \left\{ e^{-(\lambda, \nu_t^{\varepsilon,\alpha})} \right\} = \int_E \bar{w}^{(1)}(t,x;\lambda)\phi(v;x)\mu(dx\, dv), \tag{6.3}$$

where $\bar{w}^{(1)}(t,x;\lambda)$ *is the solution of the (adjoint) linear transport equation*

$$\frac{\partial \bar{w}^{(1)}}{\partial t} = L\bar{w}^{(1)}, \qquad t > 0, \quad x \in R^n, \tag{6.4}$$

$$\bar{w}^{(1)}(0,x) = \int_S \lambda(x,v)\, \tilde{\phi}(v;x)\, \gamma(dv) .$$

Remark 1. This theorem is proven by showing that when $\alpha > 2$ $w^{\varepsilon,\alpha}(t,x,v;\lambda)$ of (4.4) converges as $\varepsilon \to 0$ with $t > 0$ fixed, uniformly in $(x,v) \in E$, to $\bar{w}^{(1)}(t,x;\lambda)\phi(v;x)$. The condition $t > 0$ cannot be relaxed to $t \geq 0$ because there is an initial layer correction.

Remark 2. Since (6.4) is linear, $\bar{w}^{(1)}$ is a linear functional of λ. Thus from (6.3) the Laplace functional of $\nu_t^{\varepsilon,\alpha}$ converges to the Laplace functional of a degenerate (deterministic) law. It follows that $\nu_t^{\varepsilon,\alpha}$ converges as $\varepsilon \to 0$ vaguely to $\bar{\nu}_t$, in probability. The deterministic measure $\bar{\nu}_t$ satisfies

$$(\lambda,\bar{\nu}_t) = \int_E \bar{w}^{(1)}(t,x;\lambda)\ \phi(v;x)\ \mu(dx\, dv) .$$

This is precisely the diffusion approximation of transport theory [11]-[13] stated in adjoint form.

7. GAUSS-MARKOV AND DIFFUSION LIMIT

We continue under the hypotheses of Section 6 and in particular with $\alpha > 2$ fixed.

Since $\nu_t^{\varepsilon,\alpha} \to \bar{\nu}_t$ in probability, we may examine the limiting distribution of the signed measure $\nu_t^{\varepsilon,\alpha} - \bar{\nu}_t$, suitably scaled, as $\varepsilon \to 0$. It is convenient to regard the process $\nu_t^{\varepsilon,\alpha} - \bar{\nu}_t$ as a Markov process with state space the tempered distributions on E.

THEOREM 2. *Let $\zeta_t^{\varepsilon,\alpha}$ be the process*

$$\zeta_t^{\varepsilon,\alpha} = \varepsilon^{\frac{2-\alpha}{2}} \left(\nu_t^{\varepsilon,\alpha} - \nu_t\right) \tag{7.1}$$

regarded as a Markov process in the space of tempered distributions on E. *Then for each* $2 < \alpha < 4$ *fixed,* $\zeta_t^{\varepsilon,\alpha}$ *converges weakly to a Gaussian-Markov process* ζ_t *on the space of tempered distributions on* E *such that for any* $\lambda \in C_0^\infty(E)$ *and* $t > 0$.

$$E\left\{(\lambda,\zeta_t)\right\} = 0 \tag{7.2}$$

and

$$E\left\{(\lambda,\zeta_t)^2\right\} = \left(\bar{z}^{(2)}(t;\lambda)\phi,\mu\right) \tag{7.3}$$

where $\bar{z}^{(2)} = \bar{z}^{(2)}(t,x;\lambda)$ *is the solution of*

$$\frac{\partial \bar{z}^{(2)}}{\partial t} = L\bar{z}^{(2)} + 2g(x)\left(\bar{w}^{(1)}\right)^2, \quad t > 0, \quad x \in R^n, \tag{7.4}$$

$$\bar{z}^{(2)}(0,x;\lambda) = 0 .$$

Here L *is the operator defined by (6.1) and* $\bar{w}^{(1)}(t,x;\lambda)$ *is the solution of (6.4). The function* $g(x) \geq 0$ *is given by*

$$g(x) = \frac{1}{2}\int_S\int_S\int_S \tilde{\phi}(v;x)\ \gamma(dv)\ q(x,v)\ \pi^{(2)}(x,v,dv_1\ dv_2) \cdot \phi(v_1;x)\ \phi(v_2;x) \tag{7.5}$$

with $\pi^{(2)}$ *defined by* (2.12).

Remark 3. When $\alpha \geq 4$ we must subtract an additional $O(\varepsilon)$ deterministic term from $\nu_t^{\varepsilon,\alpha}$ in (7.1).

Note that the covariance equation for the limiting Gaussian process is a linear inhomogeneous equation driven by the solution of the mean adjoint equation (6.4). The function $g(x)$ is the new basic element necessary for characterizing the asymptotics of the fluctuation process $\zeta_t^{\varepsilon,\alpha}$.

8. CONTINUOUS-STATE BRANCHING AND DIFFUSION LIMIT

We now consider the case $\alpha = 2$, always under the hypotheses of Section 6.

THEOREM 3. *For each* $t > 0$ *and finite and for each* $\lambda(x,v) \geq 0$, *continuous and of compact support in* x,

$$- \lim_{\varepsilon \downarrow 0} \log E_{\mu^{\varepsilon,2}} \left\{ e^{-(\lambda, \nu_t^{\varepsilon,2})} \right\} = \int_E \bar{w}(t,x;\lambda) \phi(v;x) \, \mu(dx\, dv). \tag{8.1}$$

Here $\bar{w}(t,x;\lambda)$ *is the solution of the nonlinear diffusion equation*

$$\frac{\partial \bar{w}}{\partial t} = L\bar{w} - g\bar{w}^2 , \qquad t > 0, \quad x \in R^n , \tag{8.2}$$

$$\bar{w}(0,x;\lambda) = \int_S \lambda(x,v)\, \tilde{\phi}(v;x)\, \gamma(dv), \quad x \in R^n ,$$

with L *defined by* (6.1) *and* g *by* (7.5).

Remark 4. Actually we prove that the process $\nu_t^{\varepsilon,2}(\cdot \times S)$ converges weakly as a measure on $D([0,\infty); M^+(R^n))$ to a diffusion Markov process on $M^+(R^n)$, i.e. a probability measure on $C([0,\infty); M^+(R^n))$. Note that the limit process is not Gaussian. It is an infinite dimensional analog of the processes obtained in [15], [16]; it is analyzed by Dawson in [9], [10] where the term measure diffusion process is used.

<u>Remark 5</u>. The connection between Theorems 2 and 3 becomes clear from the following observation. The analysis in all cases centers around (4.4). Now when $\alpha > 2$ $w^{\varepsilon,\alpha}$ behaves for ε small and $t > 0$ like $\bar{w}^{\varepsilon}\phi$ where $\bar{w}^{\varepsilon}$ satisfies

$$\frac{\partial \bar{w}^{\varepsilon}}{\partial t} = L\bar{w}^{\varepsilon} - \varepsilon^{\alpha-2} g(\bar{w}^{\varepsilon})^2 , \qquad t > 0, \tag{8.3}$$

$$\bar{w}^{\varepsilon}(0,x;\lambda) = \int_S \lambda(x,v)\ \tilde{\phi}(v;x)\ \gamma(dv) .$$

Thus when $\alpha = 2$ we get (8.2) and when $\alpha > 2$ we get (6.4) at first and after a renormalization we get, effectively, (7.4). The case $\alpha < 2$ can, of course, be dealt with in a similar manner.

9. EXISTENCE THEOREM FOR THE BOUNDARY RENEWAL EQUATION

In the next few sections we shall give some results regarding ν^B, the boundary BTP defined in Section 3 which is a point process on E_B i.e., a probability measure on $M_p^+(E_B)$.

First we scale the process in the manner described in Section 4 and denote the scaled process by $\nu^{B,\varepsilon,\alpha}$ with $\alpha \geq 2$. Thus, (3.9) changes to

$$E_{\mu^{\varepsilon,\alpha}}\left\{e^{-(\lambda,\nu^{B,\varepsilon,\alpha})}\right\} = \exp\left\{-\left(w^{B,\varepsilon,\alpha}(\lambda),\mu\right)\right\} \tag{9.1}$$

where $\lambda \in C_0^+(E_B)$ and $w^{B,\varepsilon,\alpha}(x,v;\lambda)$ satisfies *formally* the boundary renewal equation

$$v\cdot \frac{\partial w^{B,\varepsilon,\alpha}}{\partial x} - \varepsilon q\left[\frac{e^{\varepsilon^{\alpha} w^{B,\varepsilon,\alpha}}\Phi(e^{-\varepsilon^{\alpha} w^{B,\varepsilon,\alpha}}) - 1}{\varepsilon^{\alpha}}\right] = 0, \qquad (x,v) \in E, \tag{9.2}$$

$$w^{B,\varepsilon,\alpha}(x,v;\lambda) = \lambda(x,v) , \qquad (x,v) \in E_B .$$

We shall assume in the sequel that hypotheses (5.1)-(5.6) and (6.1) hold. In addition $D \subset R^n$ is a bounded open set with smooth boundary ∂D. Only the case $\alpha \geq 2$ is considered, and *the* S *will be assumed to be compact.*

THEOREM 4. *There exists an* $\varepsilon_0 > 0$ *such that for each* ε *with* $0 < \varepsilon \leq \varepsilon_0$ *the integral equation version of the boundary renewal equation* (9.2) *has a unique bounded measurable solution* $w^{B,\varepsilon,\alpha}(x,v;\lambda)$. *This solution has the form*

$$w^{B,\varepsilon,\alpha} = \bar{w}^{B,\varepsilon,\alpha} + \tilde{w}^{B,\varepsilon,\alpha} \tag{9.3}$$

where $\bar{w}^{B,\varepsilon,\alpha}(x,v;\lambda)$ *is continuous for* $(x,v) \in E \cup E_B$ *and*

$$\lim_{\varepsilon \downarrow 0} \sup_{(x,v)\in E\cup E_B} |\tilde{w}^{B,\varepsilon,\alpha}(x,v;\lambda)| = 0 . \tag{9.4}$$

<u>Remark 6</u>. This theorem is proved by an explicit asymptotic expansion process and so it comes out of the calculations that are necessary also for the theorems that follow. Afrer the expansion has been constructed we use a straightforward contraction mapping argument for the existence.

10. HALF-SPACE TRANSPORT PROCESS

For the analysis of the boundary BTP, $\nu^{B,\varepsilon,\alpha}$, it is necessary to have available some results regarding a Markov process on a halfline which describes the local behavior of the BTP when it is near ∂D. We shall give these results in this section.

Let M be the mean operator defined by (2.4) and define the operator Q by (cf. (6.1))

$$(Qw)(x,v) = \frac{1}{\phi(v;x)} \Big(M(\phi w)\Big)(x,v) \tag{10.1}$$

where $x \in D$ is now a parameter which we shall suppress in this section. The operator Q is the infinitesimal generator of a jump

Markov process on S which we denote by $\{V(t),\ t \geq 0\}$. This process is ergodic with invariant measure $\tilde{\phi}(v)\phi(v)\gamma(dv)$ which is approached exponentially fast, uniformly with respect to the initial point. Let $z(v)$ be a continuous function from S to $[-1,1]$, say, such that

$$\int_S z(v)\tilde{\phi}(v)\phi(v)\gamma(dv) = 0. \tag{10.2}$$

Define a process $H(t)$ on R^1 by

$$H(t) = \eta + \int_0^t z(V(s))\, ds . \tag{10.2}$$

It is easily shown [13] that this process is recurrent. With $\eta < 0$ then let τ be the first time $H(t)$ equals zero; τ is a proper random variable. Let S^+ be the set of points in S for which $z(v) \geq 0$. For any Borel subset A of S let

$$P_B(\eta,v,A) = P\left\{V(\tau) \in A \mid V(0) = v,\ H(0) = \eta\right\}, \tag{10.3}$$

where $\eta < 0$. As $\eta \to -\infty$ we expect that this probability should tend to a limiting one that does not depend on the v. In fact we have the following result (proved under the hypotheses of Section 5 by Varadhan [13]).

(10.4) There is a probability measure $p_B(A)$ on S^+ and a constant $\beta > 0$ such that

$$|P_B(\eta,v,A) - p_B(A)| \leq e^{\beta\eta}, \qquad \eta < 0.$$

In [13] we discuss in some detail how one may go about computing p_B and how it is related to Chandrasekhar's H function [17] in some particular cases. The measure p_B is the limiting measure of the exit velocities of an "effective" Markov process (the one generated by $z(v)\frac{\partial}{\partial\eta} + Q$) associated with the BTP near ∂D as we move further and further into the interior of D.

11. LAW OF LARGE NUMBERS AND DIFFUSION LIMIT FOR THE BOUNDARY PROCESS

From the existence Theorem 4 we know that equation (9.1) holds and $w^{B,\varepsilon,\alpha}$ satisfies the integral equation version of (9.2). We now fix $\alpha > 2$ and $\lambda \in C_0^+(E_B)$ and assume that all hypotheses stated above Theorem 4 hold. For each $x \in \partial D$ we let $p_B(A;x)$ be the measure described (10.4) when S^+ is $S^+(x)$ and $z(v) = \hat{n}(x)\cdot v$ where $\hat{n}(x)$ is unit outward normal to ∂D at x. Recall that S is assumed to be a compact set.

THEOREM 5. *For* $\alpha > 2$ *and each fixed* $\mu \in M^+(E)$ *with* x *support contained in a compact subset of* D,

$$- \lim_{\varepsilon \downarrow 0} \log E_{\mu^{\varepsilon,\alpha}} \left\{ e^{-(\lambda,\nu^{B,\varepsilon,\alpha})} \right\} = (\bar{w}_1^B \phi, \mu)$$

$$= \int_E \bar{w}_1^B(x;\lambda)\ \phi(v;x)\ \mu(dx\ dv) \tag{11.1}$$

where $\bar{w}_1^B$ *is the solution of*

$$L\bar{w}_1^B = 0\ , \qquad\qquad , \quad x \in D \tag{11.2}$$

$$\bar{w}_1^B(x) = \int_{S^+(x)} \lambda(x,v)\ \frac{p_B(dv;x)}{\phi(v;x)} \quad , \quad x \in \partial D \tag{11.3}$$

Here L *is the diffusion operator given by* (6.1).

<u>Remark 7</u>. Since $\bar{w}_1^B$ is a linear functional of λ it follows that $\nu^{B,\varepsilon,\alpha}$ converges vaguely, in probability, as $\varepsilon \to 0$ to $\bar{\nu}^B$ which is a deterministic measure on E_B. We have that $(\lambda,\bar{\nu}^B) = (\bar{w}_1^B\phi,\mu)$ for each $\lambda \in C_0^+(E_B)$ which characterizes completely $\bar{\nu}^B$. It is easily seen that (11.2), (11.3) is just the adjoint version of the result obtained in [11] by analyzing the usual transport equations.

Remark 8. If μ has support near ∂D then a boundary layer correction is necessary. In the course of proving Theorem 5, one actually obtains all the necessary information for the more general result. The proof is similar to the ones in [13].

12. GAUSSIAN AND DIFFUSION LIMIT

As in Section 7 we may ask how the fluctuations $\nu^{B,\varepsilon,\alpha} - \bar{\nu}^B$ behave when $\varepsilon \to 0$ and $\alpha > 2$ is fixed. Let

$$\zeta^{B,\varepsilon,\alpha} = \frac{\nu^{B,\varepsilon,\alpha} - \bar{\nu}^B}{\varepsilon^{\frac{\alpha-2}{2}}} \tag{12.1}$$

which we regard as a tempered distribution-valued process.

THEOREM 6. *Under the hypotheses of Theorem* 5 *and* $2 < \alpha < 4$, $\zeta^{B,\varepsilon,\alpha}$ *converges weakly to a Gaussian process* ζ^B *such that for any* $\lambda \in C_0^\infty(E_B)$, $E\{(\zeta^B,\lambda)\} = 0$ *and*

$$E\{(\zeta^B,\lambda)^2\} = (\bar{z}_2^B\phi,\mu)$$

where $\bar{z}_2^B(x;\lambda)$ *is the solution of*

$$\begin{aligned} L\bar{z}_2^B + 2g(\bar{w}_1^B)^2 &= 0\ , & x &\in D\ , \\ \bar{z}_2^B(x;\lambda) &= 0\ , & x &\in \partial D\ , \end{aligned} \tag{12.2}$$

and g *is given by* (7.5) *while* $\bar{w}_1^B$ *is the solution of* (11.2),(11.3).

13. CONTINUOUS-STATE BRANCHING AND DIFFUSION LIMIT

In the case $\alpha = 2$, $\nu^{B,\varepsilon,2}$ has a limit that is not a Gaussian process. We have the following result.

THEOREM 7. *Under the hypotheses of Theorem* 5 *and for each* $\lambda \in C_0^+(E_B)$,

$$-\lim_{\varepsilon\downarrow 0} \log E_{\mu^{\varepsilon,2}}\left\{e^{-(\lambda,\nu^{B,\varepsilon,2})}\right\} = (\bar{w}^B\phi,\mu)$$

$$= \int_E \bar{w}^B(x;\lambda)\phi(\nu;x)\ \mu(dx\ dv) \tag{13.1}$$

where $\bar{w}^B(x;\lambda)$ *is the solution of the nonlinear elliptic boundary value problem*

$$L\bar{w}^B - g(\bar{w}^B)^2 = 0 \quad , \quad x \in D, \tag{13.2}$$

$$\bar{w}^B(x,\lambda) = \int_{S^+(x)} \lambda(x,v)\,\frac{p_B(dv;x)}{\phi(v;\ x)}\,, \quad x \in \partial D\ .$$

Remark 9. If the mean generator M is not exactly critical but is supercritical to $O(\varepsilon^2)$, i.e. the principal eigenvalue is $\varepsilon^2\theta$ where θ is a constant, say, then in all the above results the operator L must be replaced by $L+\theta$. If θ is smaller than the first eigenvalue ψ of L on D (with zero boundary conditions), then (11.2), (11.3) has a unique solution and similarly (12.2). In case (13.2), θ can actually be a bit larger than ψ if g is strictly positive. However, the limit process ν^B with log Laplace function $(\bar{w}^B\phi,\mu)$ will not have finite moments in this case (it will have finite moments if $\theta < \psi$).

REFERENCES

[1] G. I. Bell, Stochastic formulations of neutron transport, in SIAM-AMS Proceedings, Vol. 1, Transport Theory, R. Bellman, G. Birkhoff and I. Abu-Shumays, editors, Providence, R. I., 1969, pp. 181-197.

[2] J. E. Moyal, The general theory of stochastic population processes, *Acta Math.* *108* (1962), pp 1-31. See also article in the same volume as [1], pp. 198-212.

[3] T. W. Mullikin, Branching processes in neutron transport

theory, in: Probabilistic Methods in Appl. Math., Vol. 1, A. T. Bharucha-Reid, editor, Academic Press, New York, 1968, pp. 199-281.

[4] T. Harris, The Theory of Branching Processes, Springer, Berlin, 1963.

[5] N. Ikeda, M. Nagasawa and S. Watanabe, Branching Markov processes, *J. Math. Kyoto Univ. 8* (1968), pp. 233-278, 356-410, and *9* (1969), pp. 95-160.

[6] J. Kerstan, K. Matthes and J. Mecke, Unbegrenzt teilbare Punktprozesse, Akademie-Verlag, Berlin, 1974.

[7] O. Kallenberg, Random measures, Akademie-Verlag, Berlin, 1976.

[8] J. Neveu, Processes ponctuels, École d'été de Saint Flour, 1976.

[9] D. A. Dawson, Stochastic evolution equations and related measure processes, *J. Multiv. Anal. 5* (1975), pp. 1-52.

[10] D. A. Dawson, The critical measure diffusion process, *Z. Wahrscheinlichkeitstheorie verw. Gebiete 40* (1977), 125-145.

[11] E. Larsen and J. B. Keller, Solution of the steady, one-speed neutron transport equation for small mean free paths, *J. Math. Phys. 15*, (1974), pp. 299-305.

[12] M. Williams, Ph.D. dissertation, New York Univ., 1976.

[13] A. Bensoussan, J. L. Lions, and G. C. Papanicolaou, Boundary layers and homogenization of transport processes, *J. Publ. RIMS*, Kyoto Univ., to appear.

[14] A. Bensoussan, J. L. Lions and G. C. Papanicolaou, Proceedings IRIA Congress, Dec. 1977, to appear.

[15] J. Lamperti, Continuous state branching process, *BAMS 73*, (1967), pp. 382-386.

[16] M. Jirina, Diffusion branching processes with several types of particles, *Zeitschrift für Wahr. 18* (1971), pp. 34-46.

[17] S. Chandrasekhar, Radiative Transfer, Dover, New York, 1950.

STOCHASTIC PARTIAL DIFFERENTIAL EQUATION FOR THE DENSITY OF THE CONDITIONAL LAW OF A DIFFUSION PROCESS WITH BOUNDARY

E. PARDOUX (C.N.R.S.)
I.R.I.A. - LABORIA
78150 Le Chesnay
FRANCE

Let x_t be a diffusion process with boundary, as defined in STROOCK-VARADHAN [7] . Suppose we observe the process y_t given by:

$$y_t = \int_0^t h(s,x_s)ds + v_t$$

where v_t is the observation noise, which we will suppose here to be a Wiener process independent of x_t.

The filtering problem consists in characterizing the conditional law of x_t, given the observations y_s, $s \leq t$.

We will show that the conditional law is the sum of two measures : $\Sigma_t^1 + \Sigma_t^2$. Σ_t^1 has a density with respect to Lebesgue measure in R^N. Σ_t^2 is concentrated on the boundary of the subdomain of R^N where the signal process x_t is restricted to ly, and has a density with respect to Lebesgue measure on the boundary. Both densities are explicit functions of the solution of a linear stochastic P.D.E. - see Theorem 4.1.

Our proof uses Bucy formula, and the interpretation of a stochastic P.D.E. as a backward Kolmogorov equation - theorem 3.2.

This work extends our previous results [6]to more general boundary conditions, and presents a new approach to the problem. Theorem 3.2 is completely new. In the case of ordinary diffusion processes, without boundary, results similar to theorem 4.1 have been obtained by KRYLOV-ROSOVSKII [3], with different methods.

1. THE FILTERING PROBLEM.

1.1. Definition of a diffusion process with boundary.

We follow the work of STROOCK-VARADHAN [7], and introduce here all the hypotheses that will be needed in the sequel.

Let $\mathcal{O}$ be an open subset of R^N, defined as $\mathcal{O} = \{\phi(x) > 0\}$, where:

(1.1) $\qquad \phi \in C_b^2(R^N)$

ISBN 0-12-268380-3

(1.2) $\Gamma = \partial\mathcal{O} = \{\phi(x)=0\}$, and $|\nabla\phi(x)| = 1$, $\forall x \in \Gamma$

Let L_t be a second order partial differential operator :

$$L_t = \frac{1}{2}\sum_{i,j=1}^{N} a_{ij}(t,x)\frac{\partial^2}{\partial x_i \partial x_j} + \sum_{i=1}^{N} b_i(t,x)\frac{\partial}{\partial x_i}$$

where :

(1.3) $a_{ij} = a_{ji} \in C_b([0,T]\times\bar{\mathcal{O}})$, $\forall T>0$, $i,j=1,\dots N$

(1.4) $b_i \in B([0,T]\times\mathcal{O})^*$, $\forall T > 0$, $i,j=1,\dots N$

We make the following assumptions :

(1.5) $\begin{cases} \exists\alpha > 0 \text{ s.t. } \forall(t,x) \in R_+ \times \mathcal{O}, \ \forall\xi \in R^N, \\ \frac{1}{2}\sum_{i,j=1}^{N} a_{ij}(t,x)\xi_i\xi_j \geq \alpha|\xi|^2 \end{cases}$

(1.6) $\frac{\partial a_{ij}}{\partial x_i} \in L^\infty(R_+ \times \mathcal{O})$, $i,j=1,\dots N$

Let us now define the boundary operator. Let $\delta(t,x)$ be a vector field on $R_+ \times \Gamma$ tangent to Γ, with :

(1.7) $\delta \in L^\infty(R_+; C_b^1(\Gamma)) \cap C_b(R_+ \times \Gamma)$

(1.8) $\delta(t,x).\nabla\phi(x) = 0$, $\forall x \in \Gamma$, $\forall t \geq 0$

Define the vector field $\gamma(t,x)$ by :

(1.9) $\gamma_i(t,x) = \frac{1}{2}\sum_{j=1}^{N} a_{ij}(t,x)\frac{\partial\phi}{\partial x_j}(x) + \delta_i(t,x)$ $i=1,\dots N$

We suppose that $\gamma(t,x)$ is locally Lipschitz on $R_+ \times \Gamma$.

Let ρ be a locally Lipschitz real-valued function defined on $R_+ \times \Gamma$. We suppose either $\rho \equiv 0$, or :

(1.10) $0 < \underline{\rho} \leq \rho(t,x) \leq \bar{\rho} < +\infty$, $\forall(t,x) \in R_+ \times \Gamma$; $\frac{\partial\rho}{\partial t} \in L^\infty(R_+ \times \Gamma)$

---

* $B([0,T]\times\mathcal{O})$ denotes the class of bounded Borel-measurable functions on $[0,T] \times \mathcal{O}$.

Remark 1.1. : From (1.5) and (1.9), we get :

(1.11) $\gamma(t,x).\nabla\phi(x) \geq \alpha > 0$

On the other hand, the vector (ρ,γ) being defined up to a multiplicative uniformly positive constant, it is always possible, with (1.11), (1.5), to choose ρ and γ such that (1.9) and (1.8) are satisfied. $\square$

Let $\Omega_1 = C(R_+ ; R^N)$, $x_t(\omega) = \omega_1(t)$,

$$\underline{\underline{B}}_t = \sigma\{ x_s, s \leq t\} , \quad \underline{\underline{B}} = \bigvee_{t\geq 0} \underline{\underline{B}}_t$$

Under the above conditions, the following result holds (see STROOCK-VARADHAN [7]) :

Theorem 1.1 : $\forall(s,x) \in R_+ \times \bar{\mathcal{O}}$, there exists a unique probability measure P_{sx} on $(\Omega_1, \underline{\underline{B}})$ such that :

(i) $P_{sx}(x_{t\wedge s} = x) = 1$

(ii) $P_{sx}(x_t \in \bar{\mathcal{O}}) = 1 \quad \forall t \geq 0$

(iii) $\begin{cases} f(t,x_t) - \int_s^t 1_{\mathcal{O}}(f'_\sigma + L_\sigma f)(\sigma,x_\sigma)d\sigma \text{ is a } P_{sx} \text{ sub-martingale,} \\ \forall f \in C_K^{1,2}(R_+ \times R^N)^{**} \text{ such that :} \\ \rho(\sigma,x)f'_\sigma(\sigma,x) + \gamma(\sigma,x).\nabla f(\sigma,x) \geq 0, \ \forall(t,x) \in R_+ \times \Gamma \end{cases}$

Moreover, there exists a unique non decreasing, non anticipating process ξ_t^{sx} such that :

(iv) $\xi_{t\wedge s}^{sx} = 0$

(v) $\xi_t^{sx} = \int_s^t 1_\Gamma(x_\sigma)d\xi_\sigma^{sx}$

(vi) $\int_s^t 1_\Gamma(x_\sigma)d\sigma = \int_s^t \rho(\sigma,x_\sigma)d\xi_\sigma^{sx}$

** $C_K^{1,2}(R_+ \times R^N)$ denotes the class of functions which are continuous together with their first t-derivative and second x- derivative, with compact support.

$$(vii)\begin{cases} f(t,x_t) - \int_0^t 1_{\mathcal{O}}(f'_\sigma + L_\sigma f)(\sigma,x_\sigma)d\sigma - \\ \qquad -\int_0^t 1_\Gamma(\rho f'_\sigma + \gamma.\nabla f)(\sigma,x_\sigma)d\xi_\sigma^{sx} \quad \text{is a } P_{sx} \text{ martingale,} \\ \forall f \in C_K^{1,2}(R_+ \times R^N) \end{cases}$$

□

Remark 1.2. : If we compare our hypothesis with those of STROOCK-VARADHAN [7] , the restrictions we introduce are mainly hypotheses (1.5) and (1.6). This will be necessary for the study of the stochastic PDE's.

□

Let p_o be an initial probability density on $\mathcal{O}$. We suppose that :

(1.12) $$p_o \in L^2(\mathcal{O}),\ p_o(x) \geq 0 \ \text{ a.e.},\ \int_{\mathcal{O}} p_o(x)dx = 1$$

Define the probability measure P_o on $(\Omega,\underline{\underline{B}})$ by :

$$\forall B \in \underline{\underline{B}}\ ,\ P_o(x_. \in B) = \int_{\mathcal{O}} p_o(y)P_{oy}(x_. \in B)dy$$

1.2. The filtering problem.

Let $\Omega_2 = C(R_+;R^d)$, $y_t(\omega) = \omega_2(t)$, $\underline{\underline{F}}_t^o = \sigma\{y_s, s \leq t\}$, $\underline{\underline{F}} = \bigvee_{t\geq 0} \underline{\underline{F}}_t^o$, and let $\mathcal{W}$ be the Wiener measure on $(\Omega_2,\underline{\underline{F}})$. Let $\underline{\underline{F}}_t$ be F_t^o completed by the $\mathcal{W}$ null sets of $\underline{\underline{F}}$.

Let $(\Omega,\underline{\underline{G}},\underline{\underline{G}}_t,\tilde{Q}) = (\Omega_1 \times \Omega_2, \underline{\underline{B}} \times \underline{\underline{F}}, \underline{\underline{B}}_t \times \underline{\underline{F}}_t, P_o \times \mathcal{W})$.

Define $h(t,x)$:

(1.13) $$h \in B(R_+ \times \bar{\mathcal{O}}\ ;\ R^d)$$

and

$$Z_t = \exp\{\int_0^t [h(x_s),dy_s] - \frac{1}{2}\int_0^t |h(x_s)|^2 ds\}$$

where[.,.] denotes the scalar product in R^d.

Define the measure Q on $(\Omega,\underline{\underline{G}}_t)$, $\forall t \geq 0$, by :

$$\left.\frac{dQ}{d\tilde{Q}}\right|_{\underline{\underline{G}}_t} = Z_t$$

From GIRSANOV's theorem, it is easily seen that :

$$y_t = \int_0^t h(s,x_s)ds + v_t \tag{1.14}$$

where v_t is a $\underline{\underline{G}}_t$ - Q Wiener process.

v_t is also a $\underline{\underline{B}} \otimes \underline{\underline{F}}_t$ Q Wiener process, and consequently it is independent of x_t.

The filtering problem consists in computing a measure-valued process Σ_t defined as :

$$\Sigma_t f = E(f(x_t)/ \underline{\underline{F}}_t)$$

when E denotes the expectation with respect to Q. In the sequel, $\tilde{E}$ will refer to $\tilde{Q}$.

1.3. Bucy formula.

We will make use of the following result, first stated by BUCY :

Lemma 1.1.

$\forall f \in C_b(\bar{\mathcal{O}})$,

$$E(f(x_t)/ \underline{\underline{F}}_t) = \frac{\tilde{E}(f(x_t)Z_t/ \underline{\underline{F}}_t)}{\tilde{E}(Z_t/ \underline{\underline{F}}_t)} \tag{1.15}$$

Proof : Let g be any $\underline{\underline{F}}_t$ measurable random variable.

$$E\left[g \frac{\tilde{E}(f(x_t)Z_t/ \underline{\underline{F}}_t)}{\tilde{E}(Z_t/ \underline{\underline{F}}_t)}\right] = \tilde{E}\left[gZ_t \frac{\tilde{E}(f(x_t)Z_t/ \underline{\underline{F}}_t)}{\tilde{E}(Z_t/ \underline{\underline{F}}_t)}\right]$$

$$= \tilde{E}\left[g\, \tilde{E}(Z_t/ \underline{\underline{F}}_t) \frac{\tilde{E}(f(x_t)Z_t/ \underline{\underline{F}}_t)}{\tilde{E}(Z_t/ \underline{\underline{F}}_t)}\right]$$

$$= \tilde{E}\left[g\, \tilde{E}(f(x_t)Z_t/ \underline{\underline{F}}_t)\right]$$

$$= E\left[g\, f(x_t)\right]$$

□

2. STUDY OF A FORWARD STOCHASTIC P.D.E.

2.1. Notations. Statement of the result.

We go back to the domain $\mathcal{O}$ and the operators defined in 1.1.

Let $H^1(\mathcal{O})$ denote the Sobolev space of $L^2(\mathcal{O})$ functions whose first partial derivatives belong to $L^2(\mathcal{O})$.

If $u \in H^1(\mathcal{O})$, we can define its trace $\gamma_o u$ on Γ, as an element of $L^2(\Gamma)$, such that if moreover $u \in C^o(\bar{\mathcal{O}})$, then $\gamma_o u$ is identical to its usual restriction to Γ; see LIONS [4] .

We will denote by $|u|$ and (u,v) the norm and scalar product in $L^2(\mathcal{O})$, by $\|u\|$ the norm in $H^1(\mathcal{O})$. If $u,v \in H^1(\mathcal{O})$, we will make the following abuses of notation : $(u,v)_\Gamma$ will denote the scalar product of $\gamma_o u$ and $\gamma_o v$ in $L^2(\Gamma)$, and $|u|_\Gamma$ the norm of $\gamma_o u$ in $L^2(\Gamma)$.

Define a family $a(t;u;v)$ of bilinear continuous forms on $H^1(\mathcal{O})$ by :

$$a(t;u;v) = \frac{1}{2} \sum_{i,j=1}^{N} \int_{\mathcal{O}} a_{ij}(t,x) \frac{\partial u}{\partial x_i}(x) \frac{\partial v}{\partial x_j}(x)dx +$$

$$+ \sum_{i=1}^{N} \int_{\mathcal{O}} \left(\frac{1}{2} \sum_{j=1}^{N} \frac{\partial a_{ij}}{\partial x_j}(t,x) - b_i(t,x)\right)u(x) \frac{\partial v}{\partial x_i}(x)dx$$

From (1.5), $\exists\lambda$ s.t.

$$a(t,u,u) + \lambda|u|^2 \geq \alpha\|u\|^2, \ \forall u \in H^1(\mathcal{O}), \text{ p.p.t} \tag{2.1}$$

If $u,v \in \mathcal{D}(\bar{\Omega})$, we define the following bilinear form :

$$\langle D(t)u,v\rangle_\Gamma = \int_\Gamma \sum_{i=1}^{n} \frac{\partial}{\partial x_i} (\delta_i(t)u)(x)v(x)d\Gamma$$

It can be shown -see LIONS [4] - that $u,v \to \langle D(t)u,v\rangle$ can be extended to a continuous bilinear form on $H^1(\mathcal{O})$, such that, $\forall\varepsilon > o$, $\exists c(\varepsilon)$ with :

$$\langle D(t)u,u\rangle_\Gamma \leq \varepsilon|u|^2 + c(\varepsilon)\|u\|^2, \ \forall u \in H^1(\mathcal{O}) \tag{2.2}$$

Let $(\Omega,\underline{\underline{F}},\underline{\underline{F}}_t,P,w_t)$ be an R^d -valued standard Wiener process, where $\Omega = C(R_+;R^d)$, and w_t is the generic element of Ω, and

$h(t,x)$ defined by (1.13), p_o and ρ as in 1.1.

We consider the following equation :

(2.3)
$$\begin{cases} (u(t),v) + (\rho(t)u(t),v)_\Gamma + \int_o^t a(s;u(s),v)ds + \\ \quad + \int_o^t \langle D(s)u(s),v\rangle_\Gamma ds = (p_o,v) + \\ + \int_o^t [(h(s)u(s),v) + (\rho(s)h(s)u(s),v)_\Gamma, dw_s] \\ \quad \forall t > 0 \quad , \quad \forall v \in H^1(\mathcal{O}) \end{cases}$$

We will prove the following :

<u>Theorem 2.1.</u>

Equation (2.3) has a unique solution $u \in L^2(\Omega\times]0,T[;H^1(\mathcal{O}))$, $\forall T > 0$, u being $\underline{\underline{F}}_t$ adapted.

Moreover,

$$u \in L^2(\Omega;C([0,T];\ L^2(\mathcal{O}))) \text{ and}$$

$$\text{if } \rho \not\equiv 0 \quad \gamma_o u \in L^2(\Omega;C([0,T];\ L^2(\Gamma))),\ \forall T > 0$$

<u>Remark 2.1.</u> : Equation (2.3) can be formally interpreted in the following way[with additional regularity on the coefficients and the solution, the interpretation could be made rigorous]:

(2.4)
$$\begin{cases} d\,u(t,x) - \frac{1}{2}\sum_{i,j=1}^{N}(a_{ij}\frac{\partial u}{\partial x_i})(t,x)dt - \sum_{i=1}^{N}\frac{\partial u}{\partial x_j}(a_j u)(t,x)dt = \\ \quad =[(h\,u)(t,x),\ dw_t] \quad \text{in } R_+ \times \mathcal{O} \\ d(\rho(t,x)u(t,x)) + (\gamma.\nabla u)(t,x)dt + (b_o u)(t,x)dt = \\ \quad = [(\rho h\,u)(t,x),dw_t] \text{ on } R_+ \times \Gamma \\ u(o,x) = p_o(x) \end{cases}$$

where :
$$a_j = \frac{1}{2}\sum_i \frac{\partial a_{ij}}{\partial x_i} - b_j$$
$$b_o = \sum_j a_j + \sum_i \frac{\partial \delta_i}{\partial x_i}$$
$\square$

The rest of §2 will be devoted to the proof of theorem 2.1.

2.2. Existence.

Our method will be based on a discretization of the time interval $[0,T]$.

Let $0 = t_o < t_1 < \ldots < t_n = T$, with $t_{i+1} - t_i = k = \frac{T}{n}$.

Define, for $u, v \in H^1(\mathcal{O})$; $i=1,\ldots,n$:

$$(\rho^i u,v)_\Gamma = \frac{1}{k}\int_{t_{i-1}}^{t_i} (\rho(s)u,v)_\Gamma ds, \ \rho^o = \rho^1$$

$$a^i(u,v) = \frac{1}{k}\int_{t_{i-1}}^{t_i} a(s;u,v)ds$$

$$\langle D^i u,v\rangle_\Gamma = \frac{1}{k}\int_{t_{i-1}}^{t_i} \langle D(s)u,v\rangle_\Gamma \ ds$$

$$h^i(x) = \frac{1}{k}\int_{t_{i-1}}^{t_i} h(s,x)ds$$

$$\Delta w^i = w_{t_i} - w_{t_{i-1}}$$

and consider, for $i=1,\ldots,n$, equation :

$$(2.5)\quad \begin{cases} (u^i,v)+(\rho^i u^i,v)_\Gamma + k\, a^i(u^i,v) + k\langle D^i u^i,v\rangle_\Gamma = \\ = (u^{i-1},v)+(\rho^i u^{i-1},v)_\Gamma + [(h^i u^{i-1},v)+(\rho^i h^i u^{i-1},v)_\Gamma, \Delta w^i] \\ \qquad i = 1, 2, \ldots, n \quad \forall v \in H^1(\mathcal{O}) \end{cases}$$

where, for each i, u^{i-1} is given as an $H^1(\mathcal{O})$ valued random variable.

Lemma 2.1. :

$\forall i$, $\exists$ a unique $H^1(\mathcal{O})$ -valued random variable u^i, solution of equation (2.5).

Proof : For each realisation of u^{i-1} and Δw_i (2.5) is an elliptic equation, which has a unique solution $u^i \in H^1(\mathcal{O})$, if k is small enough. This is proved in LIONS [4] , and makes use of (2.1) and (2.2).

It is easy to check that :

$$(u^{i-1}, \Delta w_i) \to u^i$$

is a continuous map from $H^1(\mathcal{O}) \times R$ into $H^1(\mathcal{O})$, and so (2.5) defines a unique $H^1(\mathcal{O})$ -valued random variable u^i.

□

Choose $u^o = p_o$, and define :

$$u_n(t) = \sum_{i=1}^{n} 1_{[t_{i-1},t_i[} u^i, \text{ and } \rho_n(t),\ a_n(t,.,.),\ D_n(t)$$

and $h_n(t)$ in the same way.

Lemma 2.2. :

The sequence u_n remains in a bounded subset of $L^2(\Omega\times]0,T[;H^1(\mathcal{O}))$

Proof : choose $v = u^i + u^{i-1}$ in (2.5) :

$$(2.6)\quad \begin{cases} |u^i|^2 - |u^{i-1}|^2 + (\rho^i u^i, u^i)_\Gamma - (\rho^{i-1} u^{i-1}, u^{i-1})_\Gamma + \\ + k\ a^i(u^i, u^i + u^{i-1}) + k\langle D^i u^i, u^i + u^{i-1}\rangle_\Gamma - \\ - ((\rho^i - \rho^{i-1}) u^{i-1}, u^{i-1})_\Gamma = \\ = [(h^i u^{i-1}, u^i + u^{i-1}) + (\rho^i h^i u^{i-1}, u^i + u^{i-1})_\Gamma) \Delta w^i] \end{cases}$$

Choose $v = u^i - u^{i-1}$ in (2.5) :

$$(2.7)\quad \begin{cases} |u^i - u^{i-1}|^2 + (\rho^i(u^i - u^{i-1}), u^i - u^{i-1})_\Gamma + \\ + k\ a^i(u^i, u^i - u^{i-1}) + k\langle D^i u^i, u^i - u^{i-1}\rangle_\Gamma = \\ = [(h^i u^{i-1}, u^i - u^{i-1}) + (\rho^i h^i u^{i-1}, u^i - u^{i-1})_\Gamma) \Delta w^i] \end{cases}$$

We have of course :

$$(2.8)\quad \begin{cases} |u^i-u^{i-1}-[h^iu^{i-1},\Delta w^i]|^2 + \\ \qquad + (\rho^i(u^i-u^{i-1}-[h^iu^{i-1},\Delta w^i]), \\ \qquad\qquad u^i-u^{i-1}-[h^iu^{i-1},\Delta w^i])_\Gamma \geq 0 \end{cases}$$

(2.6)+(2.7)+(2.8) yield :

$$(2.9)\quad \begin{cases} |u^i|^2-|u^{i-1}|^2+(\rho^iu^i,u^i)_\Gamma-(\rho^{i-1}u^{i-1},u^{i-1})_\Gamma + \\ 2k\ a^i(u^i,u^i)+ 2k <D^iu^i,u^i>_\Gamma-((\rho^i-\rho^{i-1})u^{i-1},u^{i-1})_\Gamma \leq \\ \leq 2[(h^iu^{i-1},u^{i-1})+(\rho^ih^iu^{i-1},u^{i-1})_\Gamma)\Delta w^i] + \\ +|[h^iu^{i-1},\Delta w^i]|^2+(\rho^i[h^iu^{i-1},\Delta w^i],[h^iu^{i-1},\Delta w^i])_\Gamma \end{cases}$$

From (2.9), we conclude that if $u^{i-1} \in L^2(\Omega;H^1(\mathcal{O}))$, then :

$$(2.10)\quad \begin{cases} E|u^i|^2+ E(\rho^iu^i,u^i)_\Gamma + 2k\ E\ a^i(u^i,u^i) + \\ + 2k\ E <D^iu^i,u^i>_\Gamma \leq E((\rho^i-\rho^{i-1})u^{i-1},u^{i-1})_\Gamma + \\ + E|u^{i-1}|^2+ E|\sqrt{\rho^i}u^{i-1}|_\Gamma^2 + k\ E\{|h^iu^{i-1}|^2+|\sqrt{\rho^i}h^iu^{i-1}|_\Gamma^2\} \end{cases}$$

where

$$|h^iu^{i-1}|_\Gamma^2 = \sum_{k=1}^{d} |h_k^iu^{i-1}|^2$$

From (2.10), we conclude that :

$u^{i-1} \in L^2(\Omega;H^1(\mathcal{O})) \Rightarrow u^i \in L^2(\Omega;H^1(\mathcal{O}))$.

Then, by the choice of u^o, $u_n \in L^2(\Omega\times]0,T[;H^1(\mathcal{O}))$; and moreover, taking the sum of (2.10) from k=1 to k=n,

$$(2.11)\quad \begin{cases} E|u_n(T)|^2 + E(\rho^nu_n(T),u_n(T))_\Gamma + \\ + 2E\int_o^T a_n(t;u_n(t),u_n(t))dt+ 2E\int_o^T <D_n(t)u_n(t),u_n(t)>_\Gamma dt \leq \\ \leq E\int_o^T (\frac{\rho_n(t)-\rho_n(t-k)}{k} u_n(t-k),u_n(t-k))_\Gamma\ dt + \\ + |p_o|^2+ E\int_o^T\{|h_n(t)u_n(t-k)-|^2+|\sqrt{\rho_n}(t)h_n(t)u_n(t-k)|_\Gamma^2\}dt \end{cases}$$

where $u_n(t) = u_o$ for $t < 0$.

In fact, (2.10) is true when replacing T by any $t \in [0,T]$. Then, from (2.1), (2.2), (1.10), (1.13) :

$$E\{|u_n(t)|^2 + |\rho_n(t)u_n(t),u_n(t)|_\Gamma^2\} \le c_1 \tag{2.12}$$

where c_1 is independent of $t \in [0,T]$ and n.

From (2.11), (2.12), (2.1), (2.2), (1.10) and (1.13), we conclude that $\exists c_2$ independent of n such that :

$$E \int_0^T \|u_n(t)\|^2 \, dt \le c_2$$

$\square$

From lemma 2.2, we conclude that there exists a subsequence u_μ such that :

$$u_\mu \to u \text{ in } L^2(\Omega\times]0,T[;H^1(\mathcal{O})) \text{ weakly.}$$

It is easy to show that we can take the limit in (2.5), and u is a solution of (2.3). $u \in L^2(\Omega\times]0,T[;H^1(\mathcal{O}))$, and it is easy to see that u is $\underline{\underline{F}}_t$ adapted, as a limit of $\underline{\underline{F}}_t$ adapted processes.

We have shown existence.

2.3. Uniqueness.

The uniqueness will be a corollary of :

Theorem 2.2.

Let u be an $\underline{\underline{F}}_t$ -adapted process such that :

(i) $u \in L^2(\Omega\times[0,T];H^1(\mathcal{O}))$, $\forall T > 0$

(ii) u satisfies equation (2.3).

Then,

$$u \in L^2(\Omega;C([0,T];L^2(\mathcal{O}))) \ ; \text{ if } \rho \not\equiv 0,$$

$$\gamma_o u \in L^2(\Omega;C([0,T];L^2(\Gamma))) \ ;$$

and moreover :

$$(2.13)\quad \begin{cases} |u(t)|^2+(\rho(t)u(t)_\Gamma + 2\int_0^t a(u,u)ds + \\ + 2\int_0^t \langle Du,u\rangle_\Gamma ds - \int_0^t (\rho' u,u)_\Gamma\ ds = |p_o|^2 + \\ + \int_0^t \{|hu|^2+|\sqrt{\rho}hu|^2_\Gamma\ ds + 2\int_0^t [(hu,u)+(\rho hu,u)_\Gamma)\ dw_s] \end{cases}$$

Let us first prove :

Lemma 2.3.

Suppose in addition that $\rho(t,x) \equiv \rho(x)$ is independent of t.

Then, the conclusion of theorem 2.2. is valid.

Proof : From (1.10), we can consider $(\rho u,v)$ as the scalar product on $L^2(\Gamma)$.

Let :

$$V = \{(u,v) \in H^1(\mathcal{O})\times H^{1/2}(\Gamma)^{***};\ \gamma_o u = v\}$$

$$H = L^2(\mathcal{O}) \times L^2(\Gamma)$$

Then, $V \subset H$, V is dense in H and the injection is continuous.

With this frame work, the lemma is a consequence of the following proposition, proved in PARDOUX [5] :

Proposition :

Let u be an $\underline{\underline{F}}_t$ adapted process such that :

(i) $u \in L^2(\Omega\times]0,T[;V)$

(ii) $u(t) = u_o + \int_0^t v(s)ds + \int_0^t [g(s),dw_s]$

where :

(iii) $u_o \in H$

(iv) $v \in L^2(\Omega\times]0,T[;V')$ is $\underline{\underline{F}}_t$ adapted.

(v) $g \in L^2(\Omega\times]0,T[;H^d)$ is $\underline{\underline{F}}_t$ adapted and w_t is an R^d valued standard $\underline{\underline{F}}_t$ Wiener process.

*** $H^{1/2}(\Gamma)$ is a Sobolev space on Γ , equal to $\{\gamma_o u\ ;\ u \in H^1(\mathcal{O})\}$

Then, :

$$u \in L^2(\Omega;C[0,T];H)), \text{ and moreover :}$$

$$|u(t)|^2 = |u_o|^2 + 2\int_o^t \langle v(s),u(s)\rangle ds + $$
$$+ 2\int_o^t [(u(s),g(s),dw_s] + \int_o^t |g(s)|^2 ds$$

where $\langle .,.\rangle$ denotes the duality between V and V', and $|.|^2$ either the norm in H, or the norm in H^d.

Proof of theorem 2.2.: We have already proved the theorem for $\rho\equiv 0$ Suppose $\rho\not\equiv 0$. Let $0 = t_o < t_1 < \ldots < t_n = T$, with $t_{i+1} - t_i = k\frac{T}{n}$.

Define :

$$\rho_n(t) = \frac{1}{k}\int_{t_{i-1}}^{t_i} \rho(s)ds \quad , \text{ if } \quad t \in [t_{i-1}, t_i[$$

$$\rho_n(T) = \rho(T)$$

And replace ρ by ρ_n in equation (2.3).

By the result of 2.2., we can define a solution of equation (2.3) on each subinterval $[t_{i-1}, t_i]$, and so, using lemma 2.3, we can define a solution u_n of (2.3) on $[0,T]$, with the properties :

$$u_n \in L^2(\Omega\times]0,T[;H^1(\mathcal{O})) \cap L^2(\Omega;C([0,T];L^2(\mathcal{O})),$$

$$\gamma_o u_n \in L^2(\Omega;C([0,T];L^2(\Gamma))),$$

and u_n is $\underline{\underline{F}}_t$ adapted.

Moreover, $\forall t \in [0,T]$,

$$(2.14)\begin{cases} |u_n(t)|^2 + (\rho_n(t)u_n(t),u_n(t))_\Gamma + 2\int_o^t a(u_n,u_n)ds + \\ + 2\int_o^t \langle Du_n,u_n\rangle_\Gamma\, ds - \sum_{s\leq t}((\rho_n(s)-\rho_n(s-))u_n(s),u_n(s))_\Gamma = \\ = |p_o|^2 + \int_o^t \{|hu_n|^2 + |\sqrt{\rho}hu_n|_\Gamma^2\}ds + 2\int_o^t [(hu_n,u_n) + \\ \qquad + (\rho_n hu_n,u_n)_\Gamma, dw_s] \end{cases}$$

From (2.14) u_n is bounded in $L^2(\Omega\times]0,T[;H^1(\mathcal{O})) \cap L^2(\Omega;C([0,T];L^2(\mathcal{O})))$, and $\gamma_o u_n$ is bounded in $L^2(\Omega;C([0,T];L^2(\Gamma)))$.

But, using a similar relation for $u_n - u_m$, one can show that u_n is a Cauchy sequence in the spaces listed above, which enables us to take the limit in (2.14), yielding (2.13).

□

Corollary : The solution of equation (2.3) is unique in the class of $\underline{\underline{F}}_t$ adapted processes belonging to $L^2(\Omega\times]0,T[;H^1(\mathcal{O}))$.

Proof : Let u_1 and u_2 be two such solutions. Then $\bar{u} = u_1 - u_2$ is also a solution, which verifies (2.13) with p_o replaced by 0. Standard estimate yields $\bar{u} = 0$.

□

3. INTERPRETATION OF A STOCHASTIC PDE AS A BACKWARD KOLMOGOROV EQUATION.

3.1. Statement of main result.

We use here the same frame work and notations as in §2., in particular the Wiener process $(\Omega,\underline{\underline{F}},\underline{\underline{F}}_t,P,w_t)$. We shall here reverse the time, and make use of the filtration $\underline{\underline{F}}^t$, defined as :

$$\underline{\underline{F}}^t = \sigma\{w_s,\ s \geq t\}$$

Let :

$$a^*(t;u,v) = a(t;v,u)$$

$$\langle D^*(t)u,v\rangle_\Gamma = \langle D(t)v,u\rangle_\Gamma\ ,\ \forall u,v \in H^1(\mathcal{O})$$

We are given $T > 0$ and $f \in C^o_K(\bar{\mathcal{O}})$

Consider the following equation :

$$(3.1)\quad \begin{cases} (v(t),u) + (\rho(t)v(t),u)_\Gamma + \int_0^t a^*(v(s),u)ds + \\ \int_t^T \langle D^*(s)v(s)+\rho'(s)v(s),u\rangle_\Gamma ds = (f,u)+(\rho(T)f,u)_\Gamma + \\ + \int_t^T [(h(s)v(s),u)+(\rho(s)h(s)v(s),u)_\Gamma\, dw_s] \\ \qquad \forall t < T \quad , \quad \forall u \in H^1(\mathcal{O}) \end{cases}$$

Remark 3.1. : We can formally "interpret" equation (3.1) in the following way :

$$(3.2)\quad \begin{cases} dv(t,x)+L_t v(t,x)dt = -[h(t,x)v(t,x),dw_t] , \\ \qquad (t,x) \in [0,T] \times \mathcal{O} \\ \rho(t,x)dv(t,x)+\gamma(t,x).\nabla v(t,x)dt = \\ \quad -\rho(t,x)[h(t,x)v(t,x),dw_t] \quad , \ (t,x) \in [0,T]\times \Gamma \\ \qquad v(T) = f \end{cases}$$

$\sqcap$

Let $f \in C_K^o(\bar{\mathcal{O}})$

In exactly the same way as theorem 2.1, on can prove :

Theorem 2.1 .

Equation (3.1) has a unique solution $v \in L^2(\Omega\times]0,T[;H^1(\mathcal{O}))$, v being $\underline{\underline{F}}_t$ adapted.

Moreover,

$$v \in L^2(\Omega;C([0,T];L^2(\mathcal{O}))) \quad \text{and if } \rho \not\equiv 0,$$

$$\gamma_o v \in L^2(\Omega;C([0,T];L^2(\Gamma)))$$

Let $(\Omega_1,\underline{\underline{B}},\underline{\underline{B}}_s,x_s,P_{tx})$ be the solution of the sub-martingale problem starting at $x_t=x$, as stated in §1.

Let $g(\omega_1,\omega)$ be an $\underline{\underline{B}} \otimes \underline{\underline{F}}_T$ measurable random variable defined on $\Omega_1 \otimes \Omega$. We shall use the following notation :

$$\tilde{E}_{tx}^T\, g = E_{P_{tx}\otimes P}(g(\omega_1,\omega)/ \underline{\underline{F}}_T)$$

$$= \int_{\Omega_1} g(\omega_1,\omega)dP_{tx}(\omega_1)$$

We can now state the main result of this section :

<u>Theorem 3.2.</u> : The solution v of equation (3.1) satisfies :

$$(3.3)\qquad v(t,x) = \tilde{E}^{T}_{tx}[f(x_T)\exp\{\int_t^T [h(x_s,dw_s] - \frac{1}{2}\int_t^T |h(x_s)|^2 ds\}]$$

□

By the continuous dependence of P_{tx} on (t,x) [see STROOCK-VARADHAN [7]], Theorem 3.2 has the following corollary :

<u>Corollary</u> : v(t,x) is [almost everywhere equal to] a continuous function, and equality (3.3) is valid $\forall (t,x) \in [0,T]\times \bar{\mathcal{O}}$.

□

The rest of this section will be devoted to the proof of theorem 3.2.

First remark that it suffices to prove theorem 3.2 for continuous b and h. Using the Radon-Nikodym derivative of P_{tx} with respect to the solution of the submartingale problem with b=0, we can take the limit in (3.3) for $b_n \to b$ an $h_n \to h$ in measure, with the technique of STROOCK-VARADHAN [8] .

Thus, let us suppose for the rest of this section :

$$(3.4)\qquad b \in C_b([0,T]\times \mathcal{O};\ R^N)\ ,\ \forall T > 0$$

$$(3.5)\qquad h \in C_b([0,T]\times \bar{\mathcal{O}};\ R^d),\ \ \forall T > 0$$

For the same reason, we will suppose :

$$(3.6)\qquad f \in C^2_K(\bar{\mathcal{O}})$$

3.2. <u>A time discretized approximation.</u>

For m = 1,2,..., we consider the mesh $0 = t_o < t_1 < t_2 \ldots < t_m = T$ where $t_{i+1} - t_i = \frac{T}{m}$.

Define $a_m(t,x)$, $b_m(t,x)$, $h_m(t,x)$, $\rho_m(t,x)$, $\gamma_m(t,x)$ to be independent of t on each subinterval $[t_i,t_{i+1}[$, with :

$$a_m(t,x) = a^i_m(x)\ \ ,\ \forall t \in [t_i,\ t_{i+1}[\ ,$$

and similarly for b_m, h_m, ρ_m and γ_m, with the following properties :

(3.7) $\quad a_m^i \in C_b^2(\bar{\mathcal{O}};R^{N^2})$; $b_m^i, h_m^i \in C_b^2(\bar{\mathcal{O}};R^N)$; a_m^i satisfies (1.5)

(3.8) $\quad \rho_m^i \in C_b^2(\Gamma)$, $\underline{\rho} \le \rho_m^i(x) \le \bar{\rho}$

(3.9) $\quad \gamma_m^i \in C_b^2(\Gamma;R^N)$

We suppose that the following convergences hold, as $m \to +\infty$:

(3.10) $\quad a_m \to a$, $h_m \to h$ uniformly on $[0,T]\times\bar{\mathcal{O}}$, $\forall T > 0$

(3.11) $\quad \dfrac{\partial(a_m)_{ij}}{\partial x_i} \to \dfrac{\partial a_{ij}}{\partial x_i}$ in measure on $[0,T]\times\mathcal{O}$, $\forall T > 0$

$i,j=1,\ldots N$

(3.12) $\quad b_m \to b$ uniformly on each compact subset of $R_+\times\mathcal{O}$; the functions b_m being uniformly bounded.

(3.13) $\quad \rho_m \to \rho$, $\gamma_m \to \gamma$ uniformly on $[0,T]\times\Gamma$, $\forall T > 0$

Let us drop the subscript m until the end of the subsection. For $i=0,1,\ldots,m-1$, define :

$$L^i = \sum_{k,\ell=1}^{N} a_{k\ell}^i \frac{\partial^2}{\partial x_k \partial x_\ell} + \sum_{k=1}^{N} b_k^i \frac{\partial}{\partial x_k}$$

$$\Delta w^i = w_{t_i} - w_{t_{i-1}}$$

$$k = \frac{T}{m}$$

Consider the following equations :

$$(3.14)\quad \begin{cases} \frac{1}{k}v^i(x) - (L^i v^i)(x) = \frac{1}{k}v^{i+1}(x)(1+h^i(x)\,\Delta w^{i+1}) & x \in \mathcal{O} \\ \frac{1}{k}(\rho^i v^i)(x) - (\gamma^i.\nabla v^i)(x) = \frac{1}{k}(\rho^i v^{i+1})(x)(1+h^i(x)\Delta w^{i+1}), & x \in \Gamma \\ \text{for } i=0,1,\ldots m-1, \quad \text{and } v^m = f \end{cases}$$

Lemma 3.1. :

$\forall i$, v^i is a $W^{2,p}(\mathcal{O})$ -valued random variable.

Proof : From the results of AGMON-DOUGLIS-NIREMBERG [1] , the mapping :

$$(v^{i+1}, \Delta w^{i+1}) \to v^i$$

defined by equation (3.14) is continuous from $W^{2,p}(\mathcal{O}) \times R$ into $W^{2,p}(\mathcal{O})$.

□

We will use the following notation :

$$\tilde{E}^T_{ix} = E_{P^i_{ox} \otimes P}(\cdot / \underline{\underline{F}}_T)$$

where P^i_{ox} denote the solution to the submartingale problem associated with L^i, ρ^i and γ^i, starting at time o from point x.

Lemma 3.2. :

$v^i(x)$ satisfies the following relation :

$$(3.15) \qquad v^i(x) = \tilde{E}^T_{ix} \int_o^\infty \frac{1}{k} e^{-t/k} v^{i+1}(x_t)[1+h^i(x_t)\Delta w^{i+1}]dt$$

Proof : Let $\mathcal{O}_n = \{\phi(x) > -\frac{1}{n}\}$. We suppose that n is big enough such that $\Gamma_n = \partial\mathcal{O}_n = \{\phi(x) = -\frac{1}{n}\}$.

Choose $g \in W^{2,p}(\mathcal{O}_n)$, $p > N$, such that :

$$g(x) = v^{i+1}(x)[1+h^i(x)\Delta w^{i+1}], \quad \forall x \in \bar{\mathcal{O}}.$$

Let $a^i_n \in C^2_b(\bar{\mathcal{O}}_n;R^{N^2})$, $b^i_n \in C^2_b(\bar{\mathcal{O}}_n;R^N)$, a^i_n satisfying (1.5) on $\mathcal{O}_n$, such that :

$$a^i_n(x) = a^i(x), \quad b^i_n(x) = b^i(x), \qquad \forall x \in \bar{\mathcal{O}}$$

Choose $\rho^i_n \in C^2_b(\Gamma_n)$, $\gamma^i_n \in C^2_b(\Gamma_n;R^N)$

such that :

$$\exists K \ ; \ \forall n, \ \forall x \in \Gamma_n, \ y \in \Gamma,$$

$$(3.16) \qquad |\rho^i_n(x) - \rho^i(y)| + |\gamma^i_n(x) - \gamma^i(y)| \le K|x-y|$$

Let v_n be the solution of :

$$(3.17) \quad \begin{cases} v_n(x) - k(L_n^i v_n)(x) = g(x) , & x \in \mathcal{O}_n \\ \rho_n^i v_n(x) - k(\gamma_n^i.\nabla v_n)(x) = \rho_n^i(x)g(x), & x \in \Gamma_n \end{cases}$$

From (3.7), (3.8), and (3.9),

$$v_n \in W^{2,p}(\mathcal{O}_n) \cap C^2(\bar{\mathcal{O}}) \quad \text{a.s.}$$

(see for instance BENSOUSSAN-LIONS [2]).

Define :

$$F_n(t,x) = \bar{e}^{t/k} v_n(x)$$

$$\tau_r = \{\inf \ t; |x_t| \geq r\}$$

From theorem 1.1 it follows :

$$\tilde{E}_{ix}^T F_n(t\wedge\tau_r, x_{t\wedge\tau_r}) = F_n(0,x) +$$

$$+ \tilde{E}_{ix}^T \int_0^{t\wedge\tau_r} 1_{\mathcal{O}}(F_n' + L_n^i F_n)(s,x_s)ds +$$

$$+ \tilde{E}_{ix}^T \int_0^{t\wedge\tau_r} (\rho^i F_n' + \gamma^i.\nabla F_n)(s,x_s)d\xi_s^{ix}$$

$$\tilde{E}_{ix}^T F_n(t\wedge\tau_r, x_{t\wedge\tau_r}) = v_n(x)$$

$$(3.18) \quad - \tilde{E}_{ix}^T \int_0^{t\wedge\tau} \frac{1}{k} \bar{e}^{s/k} 1_{\mathcal{O}} g(x_s)ds -$$

$$- \tilde{E}_{ix}^T \int_0^{t\wedge\tau} \bar{e}^{s/k} [\tfrac{1}{k}(\rho^i v_n)(x_s) - (\gamma^i.\nabla v_n)(x_s)] \, d\xi_s^{ix}$$

But it is easy to show that :

$$v_n \to v^i \quad \text{in } W^{2,p}(\mathcal{O}) \text{ weakly}$$

$$v_n \to v^i \quad \text{in } W^{1,p}(\mathcal{O})$$

It follows that :

$$(3.19) \quad v_n(x) \to v^i(x) \text{ uniformly on each compact subset of } \bar{\mathcal{O}}.$$

$$(3.20) \qquad \left|\frac{\partial v^n}{\partial x_i}(x) - \frac{\partial v^n}{\partial x_i}(y)\right| \leq c\|v^n\|_{W^{2,p}(\mathcal{O}_n)} |x-y|^{1-\frac{N}{p}}$$

Choosing $x \in \Gamma$ and $y \in \Gamma_n$, and using (3.17), we conclude from (3.19) and (3.20) that we can take the limit in the last term of (3.18), yielding :

$$(3.21) \quad \begin{cases} v_n(x) = \tilde{E}^T_{ix}\, e^{-(t\wedge\tau_r)}\, v_n(x_{t\wedge\tau_r}) + \\ + \tilde{E}^T_{ix} \int_o^{t\wedge\tau_r} \frac{1}{k}\, e^{-s/k}\, g(x_s)[1_{\mathcal{O}} ds + \rho^i(x_s) d\xi^{ix}_s] \end{cases}$$

But,

$$1_{\mathcal{O}}(x_s)ds + \rho^i(x_s)d\xi^{ix}_s = ds$$

and on the other hand, we can take the limit in (3.21) for $r,n \to +\infty$, using Lebesgue dominated convergence theorem, and the fact that v^i is bounded a.s., by the maximum principle.

□

3.3. Proof of theorem 3.2.

We will prove (3.3) for $t = o$, to simplify the notations.

We will first express v^o in terms of $v^m = f$, using (3.15), and then take the limit, using the "invariance principle" of STROOCK-VARADHAN [7] .

First remark that, following the lines of §2.2, it is easy to prove :

$$(3.22) \quad \begin{cases} v^o \to v(0) \quad \text{in } L^2(\Omega\times\mathcal{O}) \text{ weakly} \\ \gamma_o v^o \to \gamma_o v(o) \text{ in } L^2(\Omega\times\Gamma) \text{ weakly} \end{cases}$$

Let us now introduce a few notations.

We are given a Markov chain with $\bar{\mathcal{O}}$ as its state space. The transition of this chain occur at times that are multiple of k.

Let :

$$\int_{\bar{\mathcal{O}}} f(y)\, \Pi^k_i(x,dy) = E_{P^i_{ox}} \int_o^{+\infty} \frac{1}{k}\, e^{-t/k}\, f(x_t)dt$$

denote the transition probability of this chain at time $(i+1)k$, $i=0,\ldots,N-1$.

Let P_x^k denote the measure corresponding to this chain starting from the point $x \in \bar{\mathcal{O}}$ at time 0 on the space $\Omega = D([0,T+\delta];\bar{\mathcal{O}})$. $\delta > 0$ [We want to consider transitions occuring at time T].

We will use the following notation :

$$\tilde{E}_{kx}^{T} = E_{P_x^k \times P}(\,\cdot\,/\,\underline{\underline{F}}_T\,)$$

It then follows from (3.15) :

$$v_m^o(x) = \tilde{E}_{kx}^{T}\{f(x_T)\prod_{i=1}^{m}(1+h^{i-1}(x_{ik})\Delta w^i\} \tag{3.23}$$

From the inequalities :

$$\exp\{\sum_{i=1}^{m}(\alpha_i - \frac{\alpha_i^2}{2}) \le \prod_{i=1}^{m}(1+\alpha_i) \le \exp\{\sum_{i=1}^{m}(\alpha_i - \frac{\alpha_i^2}{2} + \frac{\alpha_i^3}{3})\}$$

It can easily be shown that $v_m^o(x)$ has the same limit as :

$$\chi_m^x = \tilde{E}_{kx}^{T}[f(x_T)\exp\{\int_o^T [h(x_t, dw_t] - \frac{1}{2}\int_o^T |h(x_s)|^2 ds\}]$$

Let ψ be a $\mathcal{F}_T$ measurable real valued bounded random variable. We want to prove that :

$$E_{P_x^k \otimes P}(\chi_m^x\psi) \to E_{P_{ox} \otimes P}(\chi_m^x\psi)$$

But it is easy to check that :

$$x \to E_P(\chi_m^x\psi)$$

is continuous and bounded on $D([0,T+\delta];\bar{\mathcal{O}})$.

Then, in order to prove (3.3), it remains to show :

<u>Lemma 3.3.</u> : $P_x^k \to P_{ox}$ weakly

<u>Proof</u> : It suffices to show that we can apply the "invariance principle" of STROOCK-VARADHAN [7] .

Define :

$$\Delta^k(t,x) = \frac{1}{k}\int_{\bar{\mathcal{O}}} |y-x|^{2+\alpha}\Pi^k_{i-1}(x,dy), \quad ik \le t < (i+1)k$$

$$a^k(t,x) = \frac{1}{k}\int_{\bar{\mathcal{O}}} (y-x)\otimes(y-x)\ \Pi^k_{i-1}(x,dy)$$

$$ik \le t < (i+1)k$$

$$b^k(t,x) = \frac{1}{k}\int_{\bar{\mathcal{O}}} (y-x)\ \Pi^k_{i-1}(x,dy) \qquad ik \le t < (i+1)k$$

Simple calculations yield :

$$\Delta^k((i+1)k,x) = \frac{\alpha+2}{k}\, E^i_{ox}\int_o^{+\infty} e^{-t/k}\ 1_{\mathcal{O}}[|x_t-x|^{\alpha}\mathrm{tr}\ a^i(x_t) + \tag{3.24}$$

$$+ \alpha\frac{(a^i(x_t)(x_t-x),x_t-x)}{|x_t-x|^{2-\alpha}}]dt +$$

$$+ \frac{\alpha+2}{k}\, E^i_{ox}\int_o^{+\infty} e^{-t/k}\ 1_{\mathcal{O}}|x_t-x|^{\alpha} b^i(x_t)(x_t-x)dt +$$

$$+ \frac{\alpha+2}{k}\, E^i_{ox}\int_o^{+\infty} e^{-t/k}|x_t-x|^{\alpha}\gamma^i(x_t)(x_t-x)d\xi^{ix}_t$$

$$a^k((i+1)k,x) = \frac{1}{k}\, E^i_{ox}\int_o^{+\infty} e^{-t/k}\ 1_{\mathcal{O}} a^i(x_t)dt + \tag{3.25}$$

$$+ \frac{1}{k}\, E^i_{ox}\int_o^{+\infty} e^{-t/k}\ 1_{\mathcal{O}}[b^i(x_t)\otimes(x_t-x) +$$

$$+ (x_t-x)\otimes b^i(x_t)]dt +$$

$$+ \frac{1}{k}\, E^i_{ox}\int_o^{+\infty} e^{-t/k}[\gamma^i(x_t)\otimes(x_t-x) +$$

$$+ (x_t-x)\otimes\gamma^i(x_t)]d\xi^{ix}_t$$

$$b^k((i+1)k,x) = \frac{1}{k}\, E^i_{ox}\int_o^{+\infty} e^{-t/k}\ 1_{\mathcal{O}} b^i(x_t)dt + \tag{3.26}$$

$$+ \frac{1}{k}\, E^i_{ox}\int_o^{+\infty} e^{-t/k}\ \gamma^i(x_t)d\xi^{ix}_t$$

In order to prove the lemma, we need to state the following results :

Lemma 3.4. :

(3.27) $\sup_{(t,x)} \Delta^k(t,x) \to 0$ as $k \to 0$

(3.28) $\exists M$, s.t. $\sup_{t,x} \|a^k(t,x)\| \leq M$

(3.29) $a^k \to a$, $b^k \to b$ uniformly on compact subsets of $[0,T]\times \mathcal{O}$.

Lemma 3.5. :

In the case $\rho > 0$, the following hold :

(3.30) $\exists M$, s.t. $\sup_{t,x} |b^k(t,x)| \leq M$

(3.31) Given $t \in [0,T], y \in \Gamma$ and $\varepsilon > 0$, $\exists k_o > 0$ and $\delta_o > 0$ s.t. if $|t-s| < \delta_o, |x-y| < \delta_o$, $k < k_o$ and $(\nabla\phi(x), a^k(s,x)\nabla\phi(x)) < \delta_o$, then :

(i) $\|a^k(s,x)\| < \varepsilon$

(ii) $\|b^k(s,x) - \rho^{-1}(t,y)\gamma(t,y)\| < \varepsilon$

Lemma 3.6. :

In case $\rho \equiv 0$, the following hold :

(3.32) $\exists M$ and $c > 0$ s.t.
$\|b^k(t,x)\| > M \Rightarrow (\nabla\phi(x), b^k(t,x)) \geq c\|b^k(t,x)\|$

(3.33) $\forall\delta > 0$, $\exists M_\delta$ s.t.
$\|b^k(t,x)\| > M_\delta \Rightarrow \phi(x) < \delta$

(3.34) Given $t \in [0,T], y \in \Gamma$ and M, $\exists \delta_o > 0$ and $k_o > 0$ s.t. if $|t-s| < \delta_o, |x-y| < \delta_o$, $k < k_o$ and $(\nabla\phi(x), a^k(s,x)\nabla\phi(x)) < \delta_o$, then $\|b^k(s,x)\| \geq M$

(3.35) Given $t \in [0,T], y \in \Gamma$ and $\varepsilon > 0$, $\exists \delta_o > 0$, $k_o > 0$ and $N_o < \infty$ s.t. if $|s-t| < \delta_o$, $|x-y| < \delta_o$, $k < k_o$ and $\|b^k(s,x)\| > N_o$, then

$$\left\| \frac{b^k(s,x)}{(b^k(s,x),\nabla\phi(x))} - \frac{\gamma(t,y)}{(\gamma(t,y),\nabla\phi(y))} \right\| < \varepsilon$$

Define :

$$\tau_r^x = \inf\{t \le T \; ; \; |x_t - x| > r\}$$

The proof of the three preceding lemmas will make use of the following result :

Lemma 3.7. :

$\exists c$ and k_o s.t. if $k < k_o$,

$$\text{(i)} \ \frac{1}{k} E^i_{ox} \int_{\tau_r^x}^{+\infty} e^{-t/k} \, d\xi_t \le c \frac{k}{r}$$

$$\text{(ii)} \ \frac{1}{k} E^i_{ox} \int_{\tau_r^x}^{+\infty} e^{-t/k} dt \le c \frac{k^3}{r^2}$$

$$\text{(iii)} \ E^i_{ox}(\xi^{ix}(t))^2 \le c(t+t^2)$$

Proof : $\frac{1}{k} E^i_{ox} \int_{\tau}^{+\infty} e^{-t/k} dt = E^i_{ox}(e^{-\tau/k})$

$$= \int_o^{+\infty} e^{-t/k} \, dP^i_{ox}(\tau \le t)$$

$$= \int_o^{+\infty} hP^i_{ox}(\tau \le t) e^{-t/h} \, dt$$

From the proof of Lemma 3.3 in STROOCK-VARADHAN [7], it is easy to check that there exists a constant c_o, independent of t, i and x, such that :

$$E^i_{ox}([x_t - x]^2) \le c_o(t+t^2)$$

Moreover, using Doob's inequality, we get :

$$E^i_{ox}(\sup_{s\le t}|x_s-x|^2) \le c_1(t+t^2)$$

$$P(\tau \le t) = P(\sup_{s\le t}|x_s-x| > r) \le c_1 \frac{t+t^2}{r^2}$$

Then :

$$E^i_{ox}(\bar{e}^{\tau/k}) \le c_1 \frac{k^3+k^4}{r^2}$$

This proves the second inequality. The first one follows from :

$$\frac{1}{k} E^i_{ox}\int_\tau^{+\infty} \bar{e}^{t/k}\, d\xi^{ix}_t \le \frac{1}{k^2} E^i_{ox}\int_\tau^{+\infty} \bar{e}^{t/k}\xi^{ix}(t)dt$$

$$\le \frac{1}{k^{3/2}}(E^i_{ox}\ \bar{e}^{\tau/k})^{1/2} \times$$

$$\times\ (E^i_{ox}\int_o^{+\infty} \bar{e}^{t/h}(\xi^{ix}(t))^2dt)^{1/2}$$

and from the third inequality, which is an easy improvement of an estimate in Lemma 3.3 of STROOCK-VARADHAN [7].

□

The proof of lemma 3.4 and (3.30) follows easily from lemma 3.7.

Proof of lemma 3.5. : It remains to prove (3.31). From lemma 3.7, we can choose k_o such that, if $k < k_o$, the two last terms are less than any choosen ε.

Then, from the hypothesis $(\nabla\phi(x),a^k(s,x)\nabla\phi(x)) < \delta_o$, and lemma 3.7, we can choose r_o small enough such that, if $r < r_o$, $k < k_o$

$$\frac{1}{k} E^i_{ox}\int_o^{\tau^x_r} \bar{e}^{t/k}\ 1_{\mathcal{O}}(\nabla\phi(x),a^i(x_t)\nabla\phi(x))dt < 2\delta_o$$

Yielding, by the coercivity hypothesis on a^i :

$$\frac{1}{k} E^i_{ox}\int_o^{\tau^x_r} \bar{e}^{t/k}\ 1_{\mathcal{O}}(x_t)dt < \frac{\delta_o}{\alpha} \tag{3.36}$$

(i) then follows easily. Moreover, we deduce by similar arguments that $\forall\varepsilon'$, $\exists r_o$ and k_o s.t. if $r < r_o$, $k < k_o$:

$$|b^k(s,x) - \frac{1}{k} E^i_{ox}\int_o^{\tau_r^x} e^{-t/k} \frac{\gamma^i(x_t)}{\rho^i(x_t)} d\xi_t^{ix}| < \varepsilon'$$

And (ii) follows from (3.13).

□

Proof of lemma 3.6. : First prove(3.33), or equivalently :

$$(3.37) \qquad \phi(x) \geq \delta \Rightarrow |b^h(t,x)| \leq M_\delta$$

Indeed, $\exists\delta' > o$ s.t. $\phi(x) \geq \delta \Rightarrow d(x,\Gamma) \geq \delta'$

Then $\phi(x) \geq \delta \Rightarrow \xi^{ix}(\tau^x_{\delta'}) = 0$

The first term in $b^k(t,x)$ is bounded. The second one is equal to :

$$\frac{1}{k} E^i_{ox}\int_{\tau^x_{\delta'}}^{+\infty} e^{-t/k}\gamma^i(x_t)d\xi_t^{ix} \leq c' \frac{k}{\delta'} ,$$

from lemma 3.7. This proves (3.37).

$$(3.38) \qquad b^k(t,x) = \frac{1}{k} E^i_{ox}\int_o^{+\infty} e^{-t/k} b^i(x_t)dt + \\ + \frac{1}{k} E^i_{ox}\int_{\tau_r^x}^{+\infty} e^{-t/k}\gamma^i(x_t)d\xi_t^{ix} + \\ + \frac{1}{k} E^i_{ox}\int_o^{\tau_r^x} e^{-t/k} \gamma^i(x_t)d\xi_t^{ix}$$

The two first terms in (3.38) can be bounded by a constant depending only on $\frac{k}{r}$. Then, if $\frac{k}{r}$ is small enough, and $|b^k(t,x)| > M$, M big enough, the norm of the last term in (3.38) is larger than $\frac{1}{2}|b^k(t,x)|$. But, if moreover r is small enough,

$$\frac{1}{k} E^i_{ox}\int_o^{\tau_r^x} e^{-t/k}(\gamma^i(x_t),\nabla\phi(x))d\xi_t^{ix} \geq$$

$$\geq \Theta|\frac{1}{k} E^i_{ox}\int_o^{\tau_r^x} e^{-t/k} \gamma^i(x_t)d\xi_t^{ix}| \geq$$

$$\geq \frac{\Theta}{2} |b^k(t,x)|$$

And (3.32) then follows easily. (3.35) is easily shown with similar arguments.

But (3.34) is trivial, because, for k_o and δ_o sufficiently small,

$$(\nabla\phi(x), a^k(s,x)\nabla\phi(x)) \geq \delta_o \quad , \forall k \leq k_o$$

this is easy to check, using (1.5) and the fact that

$$\int_{s_1}^{s_2} 1_{\mathcal{O}}(x_t)dt = s_2 - s_1 \ , \ \forall\, s_1, s_2.$$

□

We have proved theorem 3.2.

<u>Remark 3.1.</u> : An alternate proof of theorem 3.2, in the case of homogeneous coefficients, would be to write (3.15) as :

$$v^i(x) = \bar{E}\,\tilde{E}^T_{ox}\, v^{i+1}(x_{k\tau_i})\{1+h(x_{k\tau_i})\Delta w^{i+1}\}$$

where $\tau_1 \ldots \tau_m$ are independent random variables, having the common probability distribution $d\bar{P}(t) = e^{-t}\, dt$. Then, (3.23) can be written as :

$$v^o(x) = \bar{E}\,\tilde{E}^T_{ox}(x_{t_m}) \prod_{j=1}^{m} (1+h(x_{t_j})\Delta w^j)$$

with

$$t_j = \frac{j}{N}\, t\, \frac{\tau_1 + \ldots + \tau_2}{j}$$

And the proof can be carried out, make use of the law of large numbers.

□

3.4. <u>A duality result.</u>

u denoting the solution of equation (2.3), and v the solution of equation (3.1), we have the following :

<u>Theorem 3.3.</u>

The continuous process :

$$(u(t),v(t)) + (\rho(t)u(t),v(t))_\Gamma$$

is P a.s. constant in time.

Proof : It is enough to prove :

$$(u(o),v(o))+(\rho(o)u(o),v(o))_\Gamma = (u(T),v(T) + (\rho(T)u(T),v(T))_\Gamma$$

with initial conditions $u(o)$ and $v(T)$ given in $H^1(\mathcal{O})$.

Consider (2.5), the time- discretization of (2.3) :

$$(3.24)\quad \begin{cases} (u^i,v)+(\rho^i u^i,v)_\Gamma + k\ a^i(u^i,v)+ k\ \langle D^i u^i,v\rangle_\Gamma \\ = (u^{i-1},v)+(\rho^i u^{i-1},v)_\Gamma + [(h^i u^{i-1},v)+(\rho^i h^i u^{i-1},v)_\Gamma,\Delta w^i] \\ u^o = u(o)\ ,\ i=1,2\ \dots\ n\quad \forall v \in H^1(\mathcal{O}) \end{cases}$$

and consider the analog discretization of equation (3.1) :

$$(3.25)\quad \begin{cases} (v^{i-1},u)+(\rho^{i-1}v^{i-1},u)_\Gamma + k\ a^{i-1}(u,v^{i-1})+k\langle D^{i-1}u,v^{i-1}\rangle_\Gamma = \\ = (v^i,u)+(\rho^i v^i,u)\ +[(h^i v^i,u)+(\rho^i h^i v^i,u)_\Gamma,\Delta w^i] \\ v^n = v(T)\ ,\quad i=n,\ n-1,\dots,1\quad \forall u \in H^1(\mathcal{O}) \end{cases}$$

Choosing $v = v^i$ in (3.24), $u = u^{i-1}$ in (3.25) and combining both equalities, we get :

$$(u^i,v^i+(\rho^i u^i,v^i)_\Gamma + k\ a^i(u^i,v^i)+ k\ \langle D^i u^i,v^i\rangle_\Gamma$$

$$= (u^{i-1},v^{i-1})+(\rho^{i-1}u^{i-1},v^{i-1})_\Gamma + k\ a^{i-1}(u^{i-1},v^{i-1}) +$$

$$+ k\ \langle D^{i-1}u^{i-1},v^{i-1}\rangle_\Gamma$$

By recurrence :

$$(3.26)\quad \begin{cases} (u^n,v(T)+(\rho^n u^n,v(T))_\Gamma + k\ a^n(u^n,v(T)) + k\ \langle D^n,u^n,v(T)\rangle_\Gamma \\ = (u(o),v^o)+(\rho^o u(o),v^o)_\Gamma + k\ a^o(u(o),v^o)+k\ \langle D^o u(o),v^o\rangle_\Gamma \end{cases}$$

where $\rho^o=\rho^1$, $a^o=a^1$, $D^o=D^1$

But ,

$$k\, a^n(u^n, v(T)) + k\, \langle D^n u^n, v(T)\rangle_\Gamma \le C \int_{T-\frac{T}{n}}^{T} \|u^n(s)\| ds$$

$$\le C\, \frac{T}{n} \|u^n\|_{L^2(0,T;H^1(\mathcal{O}))}$$

And the last term tends to zero in mean square, as $n \to +\infty$, using lemma 2.2.

Similarly, the two last terms of Equality (3.26) tend to zero, and the result follows by weak convergence.

□

4. CHARACTERIZATION OF THE CONDITIONAL DENSITY.

It suffices now to combine theorem 3.2. and 3.3, and compare with Bucy formula, to get the result.

We go back to the framework introduced in §1. Let us identify $(\Omega, \underline{\underline{F}}, \underline{\underline{F}}_t, P, w_t)$ of §2. and 3 with $(\Omega_2, \underline{\underline{F}}, \underline{\underline{F}}_t, \mathfrak{w}, y_t)$. From theorem 3.2, we deduce :

$$\int_{\mathcal{O}} p_o(x) v(o,x) dx = \tilde{E}[f(x_T) Z_T / \underline{\underline{F}}_T]$$

Comparing with theorem 3.3, we get :

$$(4.1) \quad \begin{cases} (f, u(T)) + (\rho f, u(T))_\Gamma = \tilde{E}[f(x_T) Z_T / \underline{\underline{F}}_T] \\ \forall f \in C_K^o(\bar{\mathcal{O}}) \end{cases}$$

From (4.1), we can easily show :

Lemma 4.1. :

(i) $\forall t \ge 0$, $u(t,x) \ge 0$ a.e. in $\mathcal{O}$

$\gamma_o u\,(t,x) \ge 0$ a.e. on Γ

(ii) $\int_{\mathcal{O}} u(t,x) dx + \int_\Gamma \rho(t,x) (\gamma_o u)(t,x) dx$

is a continuous process with values in $R_+ - \{o\}$.

□

We can now conclude, from (4.1) and lemma 1.1 - u being the solution of Equation (2.3), where w_t is replaced by y_t :

Theorem 4.1. :

The law of x_t conditionned by $\underline{\underline{F}}_t$, Σ_t, is the sum of two measures :

$$\Sigma_t = \Sigma_t^1 + \Sigma_t^2$$

where Σ_t^1 has a density with respect to Lebesgue measure in $\mathcal{O}$, and :

$$d\,\Sigma_t^1(x) = \frac{u(t,x)\,dx}{(u(t),1)+(\rho(t)u(t),1)_\Gamma} \quad ,$$

and Σ_t^2 is concentrated on the boundary Γ, and has a density with respect to Lebesgue measure on Γ, and :

$$d\,\Sigma_t^2(x) = \frac{(\gamma_o u)(t,x)\,d\Gamma(x)}{(u(t),1)+(\rho(t)u(t),1)_\Gamma} \quad \square$$

In other words, $\forall f \in B(\bar{\mathcal{O}})$, $\forall t \geq 0$

$$E[f(x_t)/\,\underline{\underline{F}}_t] = \frac{(u(t),f)+(\rho(t)u(t),f)_\Gamma}{(u(t),1)+(\rho(t)u(t),1)_\Gamma}$$

REFERENCES

[1] S. AGMON - A. DOUGLIS - L. NIREMBERG : "Estimates near the boundary for solutions of elliptic partial diffential equations satisfying general boundary conditions.I" Comm. on Pure and Appl. Math. XII, 623-727 (1959).

[2] A. BENSOUSSAN - J.L. LIONS : Book to appear.

[3] N.V. KRYLOV - B.L. ROZOVSKII : "On the conditional distribution of diffusion processes" . J. Ac. Sc. USSR, (1977) - in russian.

[4] J.L. LIONS : "Equations différentielles opérationnelles". Springer-Verlag, (1961).

[5] E. PARDOUX : "Equations aux dérivées partielles stochastiques non-linéaires monotones". Thèse Université Paris XI, (1975).

[6] E. PARDOUX : "Filtrage de diffusions avec conditions frontières : caractérisation de la densité conditionnelle". In : Colloque de Statistique dans les Processus Stochastiques, Grenoble. Lecture Notes in Math. Vol. 636, Springer-Verlag.

[7] D.W. STROOCK - S.R.S. VARADHAN : "Diffusion processes with boundary conditions". Comm. Pure and Appl. Math. XXIV 147-225, (1971).

[8] D.W. STROOCK - S.R.S. VARADHAN : "Diffusion processes with continuous coefficients". Comm. Pure and Appl. Math. XXII, 479-530 and 479-530, (1969).

-=-=-=-=-=-

LARGE DEVIATIONS FOR DIFFUSION PROCESSES

Mark A. Pinsky

Department of Mathematics
Northwestern University
Evanston, Illinois

Ia. INTRODUCTION

Let $(X(t),P_x)$ be a transient Markov process on a locally compact space S. We are interested in exponential estimates of the type $P_x(X(t) \in K) \sim e^{-\lambda t}$, where λ is independent of the compact set K. In order to formalize this we introduce the upper and lower escape rates:

$$-\lambda_+ = \sup_n \liminf_{t\to\infty} \frac{1}{t} \log P_x(X(t) \in K_n) \tag{1.1}$$

$$-\lambda_- = \sup_n \limsup_{t\to\infty} \frac{1}{t} \log P_x(X(t) \in K_n) \tag{1.2}$$

where $\{K_n\}_{n\geq 1}$ is an increasing sequence of compact sets whose union is S. In general $0 \leq \lambda_- \leq \lambda_+ \leq \infty$. Many known estimates can be expressed in terms of the positivity of λ_-.

These estimates can be illustrated in the simple case of a stochastic equation with constant coefficients:

$$dx = \sigma\, dw + q\, dt. \tag{1.3}$$

X(t) is normally distributed with mean $qt \neq 0$ and variance $\sigma^2 t > 0$. The known behavior of the normal distribution then shows that

$$\lim_{t\to\infty} \frac{1}{t} \log P\{|X(t)| \leq R\} = -\frac{q^2}{2\sigma^2}. \tag{1.4}$$

ISBN 0-12-268380-3

Hence $\lambda_- = \lambda_+ = \frac{q^2}{2\sigma^2}$.

We have made detailed estimations of λ_+, λ_- in two cases. The first case includes the Brownian motion on a 2-dimensional Riemannian manifold of negative curvature (2). Previously there had been obtained lower estimates for λ_- in terms of global bounds on the sectional curvature. In this work we obtain new upper bounds on λ_+ in terms of the curvature in a neighborhood of infinity.

The second case includes the class of linear stochastic equations first introduced by Hasminskii (3,4). When the origin is a stable equilibrium point we have a transient Markov process on $S = R^n \setminus \{0\}$ and we may discuss the estimates for $\lambda_\pm$. In striking contrast to the first case, these are expressed in terms of averages with respect to an invariant measure. The estimates obtained in this case are not believed to be as sharp as possible.

Ib. EXPLICIT DEFINITION OF THE ESCAPE RATES

In this section we will show that, under some regularity properties, $\lambda_\pm$ defined by (1.1)-(1.2) are well-defined independently of the starting point and the sequence of compact sets which appear in the definition.

We make the following technical hypotheses on the transition functions:

$$\exists\, t_1 > 0 \text{ such that } P_x\{X(t_1) \in U\} > 0 \text{ for each } x \in S \text{ and each open set } U \subseteq S. \tag{1.5}$$

$$\exists\, t_0 > 0 \text{ such that for each compact set } K, \text{ the functions } x \to \frac{1}{t}\log P_x\{X(t) \in K\} \text{ are equicontinuous for } t \geq t_0. \tag{1.6}$$

We now define for each compact set K

$$-\lambda_+(x,K) = \liminf_{t\to\infty} \frac{1}{t}\log P\{X(t) \in K\} \tag{1.7}$$

$$-\lambda_-(x,K) = \limsup_{t\to\infty} \frac{1}{t}\log P\{X(t) \in K\}. \tag{1.8}$$

Proposition. $\lambda_+(x,K)$, $\lambda_-(x,K)$ are independent of x.

Proof. Let $x_1 \in S$ be arbitrary and $\varepsilon > 0$. From (1.7) we have

$$\frac{1}{t} \log P_{x_1}\{X(t) \in K\} \geq -\lambda_+(x,K) - \varepsilon \qquad (t \geq t_2).$$

From (1.6), there exists a neighborhood $N_{x_1} \ni x_1$ such that

$$\frac{1}{t} \log P_y\{X(t) \in K\} \geq -\lambda_+(x_1,K) - 2\varepsilon \qquad (y \in N_{x_1}, \; t \geq t_3).$$

Now from the Chapman-Kolmogorev equation, we have

$$\begin{aligned} P_x(X(t_1 + t) \in K) &= \int_S P_x(X(t_1) \in dy) P_y(X(t) \in K) \\ &\geq \int_{N_y} P_x(X(t_1) \in dy) P_y(X(t) \in K) \\ &\geq \exp(-t(\lambda_+(x,K) + 2\varepsilon)) P_x\{X(t) \in N_y\}. \end{aligned}$$

From (1.5) the last quantity is positive. Therefore

$$\liminf_{t \to \infty} \frac{1}{t} \log P_x\{X(t_1 + t) \in K\} \geq -\lambda_+(x_1,K) - 2\varepsilon$$

for any $\varepsilon > 0$. Therefore $\lambda_+(x,K) \leq \lambda_+(x_1,K)$ for any $x \in S$, $x_1 \in S$. Thus $\lambda_+(x,K)$ is independent of x.

To prove the independence of $\lambda_-(x,K)$, we again use hypothesis (1.6): Let $\lambda_-(K) = \inf_{y \in S} \lambda_-(y,K)$. Then for each $\varepsilon > 0$ choose y_0 such that $\lambda_-(y,K) < \lambda_-(K) + \varepsilon$.

$$\frac{1}{t} \log P_y(X(t) \in K) \geq -\lambda_-(y,K) + 2\varepsilon \qquad (t = t_1 < t_2 < \ldots, y \in N_{y_0})$$

Then from the Chapman-Kolmogorev equation, we have

$$\begin{aligned} P_x(X(t_1 + t) \in K) &= \int P_x(X(t_1) \in dy) P_y(X(t) \in K) \\ &\geq e^{-t(\lambda_-(K) - 2\varepsilon)} P_x(X(t) \in N_{y_0}). \end{aligned}$$

Therefore

$$\limsup_{t \to \infty} \frac{1}{t} \log P_x\{X(t_1 + t) \in K\} \geq -\lambda_-(K) + 2\varepsilon$$

for any $\varepsilon > 0$. Therefore $\lambda_-(x,K) \leq \lambda_-(K) - 2\varepsilon$.

Hence $\lambda_-(x,K) \leq \inf_y \lambda_-(y,K)$, which proves that $\lambda_-(x,K)$ is independent of x.

We now show that $\lambda_\pm$ defined in (1.1)-(1.2) is independent of the sequence of compact sets $\{K_n\}$. For this purpose let $\{K_n\}$, $\{K'_n\}$ be two increasing sequences of compact sets whose union is S. Suppose that we can construct a subsequence $N_1 < N_2 < \cdots \to \infty$ and an increasing sequence K''_n such that

$$K''_i \subseteq K_{N_i} \subseteq K''_{i-1}, \qquad K''_i \subseteq K'_{N_i} \subseteq K''_{i+1}.$$

Then clearly $\lambda_\pm(\{K_n\}) = \lambda_\pm(\{K''_n\}) = \lambda_\pm(\{K'_n\})$.

To construct the required sequence, let N_1 be chosen so that $K_{N_1} \cap K'_{N_1} \neq \emptyset$. Let $K''_1 = K_{N_1} \cap K'_{N_1}$; now let $N_2 > N_1$ be such that $K_{N_2} \cap K'_{N_2} \supseteq K_{N_1} \cup K'_{N_1}$ and set $K''_{N_2} = K_{N_2} \cap K'_{N_2}$. Clearly $K''_2 \supset K_{N_1} \supset K''_1$, $K''_2 \supset K'_{N_1} \supset K''_1$. Proceeding in this way we can construct the required sequence.

Varadhan has shown that the equality $\lambda_+ = \lambda_-$ holds for a wide class of diffusion processes with not necessarily self adjoint generators. The key remark is to define σ as the supremum of all λ for which the equation $\mathfrak{L}u + \lambda u = 0$ has a positive solution. It is then shown that $\lambda_+ = \sigma = \lambda_-$. The details will appear elsewhere.

Ic. COMPARISON WITH LIMIT THEOREM IN STATISTICS

Let $\{Y_n\}_{n \geq 1}$ be a sequence of i.i.d. random variables with $E(Y_n) = m > 0$ and $E(e^{tY_n}) < \infty$ for some $t > 0$. Chernoff's theorem (1) asserts the existence of

$$-\lambda = \lim_{n \to \infty} \frac{1}{n} \log P\{Y_1 + \cdots + Y_n \leq n\mu\}$$

where $\mu < m$. If, for example, $\{Y_n\}_{n \geq 1}$ are normally distributed with variance $= \sigma^2$, then it is immediate that $\lambda = (\mu - m)^2/2\sigma^2$. This shows that the Chernoff estimate is sensitive to the dimension of the region under consideration.

The analogous statement in our case is

$$\lim_{n\to\infty} \frac{1}{n} \log P\{Y_1 + \cdots + Y_n \leq a\} = -\frac{m^2}{2\sigma^2}$$

which is independent of a.

II. DIFFUSION ON A MANIFOLD OF NEGATIVE CURVATURE

Let M be a two-dimensional complete simply connected Riemannian manifold of non-positive curvature. By the theorem of Cartan-Hadamard M is diffeomorphic to R^2 via the exponential mapping at some point $0 \in M$. Taking geodesic polar coordinates in the tangent space, we write the metric in the form

$$ds^2 = dr^2 + G^2(r,\theta)d\theta^2 \tag{2.1}$$

where $G(0,\theta) = 0$, $G_r(0,\theta) = 1$, $(G_{rr}/G)(r,\theta) = -K(r,\theta) \geq 0$. The Brownian motion may be constructed by solving Itô's stochastic equations

$$\begin{aligned} dr &= \sqrt{2}\, dw_1 + (G_r/G)dt \\ d\theta &= \sqrt{2}\, \frac{dw_2}{G} - (G_\theta/G^3)dt. \end{aligned} \tag{2.2}$$

Theorem. For the diffusion process (2.2), we have

$$\sqrt{4\lambda_-} \geq \inf \frac{G_r}{G} \tag{2.3}$$

$$\sqrt{4\lambda_+} \leq \inf_\theta \limsup_{r\to\infty} \frac{G_{rr}}{G_r}. \tag{2.4}$$

We give the proof in the isotropic case, i.e., $G(r,\theta) = G(r)$. Let $m_- = \inf_{r>0} G_r/G$, $m_+ = \limsup_{r\to\infty} G_{rr}/G_r$.

Proof of the lower bound. We must show that $\lambda_- \geq m_-^2/4$. To do this, we recall the stochastic integral equation

$$r(t) = r(0) + \sqrt{2}\, w_1(t) + \int_0^t \frac{G_r}{G}(r(s))ds.$$

By definition of m_-, we have $r(t) \geq r(0) + \sqrt{2}\, w_1(t) + m_- t$. Now if $K \subseteq [0,R]$ for some R, we have

$$\begin{aligned} P\{r(t) \in K\} &\le P\{r(t) \le R\} \\ &\le P\{r(0) + \sqrt{2}\, w_1(t) + m_- t \le R\} \\ &= P\left\{w_1(1) \le \frac{R - r(0) - m_- t}{\sqrt{2t}}\right\} \\ &\le e^{-(R - r(0) - m_- t)^2/4t}. \end{aligned}$$

Thus when $t \to \infty$, we have $\overline{\lim_{t\to\infty}} \frac{1}{t} \log P\{r(t) \in K\} \le -\frac{m_-^2}{4}$, for any compact set K. Hence $\lambda_- \ge m_-^2/4$ as required.

Proof of the upper bound.

Lemma 2.1. Assume that $G_{rr} \le mG_r$ for $R_0 \le r \le R_1$. Let λ_1 be the smallest eigenvalue of the problem

$$\phi_{rr} + \frac{G_r}{G}\phi_r + \lambda\phi = 0 \tag{2.5}$$

$$\phi(R_0) = 0 = \phi(R_1).$$

Then

$$\lambda_1 \le \frac{m^2}{4} + \frac{\pi^2}{(R_1 - R_0)^2}. \tag{2.6}$$

Proof. λ_1 can equivalently be characterized by the classical variational principle

$$\lambda_1 = \inf_{f \neq 0} \int_{R_0}^{R_1} f_r^2\, G\, dr \Big/ \int_{R_0}^{R_1} f^2\, G\, dr$$

where f satisfies the boundary conditions. Choose

$$f(r) = e^{-mr} \sin \frac{\pi(r - R_0)}{R_1 - R_0} \qquad (R_0 \le r \le R_1).$$

Then f satisfies the differential equation

$$f_{rr} + mf_r + \left[\frac{m^2}{4} + \frac{\pi^2}{(R_1 - R_0)^2}\right] f = 0.$$

Multiplying this by fG and integrating by parts, we have

$$-\int_{R_0}^{R_1} f_r^2\, G\, dr + \left[\frac{m^2}{4} + \frac{\pi^2}{(R_1 - R_0)^2}\right]\int_{R_0}^{R_1} f^2 G dr = -\int_{R_0}^{R_1} (G_{rr} - mG_r)\phi^2 dr.$$

But the right hand side is non-negative by hypothesis. Therefore

$$\int_{R_0}^{R_1} f_r^2 G\,dr \Big/ \int_{R_0}^{R_1} f^2 G\,dr \le \left[\frac{m^2}{4} + \frac{\pi^2}{(R_1 - R_0)^2}\right]$$

which proves the lemma.

Lemma 2.2. Assume that $G_{rr} \le mG_r$ for $R_0 \le r \le R_1$. Then

$$\liminf_{t \to \infty} \frac{1}{t} \log P\{R_0 \le r(t) \le R_1\} \ge -\left[\frac{m^2}{4} + \frac{\pi^2}{(R_1 - R_0)^2}\right].$$

To prove this, let ϕ be the solution of the equation (2.5) with $\lambda = \lambda_1$; ϕ is normalized so that $0 < \phi \le 1$ for $R_0 < r < R_1$ and $\phi'(R_0) > 0$, $\phi'(R_1) < 0$. ϕ is extended to be zero for $r < R_0$, $r > R_1$. Let ϕ_n be a non-negative C^1 function with

$$\phi_n(r) = \begin{cases} 0 & r \le R_0 - \frac{1}{n} \\ \phi(r) & R_0 + \frac{1}{n} \le r \le R_1 - \frac{1}{n} \\ 0 & r \ge R_1 + \frac{1}{n} \end{cases}.$$

In the intervals $A_n = (R_0 - \frac{1}{n}, R_0 + \frac{1}{n})$, $B_n = (R_1 - \frac{1}{n}, R_1 + \frac{1}{n})$ we interpolate ϕ_n as a quadratic polynomial. It can be verified that when $n \to \infty$, $\phi_n(r) \to \phi(r)$, $|\phi_n'(r)| \le$ const., $\phi_n''(r) \to |\phi'(r)|$ $(r = R_0, R_1)$. Applying Itô's formula to $e^{\lambda_1 t}\phi_n(r(t))$, it results that

$$E\{e^{\lambda_1 t}\phi_n(r(t))\} = \phi_n(r(0)) + E\int_0^t e^{\lambda_1 s}(\phi_n'' + \frac{G'}{G}\phi_n' + \lambda\phi_n)ds.$$

But ϕ_n satisfies the differential equation on the intervals complementary to $A_n \cup B_n$. On $A_n \cup B_n$ we have $|\phi_n'| \le$ const. Letting $n \to \infty$, we have

$$E\{e^{\lambda_1 t}\phi(r(t))\} = \phi(r(0)) + E\int_0^t e^{\lambda_1 s}|\phi'(r(s))|dA(s)$$

where $\{A(t), t \ge 0\}$ is local time for the process $\{r(t), t \ge 0\}$. Hence

$$E\phi(r(t)) \ge e^{-\lambda_1 t} E\int_0^T e^{\lambda_1 s}|\phi'(r(s))|dA(s) \qquad (t \ge T).$$

Recalling that $\phi(r) \leq I_{(R_0,R_1)}$, we see that

$$\liminf_{t \to \infty} \frac{1}{t} \log P\{R_0 \leq r(t) \leq R_1\} \geq -\lambda_1 \geq -\left[\frac{m^2}{4} + \frac{\pi^2}{(R_1 - R_0)^2}\right]$$

which proves lemma 2.2.

Proof of the upper estimate. Let $m_+ = \limsup_{r \to \infty} G_{rr}/G_r$. Given $\varepsilon > 0$, choose R_0 such that $G_{rr}/G_r \leq m_+ + \varepsilon$ for $r \geq R_0$. Then for any R_1 by lemma 2.2

$$\lambda_+((R_0,R_1)) \leq \frac{(m_+ + \varepsilon)^2}{4} + \frac{\pi^2}{(R_1 - R_0)^2} .$$

Letting $R_1 \to \infty$, we have $\lambda_+ \leq (m_+ + \varepsilon)^2/4$ for any $\varepsilon > 0$. Letting $\varepsilon \to 0$ completes the proof.

III. LINEAR STOCHASTIC EQUATIONS

In this section we estimate λ_-, λ_+ in the case of a system of linear stochastic equations. These are of the form

$$dX = AXdt + B_r X dw_r \tag{3.1}$$

where $(A, B_1, \ldots, B_\ell)$ are $n \times n$ matrices and $(w_1, \ldots, w_\ell)$ is an ℓ-dimensional Wiener process. We use the summation convention throughout.

$\{X(t),\ t \geq 0\}$ is a diffusion process on $S = R^n \setminus \{0\}$; the infinitesimal generator is

$$\mathcal{L}f = (Ax, \text{grad } f) + \tfrac{1}{2}(B_r x, \text{Hess } f\ B_r x). \tag{3.2}$$

We introduce the quantity

$$Q(x) = \frac{(Ax,x)}{|x|^2} + \frac{1}{2}\frac{|B_r x|^2}{|x|^2} - \frac{(B_r x,x)^2}{|x|^4} . \tag{3.3}$$

Clearly $Q(x) = Q(x/|x|)$. We now state the basic hypotheses:

$$(B_r x, \xi)^2 \geq \alpha |x|^2 |\xi|^2 \tag{3.4}$$

where $\alpha > 0$ and x, ξ are arbitrary vectors in R^n. We require the stability condition

$$\int_{|x|=1} Q(x)m(dx) = \nu < 0. \tag{3.5}$$

$m(dx)$ is a probability measure on the sphere S^{n-1} which is obtained in the following manner: $\mathcal{L}$ restricted to functions on S^{n-1} is a non-degenerate elliptic operator with a one-dimensional null space, consisting of constant functions. Its adjoint $\mathcal{L}^*$ has a one dimensional null space which consists of multiples of $m(dx)$. In order to state the main result we need some supplementary notation. Let

$$\bar{A}(x) = \frac{Ax}{|x|} - \frac{(Ax,x)x}{|x|^3} - \frac{|B_r x|^2}{2|x|^3} x + \frac{3}{2}\frac{(B_r x,x)^2}{|x|^5} x - \frac{(B_r x,x)^2 B_r x}{|x|^3} \tag{3.6}$$

$$\bar{B}_r(x) = \frac{B_r x}{|x|} - \frac{(B_r x,x)}{|x|^2} \tag{3.7}$$

$$\bar{\mathcal{L}}f = (\bar{A}(x), \text{grad } f) + \tfrac{1}{2}(\bar{B}_r(x), \text{Hess } f\ \bar{B}_r(x)). \tag{3.8}$$

$\bar{\mathcal{L}}$ is an operator on S^{n-1}, the unit sphere in R^n. More precisely, if f is a homogeneous function of degree zero, then $\bar{\mathcal{L}}f = \mathcal{L}f$ is also homogeneous of degree zero; if $\mathcal{L}f$ is a radial function, then $\bar{\mathcal{L}}f = 0$.

Hypothesis (3.4) implies that $\bar{\mathcal{L}}$ is a non-degenerate elliptic operator. Therefore the Fredholm alternative applies: if $g \in C^\infty(S^{n-1})$ satisfies the condition $\int_{S^{n-1}} g(x)m(dx) = 0$, then the equation $\bar{\mathcal{L}}f = -g$ has a solution $f \in C^\infty(S^{n-1})$. The solution is unique to within an additive constant. Applying this to $g = Q - \nu$ we have a function $h \in C^\infty(S^{n-1})$ which solves the equation

$$\bar{\mathcal{L}}h = -(Q - \nu). \tag{3.9}$$

We set

$$H^2 = \sup_{|x|=1} (\bar{B}_r(x), \text{grad } h + \gamma_r)^2 \tag{3.10}$$

$$s^2 = \sup_{|x|=1} \sigma_1^2(x) \tag{3.11}$$

$$n = \int_{|x|=1} \sigma_1^2(x) m(dx) \tag{3.12}$$

where $\sigma_1^2(x)$, γ_r are defined below. Then we have the following result.

Theorem. Let $(X(t),P_x)$ be the solution of (3.1), when (3.4),(3.5) are satisfied. Then

$$\lambda_+ \leq \frac{\nu^2}{2n} \tag{3.13}$$

$$\lambda_- \geq \frac{\nu^2}{2(H^2+s^2)} . \tag{3.14}$$

Proof of the upper bound. For this purpose we give an equivalent description of the process. Let $(\theta_1,\ldots,\theta_n)$ be the coordinates of a point on the unit sphere. By solving for θ_n, we may first write the stochastic equations in the form

$$\begin{aligned} d\theta_i &= \beta_{ir}(\theta)dw_r + \alpha_i(\theta)dt, \qquad (1 \leq i \leq n-1) \\ d\rho &= (B_r\theta,\theta)dw_r + Q(\theta)dt. \end{aligned} \tag{3.15}$$

From hypothesis (3.4), the following symmetric matrix is positive definite:

$$\mathfrak{A} = \left(\begin{array}{c|c} (\beta\beta^t)_{ij} & \beta_{ir}(B_r\theta,\theta), \quad 1 \leq i \leq n-1 \\ \hline X & (B_r\theta,\theta)^2 \end{array} \right)$$

By successive algebraic reduction, we may write $\mathfrak{A} = \Sigma\Sigma^t$, where Σ is triangular:

$$\Sigma = \left[\begin{array}{c|c} \tilde{\beta} & \begin{matrix} 0 \\ \vdots \\ 0 \end{matrix} \\ \hline \gamma_1(\theta)\cdots\gamma_{n-1}(\theta) & \sigma_1(\theta) \end{array} \right]$$

Let $(\tilde{w}_1,\ldots,\tilde{w}_n)$ be a new Brownian motion process. The desired representation is then

$$d\theta_i = \tilde{\beta}_{ir}(\theta)d\tilde{w}_r + A_i(\theta)dt \qquad (1 \le i \le n-1) \tag{3.16}$$

$$d\rho = \sigma_1(\theta)d\tilde{w}_n + \gamma_r(\theta)d\tilde{w}_r + Q(\theta)dt. \tag{3.17}$$

In (3.16), r is summed from 1 to n -1. $\sigma_1^2(\theta)dt$ is the conditional variance of $d\rho$, given $\tilde{w}_1(t),\ldots,\tilde{w}_n(t)$. It may be computed by the formula $\sigma_1^2(\theta) = \sigma^2(\theta) - \langle\gamma, (\tilde{\beta}\tilde{\beta}^t)^{-1}\gamma\rangle$ (5, p. 28).

It is readily verified that $(\theta(t),\rho(t)))$ is a diffusion process with generator

$$\tilde{\mathfrak{L}}f = \frac{1}{2}\sigma^2(\theta)\frac{\partial^2 f}{\partial\rho^2} + Q(\theta)\frac{\partial f}{\partial\rho} + \beta_i\frac{\partial^2 f}{\partial\rho\partial\theta_i} + (\bar{\mathfrak{L}}f)$$

which is the form of $\mathfrak{L}f$ in radial coordinates (ρ,θ). Therefore, to prove the theorem we may work with the solution of the stochastic system (3.16)-(3.17). Applying Itô's formula to $\rho + h(\theta)$, we have

$$\rho(t) + h(\theta(t)) = \rho(0) + h(\theta(0)) + \nu t + M_1(t) + M_2(t)$$

where

$$M_1(t) = \int_0^t \sigma_1(\theta(s))dw_1(s)$$

$$M_2(t) = \int_0^t [\bar{B}_r(\theta(s))\text{grad } h(\theta(s)) + \gamma_r(\theta(s))]dw_r(s)$$

$$P(t) = h(\theta(t)) - h(\theta(0)).$$

The processes $M_1(t)$, $M_2(t)$ are subject to the following limit theorems, with probability one:

$$\frac{1}{t}\langle M_1(t)\rangle \to \int_{|x|=1}\sigma_1(x)^2 m(dx) \qquad (t\to\infty)$$

$$\frac{1}{t}M_2(t) \to 0 \qquad (t\to\infty)$$

Proof of the lower bound. Thus for any $R > 0$, $\varepsilon > 0$ we have

$$
\begin{aligned}
P\{|X(t)| > R\} &= P\{\rho(t) > \log R\} \\
&= P\{M_1(t) > \nu t + \log R + P(t) - M_2(t)\} \\
&= E\left\{1 - \operatorname{erf}\left[\frac{\nu t + \log R + P(t) - M_2(t)}{\langle M_1(t)\rangle^{\frac{1}{2}}}\right]\right\} \\
&\geq E\left\{1 - \operatorname{erf}\left[\frac{\nu t + \log R + P(t) - M_2(t)}{\langle M_1(t)\rangle^{\frac{1}{2}}}\right];\right. \\
&\qquad \left.|M_2(t)| \leq \varepsilon t, |\langle M_1(t)\rangle^{\frac{1}{2}} - \nu t| \leq \varepsilon\right\}.
\end{aligned}
\tag{3.18}
$$

We use these estimates together with the inequality $\sqrt{2\pi}\,(1 - \operatorname{erf} x) \geq \frac{1}{2x} e^{-x^2/2}$ $(x \geq 2)$. Thus for any $\varepsilon > 0$ and t sufficiently large, we have

$$P\{|X(t)| > R\} \geq \frac{e^{-x^2/2}}{2x\sqrt{2\pi}} P\{|M_2(t)| \leq \varepsilon t, |(M_1(t))^{\frac{1}{2}} - nt| \leq \varepsilon t\}$$

where $x = [\nu t + 2\|h\|_\infty + \log R + \varepsilon t]/(nt(1-\varepsilon))^{\frac{1}{2}}$. Therefore

$$\liminf_{t \to \infty} \frac{1}{t} \log P\{|X(t)| > R\} \geq -\frac{(\nu + \varepsilon)^2}{2n(1-\varepsilon)}.$$

Since this is true for every $\varepsilon > 0$, we have $\lambda_+ \leq \nu^2/2n$ as required.

Proof of the lower bound. Return to equation (3.18) and use the estimate $\langle M_1(t)\rangle \leq ts^2$, $s^2 = \max_\theta \sigma^2(\theta)$. Thus

$$P\{|X(t)| \geq R\} \leq E\left\{1 - \operatorname{erf} \frac{\nu t + \log R + P(t) - M_2(t)}{s\sqrt{t}}\right\}.$$

But $M_2(t)$ is a martingale with mean zero and variance $\int_0^t |\bar{B}_r(\theta(s))\operatorname{grad} h(\theta(s)) + \gamma_r(\theta(s))|^2 ds \leq H^2 t$. Thus $P\{M_2(t) > x\} \leq e^{-x^2/2tH^2}$. Applying this above, we have

$$
\begin{aligned}
P\{|X(t)| > R\} &\leq \int_{-\infty}^{\infty} \frac{e^{-(rt + C - x)^2/2s^2 t}}{s\sqrt{t2\pi}} e^{-x^2/2tH^2}\, dx \\
&= \text{const.}(t^{\frac{1}{2}}) \exp\left(-\frac{(rt + C)^2}{2t(s^2 + H^2)}\right).
\end{aligned}
$$

This shows that

$$\limsup_{t \to \infty} \frac{1}{t} \log P(|X(t)| > R) \leq -\frac{\nu^2}{2(s^2 + H^2)}$$

as required.

REFERENCES

1. Bahadur, R. R., Some Limit Theorems in Statistics, SIAM 1971.
2. Malliavin, P., Diffusion et Geométrie Différentielle Globale, Montreal 1978.
3. Papanicolaou, G. C., and Blankenship, G., Stochastic Stability and Control for Systems with Wide Band Noise Disturbances, SIAM Journal of Mathematical Analysis, 1978.
4. Pinsky, M., Stochastic Stability and the Dirichlet Problem, Communications in Pure and Applied Mathematics 27 (1974), 311-350.
5. Anderson, T. W., Multivariate Statistical Analysis, Wiley, 1958.

IMPULSIVE CONTROL PROBLEMS DEPENDING ON A PARAMETER

Maurice Robin
IRIA
Rocquencourt - France

INTRODUCTION

We consider optimal stopping and impulsive control problems for a broad class of markov processes, when the data, namely the cost structure and the semi group, depend on a parameter. We will use the results of [6] for the characterizations of the optimal cost function and for the existence of an optimal control. The dependency w.r.t. a parameter for the impulsive control of a diffusion process was considered by BENSOUSSAN-LIONS [1], [2]. We will describe here, as one application, the approximation of the impulsive control for a diffusion process by a similar problem for a pure jump markov process.

1 - NOTATIONS AND ASSUMPTIONS

Let E be a metric locally compact space, $\Omega = D(o,\infty;E)$ the space of right continuous, left limited functions from R^+ with values in E.

Let $x_t(\omega) = \omega(t)$ for $\omega \in \Omega$, θ_t the translation operator on Ω,

$\mathcal{F}^o_t$ the σ-field generated by $\{x_t , o \leq s \leq t\}$, $\mathcal{F}^o = \mathcal{F}^o_\infty$,

$\mathcal{F}_t$, $\mathcal{F}$ the universally completed σ-field obtained from $\mathcal{F}^o_t$, $\mathcal{F}^o$.

We assume a markov process is given, depending on $\lambda \in R^+$:

ISBN 0-12-268380-3

$$X_\lambda = \{\Omega, \mathcal{F}_t, x_t, P_x^\lambda\}. \tag{1.1}$$

Let $\phi^\lambda_{(t)}$ be the corresponding semi group.

We assume that:

(i) $\phi^\lambda_{(t)}$ is a Feller semi group, that is, if C is the space of bounded continuous functions on E :

$$\left| \begin{array}{l} \phi^\lambda_{(t)} f \in C \quad \forall f \in C, \quad t > 0 \\ \lim\limits_{t \to o} \phi^\lambda_{(t)} f(x) = f(x) \quad \forall f \in C, \forall x \in E \end{array} \right. \tag{1.2}$$

and that X is quasi left continuous.

(ii) $P_x^\lambda[\sup\limits_{o \le t \le T} |x_t - x| > R] \le \gamma_T(R)$ (1.3)

where $\lim\limits_{R \uparrow \infty} \gamma_T(R) = 0$ for fixed T

$\lim\limits_{T \downarrow o} \gamma_T(R) = 0$ for fixed R

uniformly w.r.t. λ, x.

We will say that "$g^\lambda \to g$ in C_k" if g^λ converges to g uniformly on compact subsets of E.

In the following sections, we will assume also that are given $\lambda \in [0,1]$:

$$\left\{ \begin{array}{l} f^\lambda,\ \psi^\lambda \in C,\ f, \psi \in C, \quad \psi^\lambda, \psi \ge 0^{(1)},\ \|f^\lambda\|,\ \|\psi^\lambda\| \le c^t \\ f^\lambda \to f \text{ in } C_K \\ \psi^\lambda \to \psi \text{ in } C_K \end{array} \right. \tag{1.4}$$

and moreover it is assumed that :

$$\left\{ \begin{array}{l} \text{if } g^\lambda \to g \text{ in } C_K,\ ||g^\lambda|| \le \text{constant.} \\ \text{Then, for every } t \ge 0 \\ \phi^\lambda(t) g^\lambda \to \phi(t) g \text{ in } C_K. \end{array} \right. \tag{1.5}$$

---

(1) This last assumption is not essential but will simplify a little bit the proofs.

2 - OPTIMAL STOPPING PROBLEM

For any $\mathcal{F}_t$ stopping time τ, let us define :

$$\begin{cases} J_x^\lambda(\tau) = E_x^\lambda\{ \int_o^\tau e^{-\alpha s} f^\lambda(x_s)ds + e^{-\alpha\tau} \psi^\lambda(x_\tau)\} \\ J_x(\tau) = E_x\{ \int_o^\tau e^{-\alpha s} f(x_s)ds + e^{-\alpha\tau} \psi(x_\tau)\} \end{cases} \tag{2.1}$$

and :

$$u^\lambda(x) = \operatorname*{Inf}_\tau J_x^\lambda(\tau), \quad u(x) = \operatorname*{Inf}_\tau J_x^\lambda(\tau). \tag{2.2}$$

In order to study the convergence u^λ to u, we will need some results on optimal stopping problem .

Consider the "penalized problem" :

$$u^\varepsilon(x) = \int_o^\infty e^{-\alpha s} \phi(s)[f - \frac{1}{\varepsilon}(u^\varepsilon-\psi)^+]ds \tag{2.3}$$

Then we have the following:

<u>Theorem 1.</u> <u>Under the assumption of § 1,</u>

(i) <u>(2.3) as a unique solution $u^\varepsilon \in C$</u>

(ii) $u^\varepsilon(x) = \operatorname*{Inf}_v E_x \int_o^\infty e^{-\alpha s} e^{-\frac{1}{\varepsilon}\int_o^s v(r)dr} [f + \frac{1}{\varepsilon} v\, \psi]ds$

where the infinum is taken on adapted process v with values in [0,1].

(iii) $u^\varepsilon \to u$ in C_K.

<u>Proof.</u> We only give the steps of the proof which are useful for the present problem. Details and more general results can be found in [6], and in [1] for the case of diffusion processes.

For proof of (i) use the non-stationary analogue of (2.3) :

$$u_T^\varepsilon(x) = \int_t^T e^{-\alpha(s-t)}\,\phi(s-t)[f - \frac{1}{\varepsilon}(u_T^\varepsilon(s) - \psi)^+]ds \tag{2.4}$$

$$u_T^\varepsilon \in C^o([0,T];\ C).$$

(2.4) is solved by application of the fixed point theorem of contraction mappings.

Then one can show that :

$$||u_T^\varepsilon(t) - u^\varepsilon||_C \leq e^{-\alpha(T-t)}\,\frac{1}{\alpha}(||f|| + ||\psi||)$$

where u^ε is defined by (ii), and therefore $T \uparrow \infty$ shows that this u^ε is solution of (2.3). Uniqueness follows from the stochastic interpretation (ii).

Let us now outline the proof of (iii) which will give an estimate used below.

Defining $\tau^\varepsilon = \text{Inf}(s \geq 0,\ u^\varepsilon(x_s) \geq \psi(x_s))$, and :

$$v_\varepsilon(t) = \begin{cases} 0 & \text{for } t < \tau^\varepsilon \\ 1 & \text{for } t \geq \tau^\varepsilon \end{cases}$$

one has that v_ε is optimal for $J_x^\varepsilon(v)$, and moreover that $u^\varepsilon(x) = J_x^\varepsilon(v_\varepsilon) = J_x(\tau^\varepsilon)$. Therefore $u^\varepsilon \geq u$.

Let τ an arbitrary stopping time, and :

$$v_\tau(t) = \begin{cases} 0 & t < \tau \\ 1 & t \geq \tau \end{cases}$$

Now we look for an estimate of :

$$Y = J_x^\varepsilon(v_\tau) - J_x(\tau).$$

Let $h > 0$, one has $Y = Y_1 + Y_2 + Y_3$ where :

$$Y_1 = E_x \int_\tau^{+\infty} e^{-\alpha s}\, e^{-\frac{1}{\varepsilon}(s-\tau)}\, f\, ds$$

$$Y_2 = E_x \int_{\tau+h}^{+\infty} e^{-\alpha s}\, e^{-\frac{1}{\varepsilon}(s-\tau)}\, \frac{1}{\varepsilon}\, \psi(x_s)ds$$

$$Y_3 = E_x\left(\int_\tau^{\tau+h} e^{-\alpha s}\, e^{-\frac{1}{\varepsilon}(s-\tau)}\, \frac{1}{\varepsilon}\, \psi(x_s)ds - e^{-\alpha\tau}\, \psi(x_\tau)\right).$$

One has :

$$|Y_1| \le \varepsilon\, ||f||,$$

$$|Y_2| \le e^{-h/\varepsilon}||\psi||,$$

$Y_3 = Y_3' + Y_3''$ where

$$Y_3' = E_x \int_\tau^{\tau+h} e^{-\alpha s}\, e^{-\frac{1}{\varepsilon}(s-\tau)}\, \frac{1}{\varepsilon}\, [\psi(x_s) - \psi(x_\tau)]\, ds$$

$$Y_3'' = E_x\, \psi(x_\tau)\, \left[\int_\tau^{\tau+h} e^{-\alpha s}\, e^{-\frac{1}{\varepsilon}(s-\tau)}\, \frac{1}{\varepsilon}\, ds\right] e^{-\alpha\tau}.$$

Then $\psi \ge 0$ implies $Y_3'' \le 0$, and, one has also :

$$|Y_3'| \le E_x\, e^{-\alpha\tau} \sup_{\tau\le s\le\tau+h} |\psi(x_s) - \psi(x_\tau)|$$

or, by the strong markov property :

$$|Y_3'| \le E_x\, e^{-\alpha\tau} E_{x_\tau} \sup_{o\le s\le h} |\psi(x_s) - \psi(x_o)|.$$

Let $Z = \sup_{o\le s\le h} |\psi(x_s) - \psi(x_o)|$, and $T > 0$, then :

$$|Y_3'| \le E_x\, e^{-\alpha\tau}\, \chi_{\tau\le T}\, Z + E_x\, e^{-\alpha\tau}\, \chi_{\tau>T}\, Z = I + II^{(2)}.$$

$$II \le 2e^{-\alpha T}||\psi||.$$

Now if $A = \{\omega, \sup_{o\le t\le T} |x_t - x| \ge R\}$:

$$I = E_x\, e^{-\alpha\tau}\, \chi_{\tau\le T}\, \chi_A\, Z + E_x\, e^{-\alpha\tau}\, \chi_{\tau\le T}\, \chi_{\bar{A}}\, Z = III + IV.$$

By (1.5) one has :

---

(2) $\chi_{\tau\le T} = 1$ if $\tau \le T$, 0 otherwise.

$$III \leq ||\psi|| \frac{c.T}{R^2} .$$

One has also :

$$P_x(\sup_{o \leq s \leq h} |x_s - x| > \delta) \leq \frac{c.h}{\delta^2} ,$$

so, we introduce:

$$B = \{\omega \in \Omega, \sup_{o \leq s \leq h} |x_s - x_o| \leq \delta\}$$

and dividing IV with B and $\bar{B}$, one obtains :

$$IV \leq ||\psi||.c.\frac{h}{\delta^2} + E_x \; e^{-\alpha\tau} \; \chi_{\tau \leq T} \; \chi_{\bar{B}} \; Z.$$

For K compact in E, we define :

$$K_R = \{y = z + z^1 \; |z| \leq R, \; z^1 \in K\}$$

$$K_{R}, \delta_o = \{ \; y = z_1 + z_2, \; z_1 \in K_R, |z_2| \leq \delta_o\} \text{ for } \delta_o > 0$$

and :

$$\rho_\psi(\delta) = \sup_{\substack{x', x' \in K_R, \delta_o \\ |x' - x''| \leq \delta}} |\psi(x') - \psi(x'')| \text{ for } \delta \leq \delta_o.$$

Then :

$$IV \leq ||\psi||.c.\frac{h}{\delta^2} + \rho_\psi(\delta).$$

Putting together the estimates we obtain:

$$\sup_{x \in K} |u^\varepsilon(x) - u(x)| \leq \varepsilon \; ||f|| + ||\psi||.[e^{-h/\varepsilon} + 2e^{-\alpha T} + \frac{cT}{R^2} + \frac{ch}{\delta^2}] + \rho_\psi(\delta) \tag{2.5}$$

which gives (iii) of the theorem. ■

Remark. The proof in [6] is slightly different and is based on approximation of ψ by smooth functions. But it does not give the estimate (2.5) which we need in the following.

Theorem 2. Under the assumptions of § 1, $u^\lambda \to u$ in C_K.

Proof. We begin by the proof of :

$$u^{\lambda,\varepsilon} \to u^\varepsilon \text{ in } C_K. \tag{2.6}$$

First if $u_T^{\lambda,\varepsilon}$ is the solution of (2.4) for ϕ^λ, f^λ, ψ^λ, we have :

$$||u_T^{\lambda,\varepsilon}(t) - u^{\lambda,\varepsilon}|| \le e^{-\alpha(T-t)} . \frac{1}{\alpha}(||f^\lambda|| + ||\psi^\lambda||).$$

Therefore $u_T^{\lambda,\varepsilon} \to u^{\lambda,\varepsilon}$ uniformly in E, uniformly w.r.t λ,ε, when $T\uparrow\infty$.

An the other hand, $u_T^{\lambda,\varepsilon}$ is the unique fixed point of the map Π^λ from $C^o([0,T];C)$ in itself defined by :

$$w = \Pi^\lambda v = \int_t^T e^{-\alpha(s-t)} \phi^\lambda(s-t)[f^\lambda - \frac{1}{\varepsilon}(v-\psi^\lambda)^+]ds.$$

That is :

$$u_T^{\lambda,\varepsilon} = \text{Lim } v^o + \sum_{j=1}^n (v_\lambda^j - v_\lambda^{j-1}) \ , \ v_\lambda^n = \Pi^\lambda \ v_\lambda^{n-1}$$

and :

$$||v_\lambda^j - v_\lambda^{j-1}|| \le c_\varepsilon . \frac{T^{j-1}}{(j-1)!} ||v_o - \Pi^\lambda v_o||.$$

Therefore :

$$||u_T^{\lambda,\varepsilon} - v_\lambda^n|| \le C_{\varepsilon,T} . k_n$$

where $k_n \to 0$ uniformly w.r.t. λ.

Now, one shows that :

$$v_\lambda^n(t) \to v^n(t) \text{ in } C_K \text{ when } \lambda \to 0.$$

This is clear for n = 1 by (1.5), and the use of (1.5) gives this by induction.

Therefore :

$$||u^{\lambda,\varepsilon} - u^\varepsilon||_K \le ||u^{\lambda,\varepsilon} - u_T^{\lambda,\varepsilon}|| + ||u_T^{\lambda,\varepsilon}(t) - v_\lambda^n(t)||$$

$$+ ||v_\lambda^n(t) - v^n(t)||_K + ||v^n(t) - u_T^\varepsilon(t)||$$

$$+ ||u_T^\varepsilon(t) - u^\varepsilon||$$

when $\lambda \to 0$, $n \to \infty$, $T \to \infty$ successively, we obtain (2.6).

Now, for any compact K of E :

$$||u^\lambda - u||_K \le ||u^\lambda - u^{\lambda,\varepsilon}||_K + ||u^{\lambda,\varepsilon} - u^\varepsilon||_K + ||u^\varepsilon - u||_K.$$

Recalling (2.5) we have :

$$||u^\lambda - u^{\lambda,\varepsilon}||_K \le \varepsilon||f^\lambda|| + ||\psi^\lambda||[e^{-h/\varepsilon} + 2e^{-\alpha T} + \frac{cT}{R^2} + \frac{ch}{\delta^2}]$$

$$+ \rho_{\psi^\lambda}(\delta) ,$$

but :

$$\rho_{\psi^\lambda}(\delta) \le 2 \sup_{y \in k_{R,\delta_o}} |\psi^\lambda(y) - \psi(y)| + \rho_\psi(\delta).$$

Therefore :

$$||u^\lambda - u||_K \le ||u^{\lambda,\varepsilon} - u^\varepsilon||_K + 4\, ||\psi^\lambda - \psi||_{K_{R,\delta_o}}$$

$$+ 2\, c_1[e^{-h/\varepsilon} + 2e^{-\alpha T} + \frac{cT}{R^2} + \frac{ch}{\delta^2}] + \rho_\psi(\delta),$$

and $\lambda \to 0$, $\varepsilon \to 0$, $R \to \infty$, $T \to \infty$, $h \to 0$, $\delta \to 0$ give the result. ■

3 - IMPULSIVE CONTROL PROBLEM

We begin by giving some results on a type of impulsive control problem for Feller-markov processes.

Detailed proofs and others cases can be found in [6]. In addition to the data and assumptions of § 1, we take :

$$\left| \begin{array}{l} U \text{ compact in } E = \mathbb{R}^n \text{(1)} \\ f \in C,\ f \geq 0 \\ c \in C^o(U),\ c(\xi) \geq k > 0 \qquad \forall \xi \in U \end{array} \right. \tag{3.1}$$

and we set :

$$M\varphi(x) = \operatorname*{Inf}_{\xi \in U} [c(\xi) + \varphi(x+\xi)]. \tag{3.2}$$

We consider u^λ maximum solution of the set of inequalities :

$$\left| \begin{array}{l} u^\lambda \leq M u^\lambda \\ u^\lambda(x) \leq e^{-\alpha t} \phi^\lambda(t) u^\lambda(x) + \int_o^t e^{-\alpha s} \phi^\lambda(s) f(x) ds \\ u^\lambda \in C \end{array} \right. \tag{3.3}$$

and u maximum solution of the similar problem, for $\phi(t)$ instead of ϕ^λ.

It follows from [6] chapter 5 that u^λ, u exist and are the optimal cost of impulsive control problems (inventory - control - type) and we have the following additional results.

$u^\lambda(x)$ is the unique solution of the equation :

$$u^\lambda(x) = \operatorname*{Inf}_{\tau} E_x^\lambda \left[\int_o^\tau e^{-\alpha s} f^\lambda(x_s) ds + e^{-\alpha\tau} M u^\lambda(x_\tau) \right] \tag{3.4}$$

and a similar equality holds for u.

---

(1) In fact it is enough to assume that E is a vector space.

We have also (see [6] chapter 5) by adaptation of an idea of J.L. Menaldi [5].

Lemma 3.1. Define

$$u^{n,\lambda}(x) = \operatorname*{Inf}_{\tau} E_x^\lambda \left[\int_0^\tau e^{-\alpha s} f(x_s)ds + e^{-\alpha\tau} M\, u^{n-1,\lambda}(x_\tau)\right] \quad (3.5)$$

$$u^{o,\lambda}(x) = E_x^\lambda \int_0^\infty e^{-\alpha s} f(x_s)ds.$$

Then :

$$||u^{n,\lambda} - u^\lambda|| \leq \frac{C}{k.n}, \quad (3.6)$$

where c depends only on α, $||f||$.

It is now easy to show:

Theorem 3.1. Under the assumptions of § 1 and 3.1, $u^\lambda \to u$ in C_k when $\lambda \to 0$.

Proof. Let us prove first that $u^{n,\lambda} \to u^n$ in C_K for any fixed n, when $\lambda \to 0$.

We have clearly $u^{o,\lambda} \to u^o$ in C_K by (1.5). By the result of section 2 this implies :

$$u^{1,\lambda} \to u^1 \text{ in } C_K,$$

which in turn implies easily :

$$M\, u^{1,\lambda} \to M\, u^1 \text{ in } C_K.$$

Therefore, again by theorem 2, $u^{2,\lambda} \to u^2$ in C_K and the induction argument is obvious.

Now let K any compact subset of E :

$$||u^\lambda - u||_K \leq ||u^\lambda - u^{n,\lambda}||_K + ||u^{n,\lambda} - u^n||_K + ||u^n - u||_K.$$

By (3.6) :

$$||u^{\lambda} - u||_K \leq ||u^{n,\lambda} - u^n||_K + 2 . \frac{C}{K.n}$$

$\lambda \to 0$ and then $n \to \infty$ give the results. ■

Remark. One can add a dependency of $c(\xi)$ on λ assuming $c_{\lambda} \to c$ uniformly on every compact subset.

More generally if we have an operator M_{λ}, it is enough that $M_{\lambda}\varphi_{\lambda} \to M\varphi$ when $\varphi_{\lambda} \to \varphi$ in C_K to use the previous method.

4 - APPLICATIONS

We now use the results of §§2 and 3 for the approximation of the impulsive control problem of a diffusion process by a similar problem for a jump markov process.

We will begin, in fact, with a variational inequality.

Let us consider, for the sake of simplicity, the diffusion process, with value in $\mathbb{R}$, associated to the generator :

$$A = a(x) \frac{d^2}{d\, x^2} \tag{4.1}$$

where $a(x) \geq \beta > 0$ $a \in C_b^o(R)$.(One could take $a(x) \geq 0$, a uniformly lipschitz and bounded; the essential point is the uniqueness of the diffusion process associated with (4.1)).

A classical finite-difference approximation of (4.1) is given by ($h > 0$) :

$$A_h u(x) = \frac{2a(x)}{h^2}\left[\frac{1}{2} u(x+h) + \frac{1}{2} u(x-h) - u(x)\right] \tag{4.2}$$

and this can be considered as the infinitesimal generator of a jump markov process with values in R.

To solve a variational inequality for (4.1) one can use the approximate problem :

$$\begin{cases} -A_h u_h + \alpha u_h \le f & x \in R \\ u_h \le \psi \\ (-A_h u_h + \alpha u_h - f)(u_h - \psi) = 0. \end{cases} \quad (4.3)$$

Actually, one will solve this problem only for $x = i\,h$, $i \in \mathbb{Z}$ (plus a troncation in practical computations, of course).

The main point will be to check the property (1.5) when $h \to 0$.

Let us notice first that one can show that A_h generates a Feller process with values in $\mathbb{R}$ (endowed with the euclidian topology) : $X_h = (\Omega\ , \mathcal{F}_t, x_t, P_x^h)$. Moreover, using the fact that, for f bounded measurable :

$$f(x_t) - f(x) - \int_o^t A_h f(x_s) ds$$

is a P_x^h-martingale, it is easy to show (see [6] for example), that:

$$P_x^h\ [\sup_{o \le t \le T} |x_t - x| > R] \le \frac{C.T}{R^2} \quad (4.4)$$

uniformly w.r.t. h,x. (Therefore (1.3) holds.)

To check (1.5), we will proceed through several steps.

<u>First step</u> : $\{P_x^h,\ h > 0,\ x \in K\}$, where K is any compact, is relatively compact for the weak convergence on the space of bounded measures on $(\Omega,\mathcal{F})$.

For this we use the compactness criterion given by Lepeltier-Marchal [4], namely :

$\forall \epsilon$, T, $\exists \eta$ such that : (4.5)

$$P_x^h\ [\sup_{o \le t \le T} |x_t| > \eta] \le \varepsilon \quad \forall h > 0, \forall x \in K$$

$\forall \varepsilon$, η, $\exists \delta$ such that : (4.6)

$$P_x^h\ [\sup_{s \le t \le \delta} |x_t - x_s| > \eta] \le \varepsilon \quad \forall h > 0, \forall x \in K$$

$\forall \varepsilon, T, \eta, \exists \sigma > 0$ such that : (4.7)

$$P_x^h [\sup_{\sigma \le s \le t \le T} |x_s - x_t| > \eta] \le \varepsilon \quad \forall h > 0, \forall x \in K$$

$\forall \varepsilon, T, \eta \; \exists \delta$ such that : (4.8)

$$P_x^h [\sup_{\substack{t_1 \le t_2 \le T \\ |t_1 - t_2| \le \delta}} \sup_{t_1 \le t \le t_2} \min\{|x_t - x_{t_2}|, |x_t - x_{t_1}|\} > \eta] \le \varepsilon \quad \forall h > 0, x \in K$$

The details of the proof can be found in [6].

Let us say only that this proof is based on the property that, $\forall \tau$ bounded stopping time :

$$P_x^h [\sup_{0 \le s \le T} |x_\tau - x_{\tau+s}| > \eta] \le \frac{C.T}{\eta^2}$$

uniformly in h and $x \in \mathbb{R}$. This is a slight extension of (4.4) using the strong markov property.

Second step. Let $(h_n, x_n) \to (h, x)$, $u \to \infty$, h_n, $h > 0$, $x_n, x \in K$, then $P_{x_n}^{h_n}$ converges to P_x^h (weakly in the space of bounded measures on $(\Omega, \mathcal{F})$).

The proof is only that $\forall f$ bounded and continuous in R :

$$f(x_t) - f(x) - \int_0^t A_h f(x_s) ds$$

is a P martingale as soon as P is a limit point of $\{P_{x_n}^{h_n}\}_{n \in \mathbb{N}}$. Then using the fact that the martingale problem :

"Find P, probability measure on $(\Omega, \mathcal{F})$ such that $\forall f \in C_b^o(R)$:

$$f(x_t) - f(x) - \int_0^t A_h f(x_s) ds \text{ is a P martingale,}$$

$$P(x_0 = x) = 1 \text{"}$$

has the unique solution P_x^h, the result is obtained.

Third step. Let $(h_n,x_n) \to (o,x)$ x_n, $x \in K$, then $P_{x_n}^{h_n} \to P_x$ when P_x corresponds to the diffusion process associated to A.

If P is any limit point of $\{P_{x_n}^{h_n}\}$, then using the Billingsley's criterion [3] :

$\forall \varepsilon$, η, $\exists n_o, \delta$ such that :

$$P_{x_n}^{h_n}[\sup_t \sup_{t \le s_1, s_2 \le t+\delta} |x_{s_1} - x_{s_2}| \ge \varepsilon] \le \eta \tag{4.9}$$

$\forall n \ge n_o$

one obtains that $P[C(o,\infty;\mathbb{R})] = 1$.

The rest of the proof is also classical.

Fourth step : Proof of (1.5)

One has, on $[0,h_o] \times K$ ($h_o > 0$, K compact in $\mathbb{R}$), $P_{x_n}^{h_n} \to P_x^h$ when $(h_n,x_n) \to (h,x)$, (with $P_x^o = P_x$).

Therefore, if f is a real valued function on Ω, P_x^h a-s continuous, then :

$$\varphi(h_n,x_n) = \int_\Omega P_{x_n}^{h_n}(d\omega) f(\omega) \text{ converges to :}$$

$$\varphi(h,x) = \int_\Omega P_x^h (d\omega) f(\omega).$$

Hence, if f is P_x^h a.s continuous :

$\varphi(h,x)$ is continuous in $[0,h_o] \times K$

and therefore uniformly continuous.

Therefore, $\forall$f. P_x a-s continuous,

$\varphi(h,.)$ converges to $\int_\Omega P_x(d\omega) f(\omega)$ <u>uniformly on K</u>

when $h \to 0$.

It follows easily that $\forall g \in C_b^o(\mathbb{R})$:

$\phi^h(t)g \to \phi(t)g$ in C_K when $h \to 0$

if ϕ^h, ϕ are the semi group corresponding to A_h,A.

Therefore, the family of transition probabilities $\{P^{h_n}(x,t,.),\ x \in K\}$ is relatively compact, that is, $\forall \varepsilon > 0$, $\exists K_\varepsilon$ compact in $\mathbb{R}$, such that $P^{h_n}(x,t,k_\varepsilon) \geq 1-\varepsilon\ \forall n$, $\forall x \in K$.

Hence, if $g^{h_n} \to g$ in C_K when $h_n \to o$,

$$\sup_{x\in K} |\phi^{h_n}(t)g^{h_n}(x) - \phi(t)g(x)| \leq \sup_{x\in K_\varepsilon} |g^{h_n}(x) - g(x)| + \varepsilon(||g^{h_n}|| + ||g||) + \sup_{x\in K} |\phi^{h_n}(t)g(x) - \phi(t)g(x)|,$$

implying that (1.5) holds.

Therefore the result of section 2 holds.

One can also apply the results of section 3 taking in account the remark 3.1.

Actually, assuming for simplicity c = k = constant, the QVI for A_h will be :

$$\begin{cases} - A_h u_h + \alpha u_h \leq f \\ u_h \leq M_h u_h \\ (- A_h u_h + \alpha u_h - f)(u_h - M_h u_h) = 0 \end{cases}$$

where :

$$M_h u_h = k + \operatorname*{Inf}_{\xi \in U} u_h(x + [\tfrac{\xi}{h}]\, h)$$

where $[\frac{\xi}{h}]$ is the integer part of ξ/h.

This will allow to solve the QVI only with $x = ih$, $i \in \mathbb{Z}$ (and of course to keep the same convergence result).

One can generalize the previous result for multi-dimensional diffusion associated into the operator :

$$A = \sum_{ij} a_{ij} \frac{\partial^2}{\partial x_i \partial x_j} + \sum_j b_j \frac{\partial}{\partial x_j}$$

with only additional formalism.

REFERENCES

[1] A. BENSOUSSAN - J. L. LIONS - "Temps d'arrêt optimal et contrôle impulsionnel," vol. 1, Dunod, Paris 1978.

[2] A. BENSOUSSAN J. L. LIONS - "Inequations quasi variationnelles dépendant d'un paramètre," Annali Scuola Normale Superiore Pisa IV - Vol. IV, n° 2, 1977.

[3] P. BILLINGSLEY - "Convergence of probability measures," J. Wiley Pub. Co., 1975.

[4] J. P. LEPELTIER - B. MARCHAL - "Problèmes de martingale et equations différentielles stochastiques associées à un opérateur intégro différential," Annales de l'IMP, Vol. XII, n° 1, 1976, pp. 43-103.

[5] J. L. MENALDI - Thesis - To appear.

[6] M. ROBIN - "Contrôle impulsionnel des processus de markov," Doctoral Thesis - Paris IX, 1978.

[1] and [6] contain a more detailed bibliography.

STOCHASTIC DIFFERENTIAL EQUATION ASSOCIATED WITH THE BOLTZMANN EQUATION OF MAXWELLIAN MOLECULES IN SEVERAL DIMENSIONS

Hiroshi Tanaka

Department of Mathematics
Hiroshima University
Hiroshima, Japan

I. INTRODUCTION

The 3-dimensional Maxwellian gas is composed of a very large number of molecules moving in $\mathbb{R}^3$ according to the classical mechanics under the interaction due to an intermolecular repulsive force which is inversely proportional to the 5th power of the distance between molecules. As a $d(\geq 3)$-dimensional analogue of such gas we consider a model in which molecules move in $\mathbb{R}^d$ with an intermolecular repulsive force which is inversely proportional to the $(2d-1)^{th}$ power of the distance. For this model, the Boltzmann equation that governs the time evolution of the number density $u(t,x)$ has the simple form:

$$\frac{\partial u}{\partial t} = \int_{(0,\pi)\times S^{d-2}\times \mathbb{R}^d} (u'u_1' - uu_1)Q(\theta)d\theta d\phi dx_1 \tag{1.1}$$

where $u = u(t,x)$, $u_1 = u(t,x_1)$, $u' = u(t,x')$ and $u_1' = u(t,x_1')$. Here we are assuming the spatial homogeneity, and x, x_1 denote the velocities of molecules which are to change into x', x_1' after "collision." If S_{x,x_1} denotes the $(d-1)$-dimensional sphere with center $(x+x_1)/2$ and diameter $|x-x_1|$, then $S_{x,x_1} = S_{x',x_1'}$ according to the conservation law of momentum and energy. For each x and x_1 we choose a spherical coordinate system with polar axis

ISBN 0-12-268380-3

defined by the relative velocity $x - x_1$, and denote by (θ, ϕ) the spherical coordinate of $x' \in S_{x,x_1}$. Thus θ (= the colatitude of x') $\in (0,\pi)$, $\phi \in S^{d-2}$ and $d\phi$ is the surface element on S^{d-2}. The spherical coordinate of x_1' is $(\pi - \theta, -\phi)$. As in the 3-dimensional case the function Q (> 0) satisfies $Q(\theta) \sim \text{const.}\ \theta^{-3/2}$ as $\theta \downarrow 0$, and hence the measure $Q(d\theta) = Q(\theta)d\theta$ satisfies

$$\int_0^\pi Q(d\theta) = \infty, \qquad \int_0^\pi \theta Q(d\theta) < \infty. \tag{1.2}$$

For these materials see (1),(3),(7). Although the function $Q(\theta)$ in (1.1) is uniquely determined by the intermolecular force, what we have to assume for $Q(d\theta)$ in our discussions is that it is a (general) Borel measure on $(0,\pi)$ such that the second integral of (1.2) is convergent. In the case $d = 2$ an equation analogous to (1.1) can also be considered (5), but in our discussions it is assumed that $d \geq 3$.

The main objective of this paper is to associate with (1.1) and then to solve a certain <u>stochastic differential equation</u> (abbreviated: SDE). The relation between the Boltzmann equation (1.1) and our SDE is that the probability distribution $u(t)$, at time t, of a solution of the SDE gives a weak solution, that is, a probability measure solution to (1.1). The SDE we consider is similar to that of McKean (4) for diffusion case. In the present case, however, it is to be based upon a Poisson random measure, since the velocity changes discontinuously by "collision." Fundamentally, the present content is a straightforward extension of our earlier result for the 3-dimensional case (7), and some lemmas will be stated without proofs when they are essentially the same as those found in (7). However, some improvements will be made partly in the proof of the unique existence theorem for the SDE; these will further clarify the main body of our methods. We will also investigate some properties of the nonlinear semigroup arising from the solutions of the SDE.

II. STOCHASTIC DIFFERENTIAL EQUATION

Given a measure $Q(d\theta)$ on $(0,\pi)$ satisfying (1.2), we put for $\xi \in C_0^\infty(\mathbb{R}^d)$

$$(K\xi)(x,x_1) = \int_{(0,\pi)\times S^{d-2}} \{\xi(x') - \xi(x)\} Q(d\theta)d\phi,$$

and consider

$$\frac{d}{dt}\langle u,\xi\rangle = \langle u\otimes u, K\xi\rangle, \quad \xi \in C_0^\infty(\mathbb{R}^d) \tag{2.1}$$

where $C_0^\infty(\mathbb{R}^d)$ is the space of real valued C^∞-functions on $\mathbb{R}^d$ with compact supports. A solution of (2.1) may be regarded as a weak one of (1.1) by virtue of the following lemma which can be proved by the same method as in Appendix of (7).

<u>Lemma 2.1</u>. Let $\xi(x,x_1,y,y_1)$ be a continuous function on $\mathbb{R}^{4d}$ with compact support. Then, for each $\theta \in (0,\pi)$

$$\int_{S^{d-2}\times\mathbb{R}^{2d}} \xi(x,x_1,x',x_1')d\phi dx dx_1 = \int_{S^{d-2}\times\mathbb{R}^{2d}} \xi(x',x_1',x,x_1)d\phi dx dx_1.$$

The equation (2.1) makes sense not only for a function $u = u(t,x)$ but also for a (probability) measure $u = u(t,dx)$. The basic idea is to think of (2.1) as a forward equation of a certain Markov process and then to find this Markov process by solving certain SDE. In this section we derive this SDE.

First we must introduce several notations. Let W be the space of $\mathbb{R}^d$-valued right continuous functions on $\mathbb{R}_+ = [0,\infty)$ having left limits. W is equipped with the Skorohod topology; it is completely metrizable and separable, and the topological Borel field $\mathcal{B}(W)$ coincides with the usual coordinate σ-field. We also write W_T for the space of $\mathbb{R}^d$-valued right continuous functions on $[0,T]$ having left limits and $\mathcal{B}(W_T)$ for the corresponding Borel field. We think of the unit interval $(0,1)$ as a probability space with probability measure $d\alpha$ (the Lebesgue measure).

A process defined on this probability space and having paths in W (or W_T) is called an α-process for simplicity; similarly a random variable defined on this probability space is called an α-random variable. An α-process $\{Y(t)\}$ is called an α-realization of a process $\{X(t)\}$ defined on some probability space, if $\{X(t)\}$ and $\{Y(t)\}$ are equivalent in law. Let $S = (0,\pi)\times S^{d-2}\times(0,1)$ and λ be the measure on S defined by $d\lambda = Q(d\theta) \otimes d\phi \otimes d\alpha$. We put $\mathbb{R}_t^0 = (0,t]$ and $\mathbb{R}_\infty^t = (t,\infty)$. Suppose we are given a probability space $(\Omega,\mathcal{F},P)$ together with an increasing family $\{\mathcal{F}_t\}_{t\geq 0}$ of sub-σ-fields of $\mathcal{F}$. Also, suppose we are given an $\mathcal{F}_t$-adapted Poisson random measure $N(\cdot)$ corresponding to λ, that is, a counting random measure on $\mathbb{R}_\infty^0 \times S$ with the following properties:

i) for any disjoint family $\{A_1,\ldots,A_\nu\}$ from $\mathcal{B}(\mathbb{R}_\infty^0\times S)$ such that $\bar{\lambda}(A_k) = \int_{A_k} dtd\lambda < \infty$ $(1 \leq k \leq \nu)$ and for nonnegative integers $n_1,\ldots,n_\nu$, the equality

$$P\{N(A_k) = n_k,\ 1 \leq k \leq \nu\} = \prod_{k=1}^{\nu}\{e^{-\bar{\lambda}(A_k)}\frac{(\bar{\lambda}(A_k))^{n_k}}{n_k!}\}$$

holds;

ii) for each $t \geq 0$ and $A \in \mathcal{B}(\mathbb{R}_t^0\times S)$ with $\bar{\lambda}(A) < \infty$, $N(A)$ is $\mathcal{F}_t$-measurable;

iii) for each $t \geq 0$ the σ-fields $\mathcal{F}_t$ and $\sigma\{N(A): A \in \mathcal{B}(\mathbb{R}_\infty^t\times S)\}$ are independent.

Putting $a(x,x_1,\theta,\phi) = x' - x$, we can now write our SDE:

$$X(t) = X(0) + \int_{(0,t]\times S} a(X(s-),Y(s-,\alpha),\theta,\phi)dN; \tag{2.2}$$

here a solution $\{X(t)\}$ is to be found as an $\mathcal{F}_t$-adapted process under the condition that $\{Y(t)\}$ is an α-realization of $\{X(t)\}$. However, when discussing the existence problem it is more convenient to consider the SDE of the form:

$$X(t) = X(0) + \int_{(0,t]\times S} a(X(s-),Y(s-,\alpha)\theta,\mathcal{O}\phi)dN. \tag{2.3}$$

Here, by a solution of (2.3) we mean any $\mathcal{F}_t$-adapted process $\{X(t)\}$ which satisfies (2.3) for some α-realization $\{Y(t)\}$ of $\{X(t)\}$ and for some $\mathcal{F}_t$-predictable process $\mathcal{O} = \mathcal{O}(s,\alpha,\omega)$ with state space $O(d-1)$, the space of orthogonal matrices of degree $d-1$. The equation (2.3) is nothing but (2.2) with $N(\cdot)$ replaced by

$$N^*(A) = \int_{(0,\infty)\times S} \mathbb{1}_A(t,\theta,\mathcal{O}(t,\alpha,\omega)\phi,\alpha)dN, \quad A \in \mathcal{B}(\mathbb{R}^0_\infty \times S)$$

which is again an $\mathcal{F}_t$-adapted Poisson random measure corresponding to λ. The connection between (2.1) and (2.2) (or (2.3)) is, as stated in the introduction, that the probability distribution $u(t)$, at time t, of a solution of (2.2) (or (2.3)) gives a solution to (2.1).

III. SOLVING THE SDE (2.3)

In this section we assume that an $\mathcal{F}_t$-adapted Poisson random measure $N(\cdot)$ corresponding to λ is given; it is fixed throughout. The function $a(x,x_1,\theta,\phi) = x' - x$ is defined by fixing a suitable spherical coordinate system on each S_{x,x_1} so that, as a whole, $a(x,x_1,\theta,\phi)$ becomes Borel measurable in (x,x_1,θ,ϕ). In proving the existence theorem for the SDE (2.3), the following lemma will play an essential role.

Lemma 3.1. Given four points x, x_1, y and y_1 of $\mathbb{R}^d$, we can choose $\mathcal{O} = \mathcal{O}(x,x_1,y,y_1)$ from $O(d-1)$ in such a way that

$$|a(x,x_1,\theta,\phi) - a(y,y_1,\theta,\mathcal{O}\phi)| \leq \sqrt{d/4}\{|x-y| + |x_1 - y_1|\}\theta. \tag{3.1}$$

Moreover, it can be chosen so that $\mathcal{O}$ is Borel measurable in (x,x_1,y,y_1).

Proof. i) When $x = x_1$ or $y = y_1$, (3.1) clearly holds with $\mathcal{O} = I$. ii) Next we consider the case when

$$x = -x_1 = (r_1,0,0,\ldots,0)$$
$$y = -y_1 = (r_2\cos\gamma,\ r_2\sin\gamma,0,\ldots,0)$$

with r_1, $r_2 > 0$ and $0 \leq \gamma \leq \pi$. Put

$$\tilde{\mathcal{O}}_\gamma = \left(\begin{array}{cc|c} \cos\gamma & -\sin\gamma & 0 \\ \sin\gamma & \cos\gamma & \\ \hline 0 & & I_{d-2} \end{array}\right) \in O(d).$$

Then, there exists $\mathcal{O}_{r_1,r_2,\gamma} \in O(d-1)$ such that

$$r_1^{-1} r_2 \tilde{\mathcal{O}}_\gamma a(x,x_1,\theta,\phi) = a(y,y_1,\theta,\mathcal{O}_{r_1,r_2,\gamma}\phi),$$

and hence

$$\begin{aligned} &|a(x,x_1,\theta,\phi) - a(y,y_1,\theta,\mathcal{O}_{r_1,r_2,\gamma}\phi)| \\ &= |a(x,x_1,\theta,\phi) - r_1^{-1}r_2\tilde{\mathcal{O}}_\gamma a(x,x_1,\theta,\phi)| \\ &\leq \| I - r_1^{-1}r_2\tilde{\mathcal{O}}_\gamma \| |a(x,x_1,\theta,\phi)| = \| I - r_1^{-1}r_2\tilde{\mathcal{O}}_\gamma \| r_1\theta, \end{aligned} \tag{3.2}$$

where

$$\begin{aligned} \| I - r_1^{-1}r_2\tilde{\mathcal{O}}_\gamma \|^2 &= \text{the sum of the squares of the entries of } I - r_1^{-1}r_2\tilde{\mathcal{O}}_\gamma \\ &= 2\{(1-r_1^{-1}r_2\cos\gamma)^2 + (r_1^{-1}r_2\sin\gamma)^2\} \\ &\quad + (d-2)(1-r_1^{-1}r_2)^2. \end{aligned}$$

Because $r_1^2 - 2r_1r_2\cos\gamma + r_2^2 \geq (r_1 - r_2)^2$,

$$\begin{aligned} \| I - r_1^{-1}r_2\tilde{\mathcal{O}}_\gamma \|^2 r_1^2 &= 2(r_1^2 - 2r_1r_2\cos\gamma + r_2^2) + (d-2)(r_1-r_2)^2 \\ &\leq d(r_1^2 - 2r_1r_2\cos\gamma + r_2^2) \\ &= (d/4)\{|x-y| + |x_1 - y_1|\}^2. \end{aligned}$$

This combined with (3.2) proves (3.1) with $\mathcal{O} = \mathcal{O}_{r_1,r_2,\gamma}$.

iii) General case ($x \neq x_1$, $y \neq y_1$). We take $\tilde{\mathcal{O}} \in O(d)$ such that

$$\tilde{\mathcal{O}}(x - x_1)/2 = x^* = (r_1,0,0,\ldots,0)$$
$$\tilde{\mathcal{O}}(y - y_1)/2 = y^* = (r_2\cos\gamma, r_2\sin\gamma,0,\ldots,0),$$

where $r_1 = |x-x_1|/2$, $r_2 = |y-y_1|/2$ and $\cos\gamma = \dfrac{(x-x_1,y-y_1)}{4r_1r_2}$, $0 \leq \gamma \leq \pi$. Then, there exist $\mathcal{O}_{x,x_1}$ and $\mathcal{O}_{y,y_1} \in O(d-1)$ such that

$$\tilde{\mathcal{O}}a(x,x_1,\theta,\phi) = a(x^*,-x^*,\theta,\mathcal{O}_{x,x_1}\phi),$$

$$\tilde{\mathcal{O}}a(y,y_1,\theta,\phi) = a(y^*,-y^*,\theta,\mathcal{O}_{y,y_1}\phi).$$

Now we put $\mathcal{O} = \mathcal{O}^{-1}_{y,y_1}\mathcal{O}_{r_1,r_2,\gamma}\mathcal{O}_{x,x_1}$. Then we have

$$|a(x,x_1,\theta,\phi) - a(y,y_1,\theta,\mathcal{O}\phi)|$$

$$= |\tilde{\mathcal{O}}a(x,x_1,\theta,\phi) - \tilde{\mathcal{O}}a(y,y_1,\theta,\mathcal{O}\phi)|$$

$$= |a(x^*,-x^*,\theta,\mathcal{O}_{x,x_1}\phi) - a(y^*,-y^*,\theta,\mathcal{O}_{r_1,r_2,\gamma}\mathcal{O}_{x,x_1}\phi)|$$

$$\leq \sqrt{d/4}\, 2|x^* - y^*|\theta \leq \sqrt{d/4}\{|x - y| + |x_1 - y_1|\}\theta,$$

proving (3.1). Since $\tilde{\mathcal{O}}$ can be chosen to be Borel measurable in (x,x_1,y,y_1), $\mathcal{O}$ itself can be chosen to be Borel measurable.

In what follows, a process $\{X(t)\}$ is said to be <u>integrable</u> for simplicity, if $E\{\sup_{0\leq s\leq t}|X(s)|\} < \infty$ for $t \in \mathbf{R}_+$.

<u>Theorem 3.1</u>. Let X be an $\mathbf{R}^d$-valued $\mathcal{F}_0$-measurable random variable with $E|X| < \infty$ and $\{Y(t)\}$ be an integrable α-process. Then there exists an $\mathcal{F}_t$-adapted integrable process $\{X(t)\}$ which satisfies

$$X(t) = X + \int_{(0,t]\times S} a(X(s-),Y(s-,\alpha),\theta,\mathcal{O}\phi)dN \tag{3.3}$$

for some $\mathcal{F}_t$-predictable process $\mathcal{O} = \mathcal{O}(s,\alpha,\omega)$ with state space $O(d-1)$.

Proof. We put $X_0(t) \equiv X$ and

$$X_1(t) = X + \int a(X_0(s-),Y(s-,\alpha),\theta,\mathcal{O}_0\phi)dN,$$

where the integration is performed on $(0,t]\times S$ and $\mathcal{O}_0 \equiv I$. Next, assuming that $\{X_k(t)\}$ is defined for $k \leq n$, we define $\{X_{n+1}(t)\}$ by

$$X_{n+1}(t) = X + \int a(X_n(s-),Y(s-,\alpha),\theta,\mathcal{O}_n\phi)dN,$$

where $\mathcal{O}_n = \mathcal{O}(s,\alpha,\omega)$ is defined by

$$\mathcal{O}_n = \mathcal{O}(X_{n-1}(s-),Y(s-,\alpha),X_n(s-),Y(s-,\alpha))\mathcal{O}_{n-1}(s,\alpha,\omega)$$

making use of the notation $\mathcal{O}(x, x_1, y, y_1)$ of Lemma 3.1. Thus we obtain a sequence $\{X_n(t)\}_{n \geq 0}$ of processes, and by Lemma 3.1

$$|a(X_n(s-),Y(s-),\theta,\mathcal{O}_n\phi) - a(X_{n-1}(s-),Y(s-),\theta,\mathcal{O}_{n-1}\phi)|$$
$$\leq \sqrt{d/4}|X_n(s-) - X_{n-1}(s-)|\theta, \tag{3.4}$$

$$E\{\sup_{0 \leq s \leq t} |X_{n+1}(s) - X_n(s)|\} \leq \sqrt{d/4}\, E\int |X_n(s-) - X_{n-1}(s-)|\theta ds d\lambda$$
$$= c \int_0^t E|X_n(s) - X_{n-1}(s)|ds. \tag{3.5}$$

Therefore, $\{X_n(t)\}$ converges to some $\mathcal{F}_t$-adapted process $\{X(t)\}$ uniformly on each finite t-interval as $n \to \infty$ almost surely. (3.4) and (3.5) also imply that $\{a(X_n(t-),Y(t-),\theta,\mathcal{O}_n\phi)\}$ converges to some $\mathcal{F}_t$-predictable $a_\infty(t,\theta,\phi,\alpha)$ uniformly on each finite t-interval almost surely. It is then clear that $X(t-) + a_\infty(t,\theta,\phi,\alpha)$ lies on $S_{X(t-),Y(t-)}$ with colatitude θ a.s., and hence there exists a mapping $\mathcal{R}$: $S^{d-2} \to S^{d-2}$ such that

$$a_\infty(t,\pi/2,\phi,\alpha) = a(x(t-),Y(t-),\pi/2, \mathcal{R}\phi).$$

But, since

$$|a(X(t-),Y(t-),\pi/2,\mathcal{R}\phi) - a(X(t-),Y(t-),\pi/2,\mathcal{R}\phi')|$$
$$= \lim_{n\to\infty}|a(X_n(t-),Y(t-),\pi/2,\mathcal{O}_n\phi) - a(X_n(t-),Y(t-),\pi/2,\mathcal{O}_n\phi')|$$
$$= |\phi - \phi'| \lim_{n\to\infty}|X_n(t-) - Y(t-)|/2 = |\phi - \phi'||X(t-) - Y(t-)|/2,$$

we have $|\mathcal{R}\phi - \mathcal{R}\phi'| = |\phi - \phi'|$, and therefore $\mathcal{R} \in O(d-1)$ a.s. Thus $\mathcal{R}$ can be written as $\mathcal{R} = \mathcal{O}(t,\alpha,\omega)$ and it can also be proved that $a_\infty(t,\theta,\phi,\alpha) = a(X(t-),Y(t-),\theta,\mathcal{O}\phi)$ holds for $\theta \in (0,\pi)$ a.s. and that $\mathcal{O}$ is $\mathcal{F}_t$-predictable. Therefore $\{X(t)\}$ satisfies (3.3) a.s., and so the proof is finished.

In the discussions of the existence and uniqueness for solutions of (2.3), we may consider it for $0 \leq t \leq T$, T being an arbitrary positive constant. From now on we fix such a T and put $\|w\| = \sup\{|w(t)|: 0 \leq t \leq T\}$ for $w \in W_T$. Let $\mathfrak{M}$ denote the space of probability measures μ on the measurable space $(W_T, \mathcal{B}(W_T))$ such that the processes $\{w(t),\ 0 \le t \le T,\ \mu\}$ are continuous in the mean and integrable. Given μ_1 and μ_2 in $\mathfrak{M}$, we define $\bar{\rho}(\mu_1,\mu_2)$ by

$$\bar{\rho}(\mu_1,\mu_2) = \inf \int_{W_T \times W_T} \|w - w'\| \tilde{\mu}(dwdw')$$

where the infimum is taken over all probability measures $\tilde{\mu}$ on $W_T \times W_T$ such that $\tilde{\mu}(B \times W_T) = \mu_1(B)$ and $\tilde{\mu}(W_T \times B) = \mu_2(B)$ for all $B \in \mathcal{B}(W_T)$. It can be proved that $\bar{\rho}$ gives a complete metric on $\mathfrak{M}$. Similarly, we define a metric ρ_1 on $\mathcal{P}_1$, the space of probability measures on $\mathbb{R}^d$ with finite absolute moments, by

$$\rho_1(u_1,u_2) = \inf \int_{\mathbb{R}^d \times \mathbb{R}^d} |x - y| \tilde{u}(dxdy)$$

where the infimum is taken over all probability measures $\tilde{u}$ on $\mathbb{R}^{2d}$ satisfying $\tilde{u}(A \times \mathbb{R}^d) = u_1(A)$ and $\tilde{u}(\mathbb{R}^d \times A) = u_2(A)$ for $A \in \mathcal{B}(\mathbb{R}^d)$.

Now, let Δ be a partition of $[0,T]$: $0 = t_0 < t_1 < \cdots < t_n = T$ and put $\Delta(s) = t_k$ for $t_k < s \le t_{k+1}$ $(0 \le k < n)$, $|\Delta| = \max\{t_{k+1} - t_k : 0 \le k < n\}$. Let Y_k, $0 \le k < n$, be $\mathbb{R}^d$-valued α-random variables with probability distributions u_k in $\mathcal{P}_1$, and $\mathcal{O}_k$ $(\in O(d-1))$, $0 \le k < n$, be $\mathcal{B}((0,1)) \otimes \mathcal{F}_{t_k}$-measurable. Putting $u_\Delta(s) = u_k$, $Y(s) = Y_k$ and $\mathcal{O}(s) = \mathcal{O}_k$ for $t_k < s \le t_{k+1}$ $(0 \le k < n)$, we consider the SDE

$$X(t) = X + \int_{(0,t]\times S} a(X(\Delta(s)),Y(s),\theta,\mathcal{O}\phi)dN, \tag{3.6}$$

where X is an $\mathcal{F}_0$-measurable random variable with distribution f in $\mathcal{P}_1$. (3.6) is nothing but a difference equation, and hence by a method similar to that of Lemma 4.1 of (7) we can prove that the probability measure on W_T induced by a solution of (3.6) is uniquely determined by Δ, f and u_k, $0 \le k < n$. Starting with $Y_k^\#$, $0 \le k < n$, that are distributed according to $u_k^\#$ (f is the same as before), we have a similar solution whose induced probability measure on W_T is denoted by $\bar{\mu}_\Delta^\#$. Then we have the following lemma which corresponds to Lemma 4.2 of (7).

<u>Lemma 3.2.</u> We can construct a $\bar{\mu}_\Delta$-distributed process $\{X_\Delta(t)\}$

and a $\bar{\mu}^{\#}_{\Delta}$-distributed process $\{X^{\#}_{\Delta}(t)\}$ in such a way that

$$E\{\sup_{0 \leq s \leq t} |X_{\Delta}(s) - X^{\#}_{\Delta}(s)|\} \leq c \int_0^t \rho_1(u_{\Delta}(s),\ u^{\#}_{\Delta}(s))ds, \tag{3.7}$$

where c is a constant depending only on d, Q and T. In particular, we have

$$\rho_1(\bar{u}_{\Delta}(t), \bar{u}^{\#}_{\Delta}(t)) \leq c \int_0^t \rho_1(u_{\Delta}(s), u^{\#}_{\Delta}(s))ds,\ 0 \leq t \leq T. \tag{3.8}$$

Proof. Choosing Y_k and $Y^{\#}_k$ so that $E_{\alpha}|Y_k - Y^{\#}_k| = \rho_1(u_k, u^{\#}_k)$, we define $\{Y(s)\}$ and $\{Y^{\#}(s)\}$ as before. Let $\{X_{\Delta}(t)\}$ and $\{X^{\#}_{\Delta}(t)\}$ be, respectively, the solutions of (3.6) with $\mathcal{O} = I$ and the following $(3.6^{\#})$:

$$X(t) = X + \int a(X(\Delta(s)),\ Y^{\#}(s), \theta, \mathcal{O}^{\#}\phi)dN, \tag{$3.6^{\#}$}$$

where $\mathcal{O}^{\#}(s,\alpha,\omega) = \mathcal{O}(X_{\Delta}(t_k), Y_k, X^{\#}_{\Delta}(t_k), Y^{\#}_k)$ for $t_k < s \leq t_{k+1}$. Then by Lemma 3.1 the left hand side of (3.7) is dominated by

$$\text{const.}\int_0^t \{E|X_{\Delta}(s) - X^{\#}_{\Delta}(s)| + \rho_1(u_{\Delta}(s), u^{\#}_{\Delta}(s))\}ds.$$

Now an application of Gronwall's inequality yields (3.7).

The following lemma can be proved by a method similar to that of Lemma 4.4 in (7).

<u>Lemma 3.3</u>. Given an $\mathcal{F}_0$-measurable random variable distributed according to f in $\mathcal{P}_1$ and an α-process $\{Y(t)\}$ distributed according to μ in $\mathfrak{M}$, we denote by $\bar{\mu}$ the probability measure on W_T induced by any integrable solution of (3.3). We also denote by $\bar{\mu}_{\Delta}$ the probability measure on W_T for a solution of (3.6) with $Y_k = Y(t_k)$, $0 \leq k < n$. Then, for any $\varepsilon > 0$ there exists a constant $\delta > 0$ for which the following assertion holds: If $|\Delta| < \delta$, then on a suitable probability space $(\tilde{\Omega}, \tilde{\mathcal{F}}, \tilde{P})$ we can construct a $\bar{\mu}_{\Delta}$-distributed process $\{\tilde{X}_{\Delta}(t)\}$ and a $\bar{\mu}$-distributed process $\{\tilde{X}(t)\}$ in such a way that

$$E|\tilde{X}_\Delta(t) - \tilde{X}(t)| < \varepsilon, \quad 0 \leq t \leq T.$$

As an immediate consequence of this lemma we have the following theorem.

Theorem 3.2. The probability measure $\bar{\mu}$ on W_T of any integrable solution of (3.3) is uniquely determined by the initial distribution f and the probability measure μ on W_T of $\{Y(t)\}$.

By virtue of this theorem, for each fixed $f \in \mathcal{P}_1$ we can define a mapping $\Phi_f : \mathfrak{M} \to \mathfrak{M}$ by $\Phi_f \mu = \bar{\mu}$. Now we can state our main results.

Theorem 3.3. There exist an integer $n_0 \geq 1$ and a positive constant $c_0 < 1$ such that

$$\bar{\rho}(\Phi_f^{n_0} \mu, \Phi_f^{n_0} \mu^{\#}) \leq c_0 \bar{\rho}(\mu, \mu^{\#})$$

holds for any $f \in \mathcal{P}_1$ and $\mu, \mu^{\#} \in \mathfrak{M}$.

Proof. From Lemmas 3.2 and 3.3 it follows that

$$\bar{\rho}(\Phi_f \mu, \Phi_f \mu^{\#}) \leq c \int_0^T \rho_1(u(s), u^{\#}(s))ds,$$

$$\rho_1(\bar{u}(t), \bar{u}^{\#}(t)) \leq c \int_0^t \rho_1(u(s), u^{\#}(s))ds,$$

and hence (with obvious notation)

$$\bar{\rho}(\Phi_f^n \mu, \Phi_f^n \mu^{\#}) \leq c \int_0^T \rho_1(\bar{u}_{n-1}(s), \bar{u}^{\#}_{n-1}(s))ds \leq \frac{(cT)^n}{n!} \bar{\rho}(\mu, \mu^{\#}).$$

This clearly implies the assertion of the theorem.

Theorem 3.4. For an $\mathcal{F}_0$-measurable random variable X(0) with distribution f in $\mathcal{P}_1$, there exists an integrable solution of (2.3). The probability measure on W of such a solution is uniquely determined by f.

Proof. It is enough to prove the theorem for $0 \leq t \leq T$. By Theorem 3.3 there exists a unique fixed point $\bar{\mu}$ of $\Phi_f \mu = \mu$. Now an integrable solution of (2.3) is obtained by solving the SDE

(3.3) with a $\bar{\mu}$-distributed α-process $\{Y(t)\}$ and $X = X(0)$. The uniqueness in the law sense is also clear.

To conclude this section, we must remark that, if $\{X(t)\}$ is an integrable solution of (2.3), then the distribution $u(t)$ of $X(t)$ is a solution of (2.1). But, this can be verified applying the transformation formula to the stochastic differential $d\xi(X(t))$, $\xi \in C_0^\infty(\mathbb{R}^d)$.

IV. THE ASSOCIATED NONLINEAR SEMIGROUP

Given $f \in \mathcal{P}_1$, we denote by $T_t f$ the probability distribution of $X(t)$, where $\{X(t)\}$ is an integrable solution of (2.3) with initial distribution f. Then, for each $t \geq 0$ T_t is nonlinear operator from $\mathcal{P}_1$ into itself, and the uniqueness in the law sense of the solutions of (2.3) implies the semigroup property: $T_{t+s}f = T_t T_s f$ for $t,s \geq 0$ and $f \in \mathcal{P}_1$. If we denote by $\mathcal{P}_2$ the space of probability distributions on $\mathbb{R}^d$ with finite absolute second moments, then it can be proved that T_t also maps $\mathcal{P}_2$ into itself. For f_1, $f_2 \in \mathcal{P}_2$ we put

$$e(f_1,f_2) = \inf \int_{\mathbb{R}^d \times \mathbb{R}^d} |x - y|^2 \tilde{f}(dxdy),$$

$$\rho(f_1,f_2) = \sqrt{e(f_1,f_2)},$$

where the infimum is taken over all probability distributions $\tilde{f}$ in $\mathbb{R}^d$ such that $\tilde{f}(A \times \mathbb{R}^d) = f_1(A)$ and $\tilde{f}(\mathbb{R}^d \times A) = f_2(A)$ for $A \in \mathcal{B}(\mathbb{R}^d)$. Then ρ gives a metric on $\mathcal{P}_2$. Given $f \in \mathcal{P}_2$ satisfying

$$v = (1/d)\int|x - m|^2 f(dx) > 0, \quad m = \int xf(dx), \tag{4.1}$$

we write $g = g_f$ for the Gaussian distribution with density $(2\pi v)^{-d/2}\exp(-|x - m|^2/2v)$ and put $e(f) = e(f,g)$.

The result in this section is stated in the following theorem; it is again a straightforward extension of the corresponding result in (7).

Theorem 4.1. i) T_t is non-expansive with respect to ρ:

$$\rho(T_t f_1, T_t f_2) \leq \rho(f_1, f_2), \quad f_1, f_2 \in \mathcal{P}_2.$$

ii) For $f \in \mathcal{P}_2$ satisfying (4.1), $e(T_t f)$ decreases to 0 as $t \uparrow \infty$, and hence in particular $T_t f$ converges weakly to g as $t \to \infty$.

The proof of this theorem can be carried out by the same methods as those of Theorems 7.1 and 9.1 of (7). The only parts which depend upon the dimension d more or less are the following:

(A) An upper estimate for $e(\Pi_{x,x_1,\theta}, \Pi_{y,y_1,\theta})$, $\Pi_{x,x_1,\theta}$ being defined below.

(B) The d-dimensional version of a result of Ikenberry and Truesdell (2) on the time evolution of moments.

Given $x, x_1 \in \mathbb{R}^d$ $(x \neq x_1)$ and $0 < \theta < \pi$, we denote by $\Pi_{x,x_1,\theta}$ the uniform distribution on the $(d-2)$-dimensional sphere $S_{x,x_1,\theta}$ with center $\{x + x_1 + (x - x_1)\cos\theta\}/2$, of radius $\sin\theta|x - x_1|/2$ and lying on a $(d-1)$-dimensional hyperplane which is perpendicular to $x - x_1$. Then an upper bound for $e(\Pi_{x,x_1,\theta}, \Pi_{y,y_1,\theta})$ in the d-dimensional case is given by

$$\begin{aligned} e(\Pi_{x,x_1,\theta}, \Pi_{y,y_1,\theta}) &\leq \left|\frac{1+\cos\theta}{2}(x-y) + \frac{1-\cos\theta}{2}(x_1-y_1)\right|^2 \\ &+ \frac{\sin^2\theta}{4}\{|x-x_1|^2 + |y-y_1|^2\} \\ &- \frac{\sin^2\theta}{2(d-1)}\{(d-2)|x-x_1||y-y_1| + |(x-x_1, y-y_1)|\}. \end{aligned}$$

As for (B), we start with a linearly independent complete system $\{\xi_k^\ell(x): 1 \leq \ell \leq a_k\}$ of homogeneous harmonic polynomials of degree k on $\mathbb{R}^d$, a_k being $\binom{d+k-1}{k} - \binom{d+k-3}{k-2}$, and put

$$\xi_{\underline{n}}(x) = |x|^{2r}\, \xi_k^\ell(x) \quad \text{for } \underline{n} = (r,k,\ell), \quad \xi_{(0,0,0)}(x) = 1,$$

for $r = 0,1,\ldots,k = 0,1,\ldots,$ and $1 \leq \ell \leq a_k$. Then $\xi_{\underline{n}}$ is a homogeneous polynomial of degree $|\underline{n}| = 2r + k$, and any homogeneous polynomial of degree n can be expressed as a linear combination of $\xi_{\underline{\underline{n}}}(x)$ with degree $|\underline{\underline{n}}| = n$. (Other elementary properties needed here are found in Chap. IV of (6).) Now, given a probability distribution f on $\mathbb{R}^d$ with $\int |x|^\nu f(dx) < \infty$ for some integer $\nu \geq 1$, we put $\mu_{\underline{\underline{n}}}(t) = \langle T_t f, \xi_{\underline{\underline{n}}} \rangle$ for $\underline{\underline{n}}$ such that $|\underline{\underline{n}}| \leq \nu$. Then the d-dimensional version of a result of (2) is that $\mu_{\underline{\underline{n}}}(t)$ satisfies

$$\frac{d}{dt}\mu_{\underline{\underline{n}}}(t) = \sum{}' \beta^{\underline{\underline{n}}}_{\underline{\underline{n}}_1,\underline{\underline{n}}_2}\mu_{\underline{\underline{n}}_1}(t)\mu_{\underline{\underline{n}}_2}(t) - \beta_{\underline{\underline{n}}}\mu_{\underline{\underline{n}}}(t),$$

where $\sum{}'$ means the summation to be taken over all pairs $(\underline{\underline{n}}_1,\underline{\underline{n}}_2)$ such that $|\underline{\underline{n}}_1| + |\underline{\underline{n}}_2| = |\underline{\underline{n}}|$ with $|\underline{\underline{n}}_1|, |\underline{\underline{n}}_2| \geq 1$; $\beta_{\underline{\underline{n}}}$ is given by

$$\beta_{\underline{\underline{n}}} = \frac{2\pi^{d/2}}{\Gamma(d/2)}\int_0^\pi \Big\{1 - \left(\cos\frac{\theta}{2}\right)^{|\underline{\underline{n}}|} P_k^{(d-2)/2}\left(\cos\frac{\theta}{2}\right) - \left(\sin\frac{\theta}{2}\right)^{|\underline{\underline{n}}|} P_k^{(d-2)/2}\left(\sin\frac{\theta}{2}\right)\Big\} Q(d\theta),$$

where $P_k^{(d-2)/2} = k!(d-3)!/(d+k-3)! \times$ (the ultraspherical polynomial of degree k associated with $(d-2)/2$); the $\beta^{\underline{\underline{n}}}_{\underline{\underline{n}}_1,\underline{\underline{n}}_2}$ are also some constants.

Now, based upon these results for (A) and (B), the proof of the theorem goes exactly along the same line as in (7).

REFERENCES

1. Ford, G. W., and Uhlenbeck, G. E., Lectures in Statistical Mechanics, Amer. Math. Soc., Providence, Rhode Island, 1963.
2. Ikenberry, E., and Truesdell, C., J. Rat. Mech. Anal. 5 (1956).
3. McKean, H. P., Arch. Rat. Mech. Anal. 21 (1966).
4. McKean, H. P., in "Lecture Series in Differential Equations," session 7. Catholic Univ. (1967).
5. Murata, H., Hiroshima Math. J. 7 (1977).
6. Stein, E. M., and Weiss, G., Introduction to Fourier Analysis on Euclidean Spaces, Princeton Univ. Press (1971).
7. Tanaka, H., Probabilistic treatment of the Boltzmann equation of Maxwellian molecules, to appear.

POINT PROCESSES AND MARTINGALES

Shinzo Watanabe

Department of Mathematics
Kyoto University
Kyoto, Japan

I. INTRODUCTION

Motivation of the present paper arises in the study of multidimensional diffusion processes with boundary conditions. Let D be a domain in R^n (or a manifold) with a smooth boundary ∂D. A most general (smooth) conservative diffusion process on $\overline{D} = D \cup \partial D$ (a strong Markov process with continuous trajectories) is described by the following analytic data:

1) a second order differential operator on D;

$$Af(x) = \frac{1}{2} \sum_{i,j=1}^{d} a^{ij}(x) f''_{ij}(x) + \sum_{i=1}^{d} b^{i}(x) f'_{i}(x) \tag{1.1}$$

where $a^{ij}(x)$ and $b^{i}(x)$ are smooth functions on $\overline{D}$ such that $(a^{ij}(x))$ is non-negative definite,

2) a Wentzell's boundary operator ; choosing a local coordinate $x = (x^1, x^2, \ldots, x^d)$ as $x \in D$ if $x^d > 0$ and $x \in \partial D$ if $x^d = 0$,

$$Lf(x) = \frac{1}{2} \sum_{i,j=1}^{d-1} \alpha^{ij}(x) f''_{ij}(x) + \sum_{i=1}^{d-1} \beta^{i}(x) f'_{i}(x) + \mu(x) f'_{d}(x) - \rho(x)\, Af|_{\partial D}(x) \tag{1.2}$$

where $\alpha^{ij}(x)$, $\beta^{i}(x)$, $\mu(x)$ and $\rho(x)$ are smooth functions

ISBN 0-12-268380-3

on ∂D such that $(\alpha^{ij}(x))$ is non-negative definite, $\mu(x) \geq 0$, $\rho(x) \geq 0$ and $\mu(x) + \rho(x) > 0$. How can we define the notion of a diffusion process generated by given data (A, L) ?

Analytically, it is a diffusion process on $\overline{D}$ whose semigroup T_t on $C(\overline{D})$ is generated by the operator A with the domain $D(A) = C^2(\overline{D}) \cap \{f; Lf=0\}$. We refer to, e.g. Sato-Ueno [5], Bony-Courrege-Priouret [2], Taira [7] for the problem of the existence and the uniqueness of such diffusions from this analytical point of view.

Probabilistically, it may be defined as follows: it is a $\overline{D}$-valued contiuous stochastic process $X = (X_t)$ such that there exists a continuous non-decreasing process ϕ_t with $\phi_0 = 0$ and $\int_0^t I_{\partial D}(X_s)d\phi_s = \phi_t$ a.s. and the following is statisfied :

(i) for every $f \in C^2(\overline{D})$,

$$f(X_t) - f(X_0) = \text{a martingale} + \int_0^t (Af)(X_s)ds + \int_0^t (Lf)X_s)d\phi_s$$

(ii) $\int_0^t I_{\partial D}(X_s)ds = \int_0^t \rho(X_s)d\phi_s$.

If we define

$$L'f(x) = Lf(x) + \rho(x)Af|_{\partial D}(x) = \frac{1}{2}\sum_{i,j=1}^{d-1} \alpha^{ij}(x)f''_{ij}(x) + \sum_{i=1}^{d-1} \beta^i(x)f'_i(x) + \mu(x)f'_d(x), \tag{1.3}$$

then the above (i) is equivalent, under the condition (ii), to

(i)' $f(X_t) - f(X_0) = \text{a martingale} + \int_0^t (Af)(X_s)I_D(X_s)ds$

$$+ \int_0^t (L'f)(X_s)d\phi_s .$$

If we can show the existence of such a process $X = (X_t)$ such that $X_0 = x$ for each given $x \in \overline{D}$ and the uniqueness of the law of such processes, then X is a diffusion process $\overline{D}$ which may be called naturally as (A,L) - <u>diffusion process</u>. It is not difficult to see that a diffusion process which corresponds to (A,L) analytically as above is an (A,L) - diffusion process. Also, we can apply probabilistic methods of stochastic differential equations (SDE) (cf. [12], [4]) and martingale problems (Stroock-Varadhan [6], Anderson [1]) to show the existence and the uniqueness of (A,L) - diffusions.

We have proposed still another method of probabilistic construction of (A,L) - diffusions ([9], [12]): it is based on the construction of <u>excursions</u> of diffusions as a point process on a function space. Main aims of such a method are to construct (A,L) - diffusions in such a way that the very way of the construction itself clarify the probabilistic structure of diffusions and secondly, to remove the restriction of $\mu(x) \geq 0$, which was necessarily imposed in the method of SDE. By this method, we can construct a continuous stochastic process on $\overline{D}$ from the given data A and L for which we can easily prove the property (ii) above but it is not so easy to prove the property (i) (or (i)'). The main purpose of the present paper is to discuss a general result on point processes and martingales which will serve to settle such a question.

II POINT PROCESSES

Refering the details to [3] and [10], we recall here briefly what will be necessary in future discussions. Let $\{X, \mathcal{B}(X)\}$ be a measurable space. Notions of point processes on X and Poisson point processes are defined as in [3], [10]. In particular, for a given σ - finite measure $n(dx)$ on $\{X, \mathcal{B}(X)\}$, a stationary Poisson point process p on X with the characteristic measure n is a point process on X for which the corresponding counting measure $N_p(E) = \#\{s \in D_p;\ (s,p(s)) \in E\}$, $E \in \mathcal{B}(0,\infty) \times \mathcal{B}(X)$, is a Poisson random measure with the mean measure $dt\, n(dx)$. Let us recall that such p is constructed in the following way : let $U_n \in \mathcal{B}(X)$, $n = 1, 2, \ldots$, be such that they are disjoint, $0 < n(U_n) < \infty$ and $X = \bigcup_n U_n$. Let $\xi_i^{(n)}$, $n, i = 1, 2, \ldots$, be U_n - valued random variables with $P(\xi_i^{(n)} \in dx) = \frac{n(dx)}{n(U_n)}$ and $\tau_i^{(n)}$, $n, i = 1, 2, \ldots$, be non - negative random variables with $P(\tau_i^{(n)} > t) = e^{-tn(U_n)}$ such that $\xi_i^{(n)}, \tau_i^{(n)}$ are all mutually independent. Set

$$D_p = \bigcup_{n=1} \{\tau_1^{(n)}, \tau_1^{(n)}+\tau_2^{(n)}, \ldots, \tau_1^{(n)}+\tau_2^{(n)}+ \ldots +\tau_k^{(n)}, \ldots\}$$

then this is a disjoint union a.s. because of the independence of $\tau_i^{(n)}$ and we set, for $s \in D_p$ which is expressed as $s = \tau_1^{(n)}+\tau_2^{(n)}+ \ldots +\tau_k^{(n)}$, $p(s) = \xi_k^{(n)}$.

EXAMPLE 1. (Brownian excursions). For simplicity, we consider the one - dimensional case. Let $W^+(W^-) = \{w : [0,\infty) \ni t \to w(t) \in [0,\infty)$, (resp $(-\infty,0]$). continuous, $w(0) = 0$, $\exists\ \sigma(w) > 0$ such that, if $0 < t < \sigma(w)$ then $w(t) > 0$ (resp. < 0) and, if $t \geq \sigma(w)$ then $w(t) = 0\}$ and $\mathcal{B}(W^+)$ (resp. $\mathcal{B}(W^-)$) be the σ - field on it generated by Borel cylinder sets. Let

$$p^o(t,x,y) = \frac{1}{\sqrt{2\pi t}}\left(e^{-\frac{(x-y)^2}{2t}} - e^{-\frac{(x+y)^2}{2t}}\right) \qquad \begin{array}{l} t > 0 . \\ x,y \in [0,\infty), \\ \text{or } x,y \in (-\infty,0] \end{array}$$

$$K^+(t,x) = \sqrt{\frac{2}{\pi t^3}}\, x\, e^{-\frac{x^2}{2t}}, \quad t > 0 \quad x \in [0,\infty)$$

$$K^-(t,x) = \sqrt{\frac{2}{\pi t^3}}\, (-x)\, e^{-\frac{x^2}{2t}}, \quad t > 0 \quad x \in (-\infty,0].$$

Then σ - finite measures n^+ and n^- are defined on $\{W^+, \mathcal{B}(W^+)\}$ and $\{W^-, \mathcal{B}(W^-)\}$ respectively by

$$n\{ w: w(t_1) \in dx_1, w(t_2) \in dx_2, \ldots, w(t_n) \in dx_n, \sigma(w) > t_n\}$$
$$= K(t_1,x_1)\, p^o(t_2-t_1,x_1,x_2)\ldots p(t_n-t_{n-1},x_{n-1},x_n)dx_1dx_2\ldots dx_n$$
$$(0 < t_1 < t_2 < \ldots < t_n).$$

Let n be the σ - finite measure on the sum $W = W^+ \cup W^-$ such that $n|_{W^{\pm}} = n^{\pm}$. A stationary Poisson point process on W with the characteristic measure n is called a <u>Poisson point process of Brownian excursions.</u>

Let $(\Omega, \mathcal{F}, P)$ be a probability space with a right - continuous increasing family $(\mathcal{F}_t)_{t\geq 0}$ of sub σ - fields and $\{X, \mathcal{B}(X)\}$ be a measurable space. Let $\mathcal{P}$ be the smallest σ - field on $[o,\infty) \times X \times \Omega$ with respect to which all $g = g(t,x,\omega)$ with the following properties are measurable :

(i) for each $t > 0$, $(x,\omega) \to g(t,x,\omega)$ is $\mathcal{B}(X) \times \mathcal{F}_t$ - measurable

(ii) $t \to g(t,x,\omega)$ is left - continuous.

$\mathcal{P}$ is called the <u>predictable</u> σ - field and any function $f(t,x,\omega)$ measurable with respect to ρ is called <u>predictable.</u>

Let n be a σ - finite measure of $\{X, \mathcal{B}(X)\}$ and p be an $(\mathcal{F}_t)$ - stationary Poisson point process on X with the characteristic measure n, (cf. [10]). If a real function $f = f(t,x,\omega)$ on $[0,\omega) \times X \times \Omega$ is such that $\int_0^{t+} \int_X f(s,x,\omega)|N_p(ds.dx) < \infty$ a.a. ω, then $\int_0^{t+} \int_X f(s,x,\omega)N_p(ds,dx) = \sum_{s\leq t, s\in D_p} f(s,p(s),\omega)$ is well defined as an absolutely convergent sum. Let $\tilde{N}_p(E) = N_p(E) - \int_E dtn(dx)$, $E \in \mathcal{B}(0,\infty) \times \mathcal{B}(x)$. If $f(t,x,\omega)$ is predictable and

$$E[\int_0^{t+} \int_X |f(s,x,\,.)|^2\, dsn(dx)] < \infty \text{ then}$$

$\int_0^{t+} \int_X f(s,x,.)\tilde{N}_p(ds,dx)$ is defined as a martingale stochastic integral (cf. [10]). If furthermore $E[\int_0^{t} \int_X |f(s,x,.)|dsn(dx)]$

$<\infty$, then

$$\int_0^{t+}\int_X f(s,x,.)\tilde{N}_p(ds,dx) = \int_0^{t+}\int_X f(s,x,.)N_p(ds.dx) - \int_0^{t}\int_X f(s,x,.)dsn(dx)$$

holds but usually integrals in the right-hand side have no meaning separately. It is meaningful only as a <u>compensated</u> <u>sum</u>.

III MARTINGALES AND POINT PROCESSES

Let p be a Poisson point process of Brownian excursions given in Ex.1. A Brownian motion is constructed from p by the following steps. First. set $A(t) = \sum_{s\leq t, s\in D_p} \sigma(p(s)) = \int_0^{t+}\int_W \sigma(w)N_p(ds.dw)$, then $t \to A(t)$ is a strictly increasing right-continuous function such that $\lim_{t\uparrow\infty} A(t) = \infty$ a.s. Hence, if $\phi(t)$ is the inverse function of $t \to A(t)$, $t \to \phi(t)$ is a continuous increasing function a.s.. Secondly, for a given $t \in [0,\infty)$, if we set $s = \phi(t)$, i.e., $A(s-) \leq t \leq A(s)$, then the two cases can occur : (i) the case $A(s-) < A(s)$ which implies $s \in D_p$ and we set $B(t) = p(s)(t - A(s-))$; (ii) the case $A(s-) = A(s)$, then we set $B(t) = 0$. It is clear that $t \to B(t)$ is continuous a.s., and we can identify it as one-dimensional Brownian motion. There are several ways in doing this identification but we will establish the following general result by which it is immediate to verify that $B(t)$ is a Brownian motion.

Let $C = C([0,\omega) \to R)$ be the set of all real continuous functions defined on $[0,\omega)$ and $\mathcal{B}(C)$ is the σ - field generated by Borel cylinder sets. Let $\{X,\mathcal{B}(X)\}$ be a measurable space, n be a σ - finite measure on it and p be an $(\mathcal{F}_t)$ - stationary point process on X with the characteristic measure n defined on $(\Omega,\mathcal{F},P)$ with $(\mathcal{F}_t)_{t\geq 0}$. Suppose we are given the following:

(i) $(t,x,\omega) \in [0,\infty) \times X \times \Omega \to F^{t,x,\omega} \in C$

(ii) $(t,x,\omega) \in [0,\infty) \times X \times \Omega \to \sigma^{t,x,\omega} \in [0,\infty)$

(iii) $(t,\omega) \in [0,\infty) \times \Omega \to \{\mathcal{B}^{t,\omega}(u)\}_{u \in [0,\infty)}$ = an increasing family of σ - fields on X such that $\mathcal{B}^{t,\omega}(u) \subset \mathcal{B}(X)$. We suppose that the following conditions are satisfied. a) The mappings F and σ in (i) and (ii) are predictable i.e., $\mathcal{P}/\mathcal{B}(C)$ (resp. $\mathcal{P}/\mathcal{B}[0,\infty)$) - measurable and the mapping in (iii) is also predictable in the sense that for every $u \geq 0$ and $f(x) \in L^1(X, n(dx))$, $(t,x,\omega) \to E^n[f(x)|\mathcal{B}^{t,\omega}(u)](x)$ has a version which is predictable where $E^n[\,|\mathcal{B}^{t,\omega}(u)]$ is the conditional expectation with respest to the measure n.

b) For each (t,x,ω), the following hold : $F^{t,x,\omega}(0) = 0$ and $F^{t,x,\omega}(u) = F^{t,x,\omega}(u \wedge \sigma^{t,x,\omega})$ for all $u \geq 0$.

c) For each (t,ω), the following hold:

(i) $x \to \sigma^{t,x,\omega}$ is $\mathcal{B}^{t,\omega}(u)$ - stopping time,

(ii) $x \to \{F^{t,x,\omega}(u)\}$ is $\{\mathcal{B}^{t,\omega}(u)\}$-adapted,

(iii) for each $u \geq 0$, $x \to F^{t,x,\omega}(u)$ is in $L^2(X,n(dx))$ and for each $u_2 > u_1 > 0$ and $H \in L^2(X,n(dx))$ which is $\mathcal{B}^{t,\omega}(u_1)$ - measurable, $\int_X [F^{t,x,}(u_2) - F^{t,x,}(u_1)]H(x)n(dx) = 0$.

d) There exists a mapping $(t,x,\omega) \to \langle F^{t,x,\omega}\rangle \in C$ such that $F^{t,x,\omega}(0) = 0$, $\langle F^{t,x,\omega}\rangle(u) = \langle F^{t,x,\omega}\rangle(u \wedge \sigma^{t,x,\omega})$, $u \to \langle F^{t,x,\omega}\rangle(u)$ is increasing and, for each (t,ω),

$$\int_X (F^{t,x,\omega}(u_2) - F^{t,x,\omega}(u_1))^2 H(x)n(dx)$$

$$= \int_X (\langle F^{t,x,\omega}\rangle(U_2) - \langle F^{t,x,\omega}\rangle(u_1))H(x)\, n(dx)$$

holds for every $u_2 > u_1 > 0$ and $H \in L^\infty(X,n(dx))$ which is $\mathcal{B}^{t,\omega}(u_1)$ - measurable.

Let $\rho(t)$ be a non - negative $(\mathcal{F}_t)$ - adapted measurable process and $A(0)$ be a $\mathcal{F}_0$ - measurable non - negative random varibale. Set

$$A(t) = A(0) + \int_0^{t+} \int_X \sigma^{s,x,\omega} N_p(ds,dx) + \int_0^t \rho(s)ds \qquad (3.1)$$

and assume that $t \to A(t)$ is strictly increasing, $A(t) < \infty$ for $t > 0$ and $\lim_{t \wedge \infty} A(t) = \infty$ a.s.. Let $\phi(t)$ be the inverse function of $t \to A(t)$. Then $t \to \phi(t)$ is continuous a.s..

Let Δ be an extra point attached to X and we set $p(t) = \Delta$ if $t \notin D_p$. Also, we set $F^{t,\Delta,\omega}(u) \equiv 0$. Let H_t be the σ-field on Ω defined by

$$H_t = F_{\phi(t)-} \vee \sigma[F^{\phi(t),p(\phi(t)),\omega}(u-A(\phi(t)-));0 \le u \le t].$$

THEOREM Suppose, furthermore, the following condions are satisfied :

(1) for every $t > 0$, $E[\phi(t)] < \infty$,

(2) there exists a constant $M_1 > 0$ such that, for every $0 < u \le v < 1$ and t, ω

$$\int_X |F^{t,x,\omega}(u)| \, I_{[\sigma^{t,x,\omega} > v]} \, n(dx) \le M_1$$

(3) there exists a constant $M_2 > 0$ such that, for every $0 < u \le v$ and t, ω, x,

$$\langle F^{t,x,\omega}\rangle (v) - \langle F^{t,x,\omega}\rangle (u) \le M_2(v \wedge \sigma^{t,x,\omega} - u \wedge \sigma^{t,x,\omega}).$$

Set, for each $t > 0$,

$$M(t) = \int_0^{\phi(t)+} \int_X F^{s,x,\omega}(t-A(s-))\tilde{N}_p(ds,dx), \qquad (3.2)$$

Then $t \to M(t)$ has a continuous modification which is an (H_t)-martingale such that the corresponding quadratic variational process $\langle M\rangle(t)$ is given by

$$\langle M\rangle(t) = \int_0^{\phi(t)+} \int_X \langle F^{s,x,\omega}\rangle(t-A(s-))N_p(ds,dx), \qquad (3.3)$$

Proof can be given in almost the same line as that of Lemma 5 in [11].

Let us return to the process $B(t)$ defined above. In this case, $X = W, F^{t,w,\omega}(u) = w(u)$, $\sigma^{t,w,\omega} = \sigma(w)$, $\mathcal{B}^{t,\omega}(u) = \sigma\{w(s);s\le u\}$, $\langle F^{t,w,\omega}\rangle(u) = u \wedge \sigma(w)$ and $A(t) = \int_0^{t+}\int_W \sigma(w)N_p(ds,dw)$. Clearly, all the assumptions are satisfied. Therefore, $M(t) = \int_0^{\phi(t)+}\int_W w(t-A(s-))\tilde{N}_p(ds,dw)$ defines a continuous martingale such that $\langle M\rangle(t) = \int_0^{\phi(t)+}\int_W [(t-A(s-))\wedge\sigma(w)]N_p(ds,dw) = t$. Thus $M(t)$ is a Brownian motion. Now, $B(t) = M(t)$ since $M(t) = \int_0^{\phi(t)+}\int_W w(t-A(s-))N_p(ds,dw)$ becuase of $\int_0^{\phi(t)}\int_W w(t-A(s-))n(dw) = 0$.

<u>EXAMPLE</u> 3.1 Let $c_i(t)$, $t \in [0,\infty)$, $i = 1, 2, 3$, be given such that $c_i(t) \ge 0$ and $\sum_{i=1}^{3} c_i(t) \equiv 1$. Let p be an $(\mathcal{F}_t)$ - Poisson point process of Brownian excursions defined on $(\Omega,\mathcal{F},P)$ and $(\mathcal{F}_t)$. Under a certain regularity conditions on $c_i(t)$ (e. g. Lipschitz condition is sufficient), a stochastic equation for the process $A(t)$:

$$A(t) = \int_0^{t+}\int_{W^+} c_1^2(A(s-))\sigma(w)\, N_p(ds,dw) + \int_0^{t+}\int_{W^-} c_2^2(A(s-))\sigma(w)\, N_p(ds,dw) + \int_0^t c_3(A(s))ds \tag{3.4}$$

determines a unique $(\mathcal{F}_t)$ - adapted increasing process $A(t)$. For $c \ge 0$ and $w \in W$, we define $\Phi^c : W \to W$ by

$$(\Phi^c w)(t) = \begin{cases} cW(\frac{t}{c^2}) , & c > 0 \\ 0 , & c = 0. \end{cases}$$

Set

$$F^{t,w,\omega} = \begin{cases} \Phi^{c_1(A(s-))}(w) , & w \in W^+ \\ \Phi^{c_2(A(s-))}(w) , & w \in W^- , \end{cases}$$

$$\sigma^{t,w,\omega} = \begin{cases} c_1^2(A(t-))\sigma(w) & , \quad w \in W^+ \\ c_2^2(A(t-))\sigma(w) & , \quad w \in W^- \end{cases}$$

$\mathcal{B}^{t,w,\omega}(u)$ = the smallest σ - algebra on W with respect to which all $[w \to F^{t,w,\omega}(v)]_{v \le u}$ are measurable,

and

$$\langle F^{t,w,\omega}\rangle(u) = u \wedge \sigma^{t,w,\omega}.$$

Let $\rho(s) = c_3(A(s))$ and hence $A(t)$ coincides with that give in (3.4). Set

$$X(t) = \int_0^{\phi(t)+} \int_W F^{s,w,\omega}(t-A(s-))N_p(ds,dw)$$

$$= \begin{cases} F^{\phi(t),p(\phi(t)),\omega}(t-A(\phi(t)-)), & \text{if} \quad \phi(t) \in D_p \\ 0 & , \text{if} \quad \phi(t) \notin D_p. \end{cases}$$

Now, by the theorem,

$$M(t) = \int_0^{\phi(t)+} \int_W F^{s,w,\omega}(t-A(s-))\tilde{N}_p(ds,dw)$$

defines a continuous martingale with

$$\langle M\rangle(t) = \int_0^{\phi(t)+} \int_W [(t-A(s-)) \wedge \sigma^{s,w,\omega}]N_p(ds,dw)$$

$$= \int_0^t I_{[X(s)\neq 0]}ds. \tag{3.5}$$

Also it is eary to see that

$$\int_0^t I_{[X(s)=0]}ds = \int_0^t c_3(s)d\phi(s).$$

Then

$$M(t) = \int_0^{\phi(t)+} \int_W F^{s,w,\omega}(t-A(s-))\tilde{N}_p(ds,dw)$$

$$= \int_0^{\phi(t)+} \int_W F^{s,w,\omega}(t-A(s-))N_p(ds,dw)$$

$$- \int_0^{\phi(t)} ds \int_W F^{s,w,\omega}(t-A(s))n(dw)$$

$$= X(t) - \int_0^{\phi(t)} [c_1(A(s))-c_2(A(s))]ds$$

$$= X(t) - \int_0^t [c_1(s)-c_2(s)]d\phi(s) \qquad (3.6).$$

Also, we can conclude that, if $f(t,x)$ is continuous in $(t,x) \in [0,\infty) \times R$ which is smooth on $[0,\infty) \times [0,\infty)$ and $[0,\infty) \times (-\infty,0]$ respectively,

$$f(t,X_t) - f(0,0) = \text{a martingale}$$

$$+ \int_0^t [\frac{\partial f}{\partial t}(s,X_s) + \frac{1}{2}\frac{\partial^2 f}{\partial x^2}(s.X_s)I_{\{X_s \neq 0\}}]ds$$

$$+ \int_0^t [c_1(s)\frac{\partial f}{\partial x}(s,0+)-c_2(s)\frac{\partial f}{\partial x}(s,0-)]d\phi(s).$$

In this way, we can identify the process $X(t)$ with a one dimensional Brownian motion with a time -dependent Feller's two sided boundary condition $c_1(t)u'(0+) - c_2(t)u'(0-) - \frac{1}{2}c_3(t)u''(0) = 0$ at $x = 0$.

A multidimensional version of our theorem can be given in a obvious manner and we can apply it for the processes constructed in [12] (cf. also [9]). But we do not dicuss about it here.

REFERENCES

[1] R. Anderson : On diffusion processes with second order boundary contitions I.II, Indiana Univ. Math. J., 25(1976), 367-395 ; 403-441.

[2] J.M. Bony, P.Courège et P. Priouret : Semi-groupes de Feller sur une variété a bord compacte et problèmes aux limites intégro-différentiels du second ordre donnant lieu au

principe du maximum, Ann. Inst. Fourier, 18(1968), 369-521

[3] K. Itô : Poisson point processes attached to Markov processes, Proc. Sixth Berkeley Symp. Vol.III, Univ. California Press, Berkeley, 1972, 225-239.

[4] S. Nakao and T. Shiga : On the uniqueness of solutions of stochastic differential equations with boundary condition, J. Math. Kyoto Univ., 12(1972), 451-478.

[5] K. Sato and T. Ueno : Multi-dimensional diffusion and the Markov process on the boundary, J. Math. Kyoto Univ., 14(1965), 529-605.

[6] D.W. Stroock and S.R.S. Varadhan : Diffusion processes with boundary conditions, Comm. Pure Appl. Math., 24(1971), 147-225.

[7] K. Taira : Sur l'existence des processus de diffusion satisfaisant aux conditions aux limites de Ventcel', C. R. Acad. Sc. Paris, 284, ser.A (1977), 1133-1136.

[8] S. Watanabe : On stochastic differential equations for multidimensional diffusion processes with boundary conditions, I. II, J. Math. Kyoto Univ. 11(1971), 169-180 ; 545-551.

[9] S. Watanabe : Construction of diffusion processes by means Poisson point process of Brownian excursions, Proc. Third Japan-USSR Symp. on Probability Theory, Lecture Notes in Math., 550 Springer (1976), 650-654.

[10] S. Watanabe : Poisson point process of Brownian excursions and its applications to diffusion processes, Proc. Symp. in Pure Math. AMS., 31(1977), 153-164.

[11] S. Watanabe : Excursion point process of diffusion and stochastic integral, Proc. International Symp. on SDE, Kinokuniya (1978).

[12] S. Watanabe : Construction of diffusion processes with Wentzell's boundary conditions by means of Poisson point processes of Brownian excursions, Proc. Semester on Prob., Banach Center, Warsaw, to appear

C^k HYPOELLIPTICITY WITH DEGENERACY, Part II

Paul Malliavin

3.3 Some lemmas

3.3.1 Lemma

Denote by Q_a *the set of positive quadratic forms on* R^m *having all their eigenvalues* $\leq a$. *We can find* $N = N(a,\varepsilon)$ *points* $\{\xi_k\}$ *of the unit sphere of* R^m *such that*

$$N(a,\varepsilon) < \left(\frac{ca}{\varepsilon}\right)^{m-1}$$

$$q(\xi_k) \geq \varepsilon \qquad (q \in Q_a,\ 1 \leq k \leq N(a,\varepsilon))$$

will imply that all of the eigenvalues of q *are* $> \varepsilon/2$. (c is a numerical constant.)

Proof. The length of the gradient of $q(\xi)$ on the unit sphere is bounded by a. We have therefore to choose the ξ_k such that the ball of center ξ_k and radius $\varepsilon/2a$ covers the sphere.

3.3.2 Lemma (Estimates of the variance of the brownian motion).

Let $b_\omega(\xi)$ *a scalar valued brownian. Consider its variance* $\sigma_{[0,a]}(b)$, *defined by*

$$\sigma^2_{[0,a]}(b) = \int_0^a b^2_\omega(\xi)\frac{d\xi}{a} - \left(\int_0^a b_\omega(\xi)\frac{d\xi}{a}\right)^2.$$

Then

$$P(\sigma_{[0,a]} < \varepsilon) < 2\exp\left(-\frac{a^{\frac{1}{2}}}{8\varepsilon}\right).$$

Proof: We remark that using the rescaling in space-time $b_\omega(\xi) = a^{\frac{1}{2}} b_\omega(\xi/a)$ it is sufficient to prove the lemma for $a = 1$.

Consider on $[-\frac{1}{2},\frac{1}{2}]$ the orthonormal system 1,

ISBN 0-12-268380-3

$2^{1/2}\cos 2\pi kt$, $2^{1/2}\sin 2\pi kt$. In this system the white noise W has for expansion, G_s being independent normal variables,

$$W = G_0 + 2^{1/2} \sum_{k=1}^{+\infty} (G'_k \cos 2\pi kt - G_k \sin 2\pi kt).$$

We integrate in t and we shall get the following expression of the brownian $b_\omega(t)$, $t \in [-\frac{1}{2}, +\frac{1}{2}]$:

$$b_\omega(\tfrac{1}{2}+t) = c + tG_0 + 2^{1/2} \Sigma \frac{1}{2\pi k} (G_k \cos 2\pi kt + G'_k \sin 2\pi kt)$$

where c is a constant. It turns out that the mean value of the right hand side is equal to c; therefore the variance σ^2 is the L^2 norm of the remaining sum. We compute this L^2 norm expanding t in a sine series and then applying the Plancheral theorem. Neglecting the sine terms contribution we get

$$\sigma^2 \geq \frac{1}{(2\pi)^2} \sum_{k=1}^{+\infty} \frac{1}{k^2} G_k^2.$$

Using now the χ^2-law, z being a real parameter

$$E(\exp(-2(\pi)^2 z^2 \sigma^2)) \leq \prod_{k=1}^{+\infty} (1+z^2 k^{-2})^{-1} = \frac{1}{\cosh \pi z}$$

which implies

$$P(\sigma<\varepsilon) \leq 2 \inf_{z>0} \exp(2\pi^2 z^2 \varepsilon^2 - \pi z).$$

3.3.3 Lemma

Denote by τ the exit time of $[-h,+h]$ for the brownian starting at zero, then

$$P(\tau<ah^2) \leq 2 \exp(-\frac{1}{a}).$$

Proof: We have by a classical computation

$$E(e^{-\lambda\tau}) = \frac{1}{\mathrm{ch}(2\sqrt{\lambda})} \qquad (h=1,\ \lambda>0)$$

3.3.4 Lemma

Introduce on V a riemannian metric of reference. Define the guage $|||r|||$ of a linear frame as

$$|||r||| = \sup_{\substack{z\in T(V)\\ \|z\|=1}} \|r(z)\|_{R^n} + \sup_{\xi\in R^n} \|r^{-1}(\xi)\|_{T(V)}.$$

Then,

$\{r\in GL(V);\ \|r\|=1\}$ *is the bundle of orthonormal frames,*
$\{r\in GL(V);\ \|r\|\leq M\}$ *is a compact part of* GL(V),

$$\left(\sup_{|||r_0|||=1}\ \sup_{t\in[0,1]} |||r_\omega(t)|||\right) \in L^p(\Omega) \text{ for all } p < +\infty.$$

Proof: Everything is obvious except the last assertion, which results from the martingale maximal theorem, combined with uniform estimates of Itô's invariant of the diffusion $r_\omega(t)$ and [8], p. 244 (cf. Proof of Lemma 3.1).

3.3.5 Lemma

Suppose that there exists $t_0(\omega)$, $t_1(\omega)$, $(0 \leq t_0(\omega) < t_1(\omega) < 1)$ *such that*

$$\Sigma \int_{t_0(\omega)}^{t_1(\omega)} (f^i_{A_k} f^j_{A_k})(\tilde{r}_\omega(\xi))\,d\xi = \tilde{\alpha}$$

where $\tilde{r}_\omega(\xi)$ *is the diffusion starting from*

$$|||\tilde{r}_\omega(t_0(\omega))||| = 1,\ p(\tilde{r}_\omega(t_0)) = p(r_\omega(t_0)).$$

Suppose that

$$E(\|\tilde{\alpha}^{-1}\|^p) < +\infty \qquad \text{for all } p < +\infty$$

then this will imply:

$$\|\gamma\| \in L^p \quad \textit{for all } p < +\infty.$$

Proof: The quadratic form obtained by integration on a smaller interval is obviously smaller than the form obtained by integration over [0,1]. We have only to take care of the *renormalization procedure* by the introduction of $|||\tilde{r}_\omega(t_0(\omega))||| = 1$; this is done by 3.3.4.

3.4 *Proof of the Theorem 3.2*

3.4.0 We shall consider the sequence

$$\varepsilon_q = 2^{-q}.$$

For fixed q, we shall construct an event A_q such that

3.4.0.1 $$\|\tilde{\alpha}(\omega)\| \leq 2^{qc_0}, \omega \notin A_q$$

where c_0 is a fixed constant. Then

3.4.0.2 $$\sum_q 2^{qc_0 p} P(A_q) < +\infty \quad \text{for all } p$$

will imply the theorem.

3.4.1 The manifold V is supposed to be compact; this is not inherited by GL(V); however, limiting ourself to the r which satisfies $|||r||| \leq 2$, we get by 3.3.4 a compact manifold $\widetilde{GL}(V)$ with boundary. We shall denote by M a universal bound for the Itô invariants on $\widetilde{GL}(V)$ of all the functions that we will consider below. As these functions are finite in number we have by compactness of $\widetilde{GL}(V)$ that M is finite.

3.4.2 We shall take $t_0^q(\omega)$ as the first time between $[0,\frac{1}{2}]$ where $\Phi_s(v_\omega(t))$ reaches the value ε_q. Renormalizing r_ω such that $|||\tilde{r}_\omega(t_0^q(\omega))||| = 1$ we shall denote by

$\tilde{T}$ the stopping time where $\tilde{r}_\omega$ leaves $\widetilde{GL}(V)$. Our estimates will be for $t \in [t_0^q(\omega), \tilde{T}\wedge 1]$ and on this range 3.4.1 could be applied.

In particular using 3.4.1, 3.3.3 and the method of equations of comparison ([9]) for a riemannian metric of reference on GL(m,R)/O(m;R) we have

3.4.2.1 $$P(\tilde{T} - t_0^q < \lambda) < \exp(-\frac{c_1}{\lambda})$$

where $\lambda < 1$, c_1 will denote constants depending only upon M and V.

3.4.3 Using 3.4.1 and a comparison equation for Φ_s denoting by t_2^q the first time after t_0^q where $\Phi_s(v_\omega(t))$ takes the value ε_{q+1}, we have by 3.3.3

3.4.3.1 $$P(t_2^q - t_0^q < \lambda\varepsilon_q^2) < \exp(-\frac{c}{\lambda}),\ q > c.$$

3.4.4 Consider the event

$$A'_q = \{t_0^q > \tfrac{1}{2}\} \cup \{\tilde{T} - t_0^q < 2\varepsilon_q^3\} \cup \{t_2^q - t_0^q < 2\varepsilon_q^3\}$$

then

3.4.4.1 $$P(A'_q) \leq u(\varepsilon_q) + \exp(-c\varepsilon_q^{-1}).$$

Denote by e_q the interval $[t_0^q, t_0^q + 2\varepsilon_q^3]$; then

3.4.4.2 $$\int_{e_q} \Phi_s(v_\omega(\xi))\,d\xi > \varepsilon_q^4 \quad \text{if } \omega \notin A'_q.$$

3.4.5 We denote by Σ_j^* the tangent vectors to V which are exactly the brackets of j vectors taken amongst the A; then

$$\Sigma_s = \bigcup_{1\leq j\leq s} \Sigma_j^*.$$

Given $\ell \in (R^m)^*$, $\|\ell\|_1 = 1$ we define

$$I_j(\omega,\ell) = \int_{e_q} \sum_{Z\in\Sigma_j^*} \ell^2(f_Z(\tilde{r}_\omega(\xi)))d\xi$$

Denote by u the unit vector of $\mathbb{R}^m$ verifying $\ell(u) = 1$; given $v \in V$ there exists $B_1,\ldots,B_m \in \Sigma_s$ and scalars α_r, $|\alpha_r| \leq \det(B)$ such that

$$u = \alpha_1 B_1 + \ldots + \alpha_m B_m.$$

Applying ℓ to the two members and applying to the right hand side the Schwarz's inequality we find

$$(\det B)^2 \leq m \, \Sigma \, \ell^2(B_r).$$

Therefore 3.4.4.2 imply

3.4.5.1 $$\sum_{1\leq j\leq s} I_j(\omega,\ell) > m^{-1}\varepsilon_q^4 \, , \quad \omega \notin A'_q.$$

3.4.6 Main Lemma

Let $\beta_j \geq 5$ $(1 \leq j \leq s)$. Consider the event

$$A^j_{q,\beta_j}(\ell) = \{I_j(\omega,\ell) \geq \varepsilon_q^{\beta_j}\} \cap \{I_{j-1}(\omega,\ell) < \varepsilon_q^{4\beta_j}\}.$$

Then

$$P(A^j_q(\ell,\omega) \cap (A'_q)^*) < \exp(-c\varepsilon_q^{-1/2})$$

(*where* A* *denotes the complement of* A).

Proof: Consider

$$Y_U(\xi) = \ell(f_U(\tilde{r}_\omega(\xi))), \quad U \in \Sigma^*_{j-1}.$$

Using the fact that

$$\mathcal{L}_{\tilde{A}_k} f_U = f_{[A_k,U]}$$

we have that the increasing process ψ_U associated to the martingale part of the stochastic differential of Y_U is given by

$$d\psi_U = \sum_k \ell^2(f_{[A_k,U]}(\tilde{r}_\omega(\xi)))d\xi.$$

Therefore

$$\int_{e_q} \sum_{U\in\Sigma^*_{j-1}} d\psi_U = I_j(\omega,\ell) \geq \varepsilon_q^{\beta_j}.$$

We can therefore find $U_0 \in \Sigma^*_{j-1}$ such that

$$\int_{e_q} d\psi_{U_0} \geq c\varepsilon_q^{\beta_j}.$$

We shall write $\psi_{U_0}(\xi)$ as the sum of a brownian b_π with the change of time ψ_{U_0} and of a Lebesgue integral:

$$Y_U(\xi) = b_\pi(\psi_U(\xi)) + \int^\xi \theta(\xi)d\xi \text{ with } |\theta| < M.$$

We shall split the interval e_q in $[\varepsilon_q^{5-\beta_j}]$ equal intervals. We can therefore find at least one such interval, call it e_q', which will satisfy

$$\int_{e_q'} d\psi_{U_0} \geq c\ \varepsilon_q^{2\beta_j-5}.$$

We have

$$\int_{e_q'} |Y_{U_0}|^2 d\xi \geq M^{-1} \int_{e_q'} |Y_{U_0}|^2 d\psi_{U_0}$$

$$\int_{e_q'} |Y_{U_0}|^2 d\psi_{U_0} = \int_{e_q''} |b_\pi(\lambda) + G(\lambda)|^2 d\lambda$$

where $e_q'' = \psi_{U_0}(e_q')$, $G(\lambda) = \int^{\psi_{U_0}^{-1}(\lambda)} \theta d\xi$.

Therefore
$$\int_{e'_q} |Y_{U_0}|^2 d\psi_{U_0} \geq |e''_q| \sigma^2_{e''_q}(b\pi+G).$$

Now by Minkowski's inequality, α being a constant,

$$\|b_\pi+G-\alpha\|_{L^2} \geq \|b_\pi-\alpha'\|_{L^2} - \|G-\alpha''\|_{L^2} \qquad \alpha = \alpha' + \alpha''$$

Choose α such that the first member is $\sigma(b_\pi+G)$ and $\|G-\alpha''\|_{L^2}$ is $\sigma(G)$. We get

$$\sigma_{e''_q}(b_\pi+G) \geq \sigma_{e''_q}(b_\pi) - \sigma_{e''_q}(G).$$

This inequality is of interest only when the right hand side is positive. Furthermore the oscillation of G on e''_q is smaller than $M|e'_q|$, therefore

$$\sigma_{e''_q}(G) \leq \frac{1}{2} M|e'_q|.$$

By Lemma 3.2

$$P(\sigma_{e''_q}(b_\pi) < M|e'_q|) \leq \exp\left(- \frac{|e''_q|^{1/2}}{8M|e'_q|}\right).$$

We have

$$|e''_q|^{1/2} \geq c\ \varepsilon_q^{\beta_j-5/2}, \quad |e'_q| = 2\varepsilon_q^{\beta_j-2}.$$

Therefore with a probability greater than $1-\exp(-c\ \varepsilon_q^{-1/2})$ we have

$$\sigma_{e''_q}(b_\pi+G) \geq \frac{1}{2} M|e'_q| \qquad \text{which leads to}$$
$$\int_{e'_q} |Y_{U_0}|^2 d\xi > \frac{1}{4} M|e'_q|^2|e''_q| > c\ \varepsilon_q^{4\beta_j-9} \geq \varepsilon_q^{4\beta_j}$$

which proves the lemma.

3.4.7 Conclusion of the Proof

By 3.4.5.1 for every $\omega \notin A'_q$ there exists $j_0(\omega)$ such that

$$I_{j_0(\omega)}(\omega,\ell) \geq \varepsilon_q^5 \qquad 1 \leq j_0(\omega) \leq s$$

We apply then the Lemma 3.4.6:

$$I_{j_0(\omega)-1}(\omega,\ell) > \varepsilon_q^{4\times 5} \qquad \omega \notin A_q^{j_0}(\ell,\omega).$$

We iterate the application of the Lemma 3.4.6. As we have to consider only a fixed number of cases $j_0(\omega) = 1$, $j_0(\omega) = s-1,\ldots,j_0(\omega) = 2$, and so on. For each of these cases we have a well-defined sequence of β_j to introduce

$$\beta_j = 5\cdot 4^{j_0-j} \qquad j < j_0.$$

We define a finite union of sets introducing

$$A''_q(\ell) = \bigcup_{j,\beta_j} A^j_{q,\beta_j}(\ell)$$

and defining

$$c_0 = 5\cdot 4^s, \qquad \text{we have}$$

$$I_1(\ell,\omega) \geq \varepsilon_q^{c_0} \qquad \omega \notin A''_q(\ell) \cup A'_q \text{ with}$$

$$P(A''_q(\ell)) \leq \exp(-c\,\varepsilon_q^{-1/2}).$$

Now remark that the quadratic form

$$q(\ell) = \int_{e_q} \sum_k \ell^2(f_{A_k}(\tilde{r}_\omega(\xi)))d\xi$$

is uniformly bounded, therefore by the Lemma 3.3.1 it is sufficient to minorize $q(\ell)$ on a net R_q on the unit sphere with

$$\text{cardinal}(R_q) < c\, \varepsilon_q^{-mc_0}$$

Define

$$A''_q = \bigcup_{\ell \in R_q} A''_q(\ell).$$

Then

$$P(A''_q) \leq \varepsilon_q^{-mc_0} \exp(-c\, \varepsilon_q^{-1/2}).$$

Define

$$A_q = A'_q \cup A''_q.$$

Then A_q will satisfy 3.4.0.1 and 3.4.0.2 and the Theorem 3.2 will follow.

4. APPLICATION TO A WEAKLY PSEUDO-CONVEX DOMAIN

4.0 Notation. We shall recall some known facts (cf. [2], [3], [5]) and notations. D will denote a bounded domain in C^n with a boundary which is a C^∞ hypersurface which will be denoted by V. If $v \in V$, $T_v(V)$ denotes the real hyperplane tangent to V, $T_v^{\mathbb{C}}(V) = T_v(V) \cap \sqrt{-1}\, T_v(V)$ the complex hyperplane which is contained in $T_v(V)$. Choosing a basis of complex vectors of $T_v^{\mathbb{C}}(V)$, $Z_1, \ldots, Z_{n-1}$:

$$Z_k = \alpha_k^s \frac{\partial}{\partial z^s},$$

then the Kohn laplacian Δ^K is defined by

$$\Delta^K = \frac{1}{2} \Sigma (Z_k \overline{Z}_k + \overline{Z}_k Z_k).$$

Denoting by $\vec{n}$ the inner normal to D, the Levi matrix $c_{i,j}$ is defined by the scalar product

$$c_{i,j} = (\sqrt{-1}\, \vec{n} \,|\, [Z_i, \overline{Z}_j])_{\mathbb{R}^{2n}}.$$

The hypothesis of weak pseudo-convexity for D means that $c_{i,j}$ is a positive semi-definite matrix. We shall denote by

$$\chi(v) = \text{Trace}(c_{i,j}) = \sum_i c_{i,i} \geq 0.$$

4.1 Theorem

Denote

$$O_\varepsilon = \{v \in V, \ \chi(v) < \varepsilon\}.$$

Given $v_0 \in V$, r *being chosen* > 0, *denote*

$$B_r = \{v \in V; \text{ distance } (v,v_0) < r\}.$$

Denote by Π *the orthogonal projection of* $\mathbb{C}^n$ *on* $T^{\mathbb{C}}_{v_0}(V)$. *Let*

$$O_{\varepsilon,v_0} = \Pi(O_\varepsilon \cap B_r)$$

$$g(\varepsilon) = (2n-2) \text{ Lebesgue measure of } O_{\varepsilon,v_0}.$$

Suppose that

$$g(\varepsilon)[\log \frac{1}{\varepsilon}]^{2n-2} \to 0. \qquad (\varepsilon \to 0)$$

Then the fundamental solution $p_t(v_0,v)$ *of the heat equation*

$$\frac{\partial}{\partial t} - (\Delta^K)^*$$

(*where* $(\Delta^K)^*$ *is the adjoint of* Δ^K)

with pole at $t = 0$, $v = v_0$ is C^∞ *in* v *for all* $t > 0$.

4.2 *Proof of 4.1*:

The introduction of B_r will be taken into account by the stopping time T'_r where the process exits B_r. As trivially

$$\int_0^1 \ell^2(f_{A_k})\,d\xi \geq \int_0^{1\wedge T'} \ell^2(f_{A_k})\,d\xi$$

it is sufficient to proceed to the same estimates as in the proof of 3.2 but dealing on $[0,1\wedge T'_r]$, and therefore using 3.2 with s = 1 it is sufficient to estimate

$$u(\varepsilon) = P(\chi(v_\omega(\xi)) < \varepsilon,\ 0 < \xi < T'_{r/2}\wedge\tfrac{1}{2}).$$

Denoting

$$\tilde{v}_\omega(\xi) = \Pi(v_\omega(\xi))$$

we have

$$d\tilde{v}^i_\omega = \alpha^i_k(v_\omega)\,db^k_\omega + \beta^i(v_\omega)\,d\xi.$$

We have furthermore

$$\|\alpha\| < c, \quad \|\beta\| < c.$$

Furthermore if we have chosen r sufficiently small we have also

$$\det(\alpha) > c_1.$$

Define

$$\gamma(\varepsilon,\delta) = \sup_{v_1 \in O_\varepsilon \cap B_r} P_{v_1}(\chi(v_\omega(\xi)) < \varepsilon \text{ for } \xi \in [0,\delta]).$$

Let τ be the first exit time of $\pi(v_\omega(\xi))$ from the ball of center $\pi(v_1)$ and of radius $\delta^{1/2}$, then

$$\gamma(\varepsilon,\delta) \leq E\Big(\int_0^{\delta\wedge\tau} I_{O_{\varepsilon,v_0}}(\tilde{v}_\omega(\xi))\,\frac{d\xi}{\delta\wedge\tau}\Big).$$

Using Schwarz's inequality

$$\leq \left(E\left(\int_0^{T'} I \, d\xi\right)\right)^{1/2} \left(E\left((\delta \wedge \tau)^{-1}\right)\right)^{1/2}.$$

The first integral is evaluated by Krylov's inequality [6], the second using comparison lemmas [9] to minorize τ. We get

$$\gamma(\varepsilon,\delta) \leq c[(g(\varepsilon))^{\frac{1}{2n-2}}\delta^{-1}]^{1/2}.$$

We shall take δ such that the first member is smaller than $\frac{1}{2}$. This value of δ being fixed we evaluate $u(\varepsilon)$ by iterating, via the Markov property, this estimate and we get

$$u(\varepsilon) < \exp(-c[g(\varepsilon)]^{\frac{-1}{2n-2}}).$$

Now the hypothesis

$$g(\varepsilon)[\log \frac{1}{\varepsilon}]^{2n-2} \to 0$$

implies that

$$\varepsilon^{-k} u(\varepsilon) \to 0$$

and Theorem 3.2 gives the conclusion.

Bibliography

[1] D. L. Burkholder and R. F. Gundy: Extrapolation and interpolation for quasilinear operators on martingales. Acta Mathematica 1970 (120) pp. 249-304.

[2] K. Diedrich and J. Fornaess: Pseudo-convex domains with real analytic boundary. Annals of Math. 1978, pp. 371-385.

[3] Ju. V. Egoroff: On the subellipticity of the $\overline{\partial}$

Neumann-problem. Dokl. Akad. Nauk (Soviet Mathematics) vol. 18, p. 1078 (1977).

[4] L. Hörmander: Hypoelliptic second order differential equations. Acta Mathematica 1967, pp. 147-171.

[5] J. J. Kohn: Sufficient condition for subellipticity on a weakly pseudo-convex domain. Proc. Nat. Acad. 1977, pp. 2214-2218.

[6] N. V. Krylov: An inequality in the theory of Stochastic integrals. Theory of probability and its applications 1971 (16), pp. 438-448.

[7] H. Kunita and S. Watanabe: On square integrable martingales. Nagoya Math. Journal 1967 (30), pp. 209-246.

[8] P. Malliavin: Stochastic calculus of variation and hypoelliptic operators. Proc. of the International Symposium on Stochastic Differential Equations (Kyoto 1976) Tokyo, 1978.

[9] P. Malliavin: Asymptotics of the Green function of a riemannian manifold and Itô's stochastic integrals. Proc. Nat. Acad. 1974 (71), pp. 381-383.

A 8
B 9
C 0
D 1
E 2
F 3
G 4
H 5
I 6
J 7